Water, Flood Management and Water Security Under a Changing Climate

Anisul Haque • Ahmed Ishtiaque Amin Chowdhury
Editors

Water, Flood Management and Water Security Under a Changing Climate

Proceedings from the 7th International Conference on Water and Flood Management

Springer

Editors
Anisul Haque
Institute of Water and Flood Management (IWFM)
Bangladesh University of Engineering and Technology (BUET)
Dhaka, Bangladesh

Ahmed Ishtiaque Amin Chowdhury
Institute of Water and Flood Management (IWFM)
Bangladesh University of Engineering and Technology (BUET)
Dhaka, Bangladesh

ISBN 978-3-030-47785-1 ISBN 978-3-030-47786-8 (eBook)
https://doi.org/10.1007/978-3-030-47786-8

This Springer imprint is published by the registered company Springer Nature Switzerland AG.
The registered company address is: Gewerbestrasse 11, 6330 Cham, Switzerland

Contents

An Overview on Isotopes in Precipitation of Bangladesh

Abdul Hadi Al Nafi Khan, Md. Abdul Quaiyum Bhuyian, Md. Ariful Ahsan, Farhana Islam, Md. Masud Karim, and Md. Moniruzzaman

1 Introduction

Precipitation is an important part of the hydrological cycle, and spatiotemporal variations of precipitation isotopes may lead to conceptualize the processes involved in hydrometeorological and ecological events (Craig 1961; Dansgaard 1964; Liu et al. 2014). The precipitation stable isotopes (^{18}O and^{2}H) have been established as an important tool in Isotope Hydrology which is frequently applied in studying not only the source and dynamics, but also in atmospheric circulation and reconstruction of palaeoclimatic events (Araguas-Araguas et al. 1998). Variation of $\delta^{2}H$ and $\delta^{18}O$ with respect to space can reveal the source of vapour/moisture and the recycling events that have occurred during the transportation (Bowen and Revenaugh 2003).

Temperature, rainfall amount, altitude, humidity, vapour pressure, latitude, continentality etc. are globally known as environmental and geographic controls over isotopic variations in precipitation (Rozanski et al. 1993; Vuille et al. 2005; Chen et al. 2017). However, in local scale the environmental controls may act differently as moisture recycling, transient eddies and changing moisture sources play roles in relatively smaller domains (Dublyansky et al. 2018).

Global Network of Isotopes in Precipitation (GNIP) programme was started in 1958 with the collaboration of World Meteorological Organization (WMO) to generate better understanding on spatiotemporal climate variation, atmospheric circulation and studying interaction of water between atmosphere and biosphere. From 1961, more than 1000 locations spread across approximately 125 countries

A. H. A. N. Khan (✉) ·
M. A. Q. Bhuyian · M. A. Ahsan · F. Islam · M. M. Karim · M. Moniruzzaman
Institute of Nuclear Science and Technology, Bangladesh Atomic Energy Commission, Dhaka, Bangladesh

A. Haque, A. I. A. Chowdhury (eds.), *Water, Flood Management and Water Security Under a Changing Climate*, https://doi.org/10.1007/978-3-030-47786-8_1

have produced monthly precipitation samples in connection with GNIP programme (IAEA 2019). While somewhat late, GNIP programme was started in Bangladesh since 2009 initially with two stations (Savar at central Bangladesh and Sylhet at northeast of Bangladesh) and currently there are a total of eight GNIP stations in different parts of the country to collect monthly samples.

Geographically, Bangladesh is a small delta which shares its south border with the Bay of Bengal, which contributes significantly to the incidence of summer monsoon in South Asia. This monsoon climate affects the lives of the huge number of population inhabiting this area. Vapour source and moisture transportation patterns are very important factors for understanding and simulating the seasonal climatic condition and variability of monsoons (Aggarwal et al. 2004). Thus, as a tracer (Liu et al. 2014), stable isotopes in precipitation in a local scale can bear significant information on moisture source and its transportation and it can also be important input for regional and global climate models (Bowen and Revenaugh 2003; Hughes and Crawford 2012). Moreover, detailed knowledge on stable isotopes in precipitation can facilitate better understanding of groundwater recharge (Gonfiantini et al. 1998). The current study aims to investigate the following areas: (1) Basic characteristics of stable isotopes (^{18}O and ^{2}H) and their spatiotemporal variation across Bangladesh; (2) Source of moisture and its transportation throughout the year.

2 Study Area and Methodology

Bangladesh is located within a subtropical monsoon climate region and experiences extensive seasonal variations in terms of temperature, humidity and rainfall amount. Four noticeable seasons are observed in Bangladesh: northeast monsoon or winter (December–February), pre-monsoon or summer (March–May), southwest monsoon (June–September) and post-monsoon or autumn (October–November) (Ahasan et al. 2010; Shahid 2010; Khatun et al. 2016; Mullick et al. 2019). The hottest months are April and May during the summer season, when monthly normal temperature occurs in the range of 21.1–36 °C throughout the country. The coldest month is January with a monthly normal temperature range of 10.4–27.4 °C across different parts (Mullick et al. 2019). Heavy rainfall is observed in most regions with some exceptions in the dry, western part of the country covering the Rajshahi division. About 75% of the total rainfall amount in a year is recorded during the monsoon period (Shahid 2010). Relatively lower amount of rainfall is experienced by the west-central, central and east-central parts of Bangladesh. Annual rainfall varied in the range of 1200–1400 mm in those regions for the period of 1961–2010. For the same period, the range was 2400–2600 mm in the northeastern region and 3200–3400 mm for southern and southeastern parts (Ahasan et al. 2010).

Bangladesh maintains a total of eight GNIP stations for precipitation sample collection aiming isotopic analysis. The stations are located in different climatic and geographic zones of the country namely, Dhaka (SR), Chuadanga (CN), Dinajpur (DN), Barishal (BS), Satkhira (KH), Sylhet (HT), Bandarban (BN) and Cox's Bazar (XB). Dhaka (SR) is in the central part of Bangladesh; BS, KH and XB are three

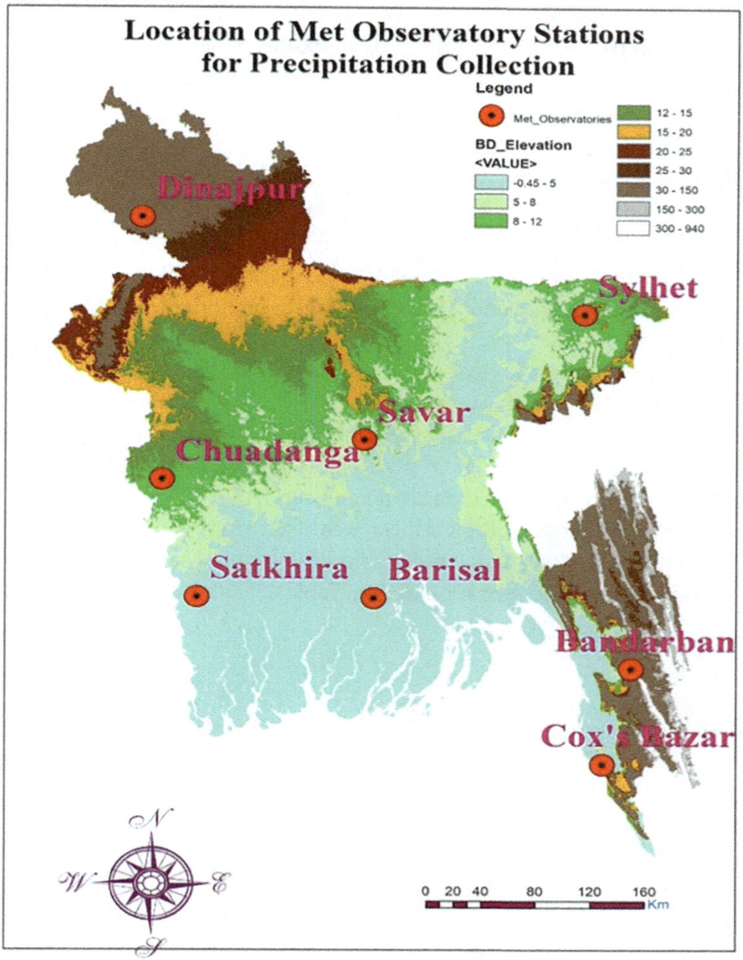

Fig. 1 Location Map of GNIP stations in Bangladesh

coastal stations; DN is in the northwestern part of the country; CN is located in the south central part of Bangladesh. Location of the current GNIP stations in Bangladesh is portrayed in Fig. 1.

For the sample collection process, mainly two types of rainfall collectors are used in Bangladesh. The old setup is comprised of a High Density Polyethylene (HDPE) bottle and a funnel while a Ping-Pong ball is placed at the funnel mouth to prevent evaporation during rainfall. Recently three of the stations (XB, HT, SR) are using PALMAX (Croatia) rainfall collector. Whichever the type, the rain collector is placed in an open space to collect the rain. Collected rainwater is immediately transferred after each rainfall event to a bottle which is kept sealed and stored inside a room. At the end of the month, monthly composite sample is prepared from the collected event based samples (IAEA 2014). Analysis was done by Liquid Water Isotopic Analyzer (LWIA) in Isotope Hydrology Laboratory of IAEA.

The isotopic results are expressed as δ-values and in ‰ (per mil.) unit, which are relative to VSMOW (Vienna Standard Mean Ocean Water) on a normalized scale.

$$\delta\ (‰) = \left(R_{sample} - R_{standard}\right)/R_{standard} \times 1000,$$

where R refers to the $^{18}O/^{16}O$ or $^{2}H/^{1}H$ ratio.

Results of the isotopic analysis and meteorological data were collected from WISER database of IAEA. Total 297 monthly samples have been used in this study from those stations having the collection period of 2009–2017. Most of the stations were operational since 2014 while observation of HT and SR stations were started from 2009.

Observation of meteorological variables, including the precipitation amount (P, mm), temperature (T, °C), relative humidity (RH, %) and vapour pressure (Vp, hPa) were available at each GNIP stations except in BN. BN shares the meteorological data from its nearest meteorological observatory station in Rangamati. All isotopic and most of the meteorological data are taken from IAEA-WISER database (WISER 2018). A portion of meteorological data was used from Reliable Prognosis (RP5 2018). A summary of two major meteorological controls, temperature and rainfall for the GNIP stations has been provided in Table 1.

3 Results

3.1 Variation in δ^2H and $\delta^{18}O$

The overall isotopic composition in precipitation (maximum, minimum and average) has been summarized in Table 2. The patterns of stable isotopic composition has been found to be different in different seasons in Bangladesh. The usual isotopic trend for June–September is: XB > CN > DN > SR > BN > KH > HT > BS. The major portion of rain occurs for a year during this period in Bangladesh. For the rest of the year the trend changes into: DN > KH > SR > BS > CN > BN > HT > XB. No distinct difference is found among the coastal and other stations in both the seasons. The mean monthly isotopic variation occurs within a very small range ($\delta^{18}O = -6.998$‰ to -5.58‰, arithmetic average) during June–September. However, Cox's Bazar shows the most enriched signature during the monsoon season while the most depleted during the non-monsoon months. This reflects the effect of seasonality and change of moisture source in this station. Satkhira and Barisal are the other two coastal stations, but showing relatively depleted $\delta^{18}O$ values, especially during monsoon months. The weighted δp values show a somewhat surprising order: HT > XB > SR > CN > DN > KH > BN > BS. The reason for Sylhet station to be in the first position may be attributed to high rainfall during the months of April and May, which is relatively enriched. The reason of KH and BS stations to be depleted in isotopes may be a result of mixing with recycled vapour.

Table 1 Summary data of rainfall and temperature of the eight rainfall stations of Bangladesh (Climate data 2019)

Rainfall station name	Yearly avg. temp. (°C)	Yearly rainfall avg. (mm)	Lowest rainfall avg. (mm)	Lowest rainfall avg. month	Highest rainfall avg. (mm)	Highest rainfall avg. month	Lowest temp. avg. (°C)	Highest temp. avg. (°C)	Temp. variation over the year (°C)
Dhaka	25.8	1990	6	December	372	July	18.8	28.9	10.1
Sylhet	24.8	3876	8	December	780	June	18.5	28.1	9.6
Chuadanga	26.2	1422	3	December	287	July	18.7	30.4	11.7
Satkhira	26.2	1655	6	December	347	July	19.1	30.2	11.1
Dinajpur	25.0	1728	1	December	402	July	17.4	28.8	11.4
Barisal	25.9	2184	10	January	444	July	19.2	30.3	11.1
Cox's Bazar	25.6	3770	7	January	1060	July	20.5	28.5	8.0
Bandarban	25.9	2528	6	January	570	July	20.5	28.8	8.3

Table 2 Summary of stable isotopic data in precipitation of different stations of Bangladesh

	$\delta^{18}O$ (‰)			$\delta^{2}H$ (‰)		
Station ID	MIN	MAX	AVG	MIN	MAX	AVG
BN	−10.72	−0.33	−5.34	−75.70	6.50	−32.42
BS	−13.28	0.19	−5.54	−97.80	10.50	−33.25
CN	−12.07	0.83	−4.43	−87.60	14.60	−25.50
DN	−11.35	1.13	−4.67	−80.66	25.60	−25.44
HT	−12.66	1.23	−4.76	−92.60	25.30	−27.34
KH	−10.93	1.24	−4.32	−76.99	12.32	−23.94
SR	−14.35	1.97	−4.68	−107.36	20.90	−28.13
XB	−10.00	−0.18	−5.35	−67.40	8.60	−32.52

3.2 Seasonal Variation

Figure 2 describes the monthly $\delta^{18}O$ variation for different years in the observed stations. The first 5 months of the year, the samples are found to be more enriched than the monsoon rain samples. DN, SR and BN stations show a linear decreasing trend from the start to end of the year in these stations. In the other five stations, not much variation occurs within the months of January to May, while a marked depletion occurs just after May. From June to September in these five stations, the $\delta^{18}O$ values are rather stable showing no distinct trends and mostly varying between −10‰ and −6‰ except Cox's Bazar. Cox's Bazar gives relatively enriched value (−8‰ to −2.76‰ usually) in that time period.

3.3 Construction of Local Meteoric Water Lines

Craig (1961) constructed a linear $\delta^{2}H$-$\delta^{18}O$ relationship on the basis of 400 water samples collected from rivers, lakes, and precipitation which is popularly known as Global Meteoric Water Line, GMWL ($\delta^{2}H = 8 \times \delta^{18}O + 10$) (Craig 1961). Collected precipitation samples in connection to this study were plotted against Craig's GMWL as shown in Figs. 3 and 4. In both marine and non-marine stations, the monsoon samples were very close to the GMWL (Fig. 3), while some divergence is observed in the non-monsoon season (Fig. 4). Local Meteoric Water Lines (LMWLs) were constructed for each stations. LMWL for each station has been presented in Table 3. The local meteoric lines are much similar to GMWL with the slope very close to 8, while DN and XB show some deviations in slope and intercept respectively. The positioning of the samples against GMWL and LMWLs clearly demonstrates the equilibrium condensation of vapour in cloud (Dansgaard 1964; Gat 1996; Managave et al. 2015).

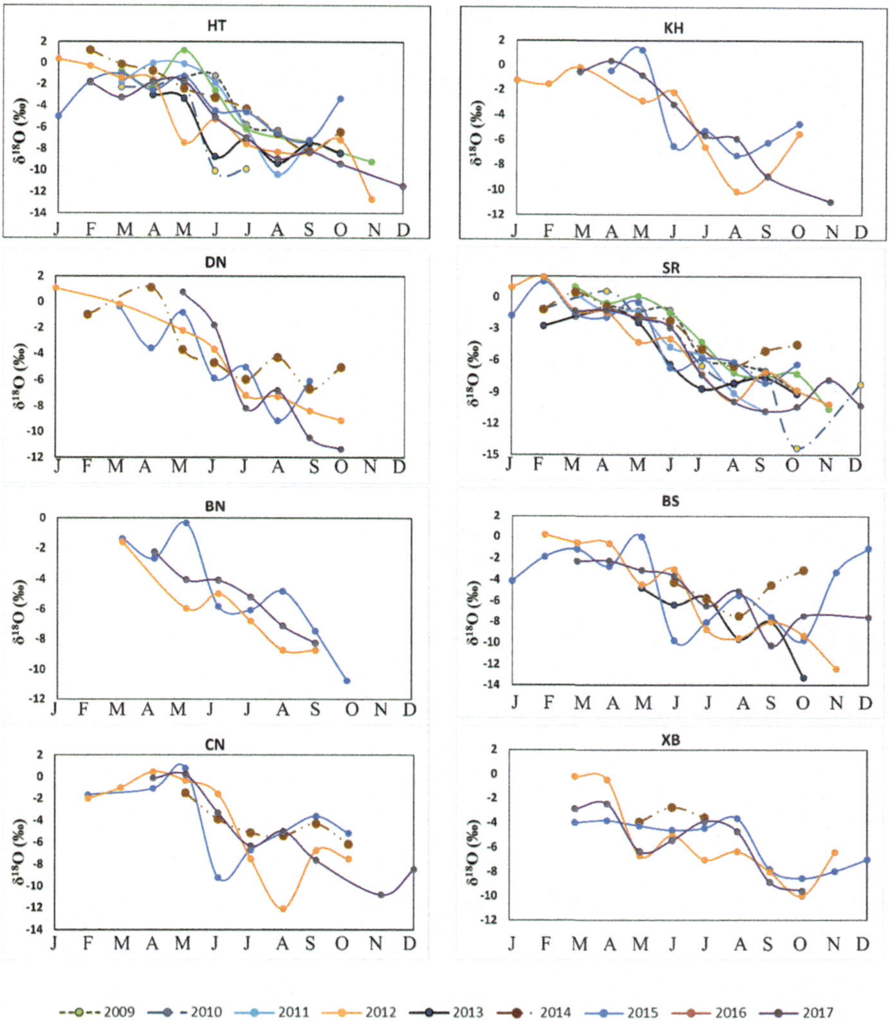

Fig. 2 Monthly Stable Isotopic trend in GNIP stations of Bangladesh

3.4 Deuterium Excess

The deuterium excess (or d-excess), defined by d-excess (‰) = δ^2H-8· δ^{18}O (Dansgaard 1964) is an important parameter which is specifically sensitive to the conditions in determining moisture source conditions and the effect of evaporation on particular water types (Pfahl and Sodemann 2014). The distribution of d-excess (Fig. 5) over different months of a year show wide dispersion from January to April and smaller variation during the monsoon months. Some samples are falling below the line maintaining d-excess = 10‰, mostly during the first half of the year, while

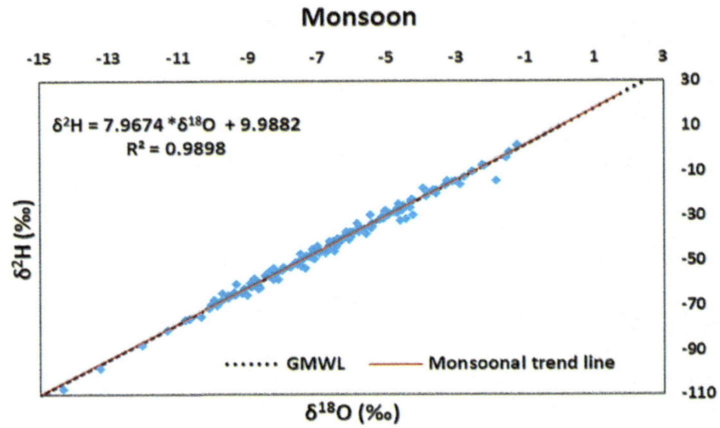

Fig. 3 Monsoon samples against GMWL

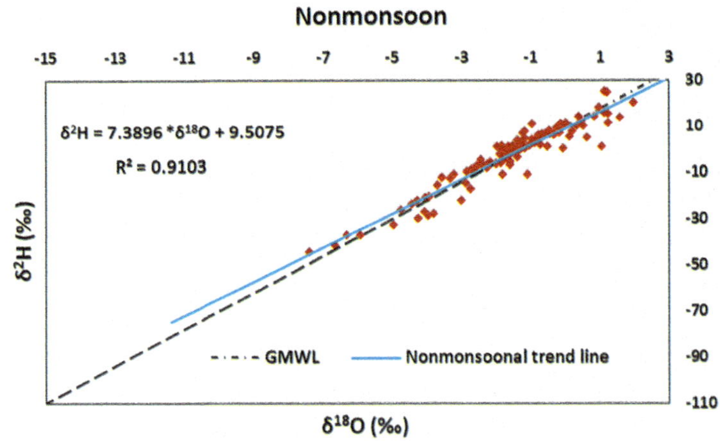

Fig. 4 Non-monsoon samples against GMWL

Table 3 Local Meteoric Water Lines of GNIP stations of Bangladesh

Station	LMWL equation
Satkhira (KH)	$\delta^2H = 7.94 \times \delta^{18}O + 10.41$
Cox's Bazar (XB)	$\delta^2H = 7.68 \times \delta^{18}O + 8.56$
Barishal (BS)	$\delta^2H = 8.06 \times \delta^{18}O + 11.33$
Dhaka (SR)	$\delta^2H = 7.82 \times \delta^{18}O + 8.19$
Sylhet (HT),	$\delta^2H = 8.05 \times \delta^{18}O + 10.95$
Dinajpur (DN),	$\delta^2H = 8.23 \times \delta^{18}O + 13.29$
Bandarban (BN)	$\delta^2H = 8.19 \times \delta^{18}O + 11.29$
Chuadanga (CN)	$\delta^2H = 8.01 \times \delta^{18}O + 10.05$
Combined (all stations)	$\delta^2H = 7.97 \times \delta^{18}O + 10.15$

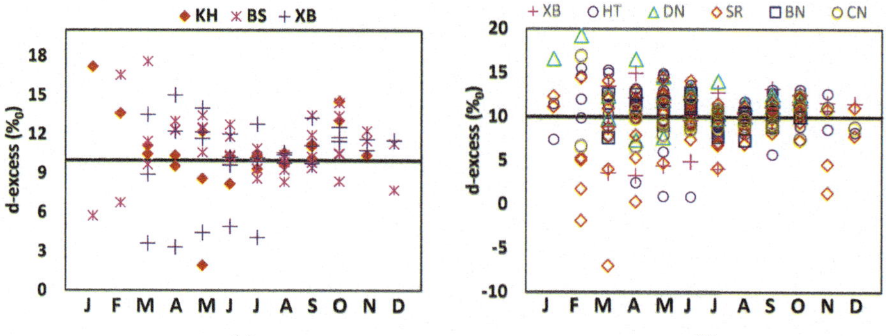

Fig. 5 Deuterium excess variation over the year (**a**) d-excess variation in coastal stations (**b**) d-excess variation in nonmarine stations (XB is included to compare)

the major portion falls above the line. Precipitation with higher d-excess values (>10‰) indicates intense evaporation (Managave et al. 2015; Dublyansky et al. 2018) while smaller d-excess values result from recycling events (Datta et al. 1991). Therefore, in non-monsoonal rain, mixture from evaporation and recycled rain can be projected from deuterium excess values. In the period of July to September, when the maximum share of the monsoonal rain occurs, the d-excess values are quite similar to the XB station or other marine stations, confirming the similar source of vapour without significant recycling events or evaporation.

3.5 *Meteorological Controls*

All eight GNIP stations in Bangladesh are located in the tropical climate region. The maximum monthly average temperature is 26.2 °C in Chuadanga and Satkhira regions, while the yearly highest average rainfall is 3876 mm in Sylhet region. The lowest precipitation is observed in December and January months, with the lowest average of 1 mm in Dinajpur. The month of July sees the highest amount of precipitation with the highest average of 1060 mm recorded in Cox's Bazar. The maximum monthly average temperature is found in Chuadanga as 30.4 °C during May. In January, the lowest monthly average temperature is recorded as 17.4 °C in Dinajpur. The difference between maximum precipitation (during wettest months) and minimum precipitation (during driest months) are 1053 mm in Cox's Bazar and 284 mm in Chuadanga respectively. The highest and lowest average temperatures vary by ~11.7 °C (Chuadanga) and ~8.0 °C (Cox's Bazar) throughout the year. High humidity is the common feature of monsoon period while relatively lower humidity is evidenced in pre or post-monsoon seasons (Khatun et al. 2016; Mullick et al. 2019). RH range during the data collection period at different stations vary from 55% to 90%. Stable isotopic relation with temperature and precipitation is shown in

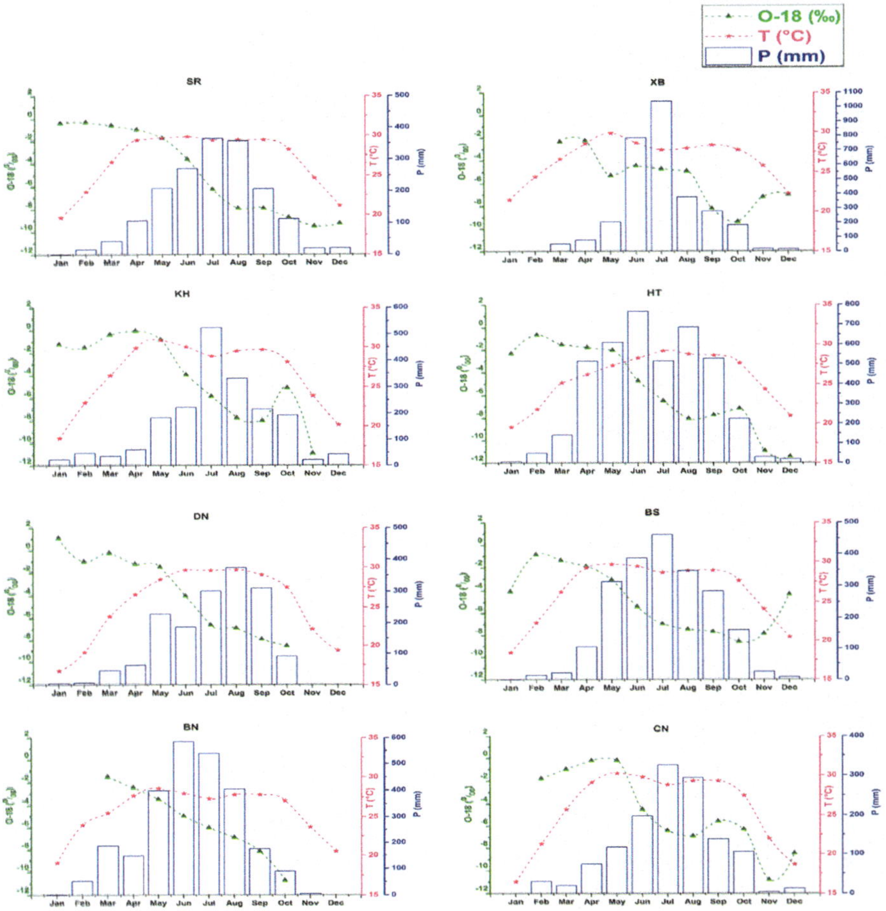

Fig. 6 Combined plot of temperature, precipitation and $\delta^{18}O$

Fig. 6. Figure 6 indicates that change in isotopic signature does not maintain the same phase all throughout the year. Hence, correlation coefficients were considered differently for January–May and June–October for Table 4, where isotopic correlations with meteorological controls had been determined. Meteorological data was considered only for the isotopic sampling period for this table. Values higher than 0.3 were considered significant and marked bold. Table 4 shows wide variations in the signs and values of correlation coefficients according to stations and seasons. Details on correlation coefficients have been discussed in sections below.

Table 4 Correlation coefficients between $\delta^{18}O$ and meteorological factors

Station ID	January to May				June to October			
	Temp.	Rainfall amount	Relative humidity	Vapor pressure	Temp.	Rainfall amount	Relative humidity	Vapor pressure
KH	0.26	**−0.70**	0.26	0.23	−0.11	−0.13	**−0.30**	−0.23
BS	0.17	**−0.71**	**−0.59**	−0.07	**0.44**	**−0.40**	**−0.42**	0.18
HT	−0.05	**−0.34**	**−0.34**	−0.23	−0.22	0.14	−0.14	**−0.31**
DN	**−0.42**	**−0.78**	−0.19	**−0.53**	**0.31**	**−0.36**	**−0.53**	−0.21
SR	−0.19	**−0.42**	**−0.48**	−0.29	**0.36**	−0.11	**−0.50**	−0.12
XB	−0.24	**−0.67**	−0.28	−0.27	0.07	**0.34**	−0.03	**0.39**
BN	**−0.31**	**−0.44**	**−0.38**	**−0.37**	0.19	**0.41**	0.18	0.03
CN	**0.64**	**0.30**	−0.07	**0.68**	0.12	**−0.53**	−0.24	0.06

3.5.1 Temperature Effect

Dansgaard (1964) found strong correlation for the plains stations between the annual mean $\delta^{18}O$ and temperature on a global scale, with an average change of 0.7‰/°C. However, the relationship between temperature and isotopic variation depends on the way how the values of temperature are averaged, i.e. weekly mean, monthly mean etc. (Ichiyanagi 2007).

Monthly mean temperature data was used to find the correlation with $\delta^{18}O$ in this study. The mean monthly temperature variation occurs in the range of approximately 16–30 °C during the sampling periods and inter station variation is rather negligible. Temperature was found to be least significant in most cases. For the period of January to May, maximum temperature dependence (41%) was observed in CN station showing positive correlation. DN station shows negative correlation and around 18% of the isotopic variation is associated with temperature during that period. The other stations have mostly negative correlation and also insignificant temperature dependence. Lower temperature dependence is observed during the monsoon period (June–October). Mean monthly temperature varies between approximately 27 and 31 °C during this time period. Most of the stations show positive correlation during this time period unlike non-monsoon season. Maximum temperature influenced isotopic variation occurs in BS (19%).

3.5.2 Amount Effect

The potential influence of monthly rainfall amount to change its isotopic composition is termed as "amount effect" which was first observed by Dansgaard (1964). For the first quarter of the year, average monthly total rainfall ranges from 0 to 55 mm in all stations except in HT where it reaches up to ≈145 mm. The precipitation amount shifts to higher range in later months up to October while the maximum monthly rainfall occurs in XB during July. Amount effect is much stronger during non-monsoon months. All the stations maintain negative correlation with precipitation amount except CN during January to May. Significant and high amount effect ($r \approx -0.7$) is observed in DN, KH, BS and XB stations, while relatively lower amount effect ($r \approx -0.5$ to -0.3) in the other ones. In June to October the amount effect seems to be less significant. The major isotopic variation ($r = -0.53$) associated to rainfall amount occurs in CN station. However "anti-amount effect" (Tian et al. 2018) is also observed in some stations which includes XB ($r = 0.34$), BN ($r = 0.41$) and HT ($r = 0.14$) stations.

3.5.3 Humidity Effect

Relative humidity alongside sea surface temperature, were found to be important factors that influence the isotopic signature of precipitation (Dansgaard 1964; Merlivat and Jouzel 1979; Clark and Fritz 1997; Lachniet 2009; Duy et al. 2018). High d-excess values are likely to occur in lower humid conditions, when rapid evaporation occurs in combination with stronger kinetic isotope effects at that time. On the other hand low d-excess values occur during high humid conditions as a result of slow evaporation (Lachniet 2009; Dublyansky et al. 2018).

Among all the parameters presented here, relative humidity seems to have the largest effect on the spatiotemporal variation of stable isotopic composition of precipitation as seen from Eq. (1). However, monsoonal and non-monsoonal effect due to RH has been presented in Table 4. RH makes mostly negative correlation with $\delta^{18}O$ except two instances. BS and SR stations show significant correlation between RH and $\delta^{18}O$ in both seasons.

$$\delta^{18}O = -1.293 * RH + 73.73 \qquad r = -0.5 \tag{1}$$

Almost no correlation ($r = -0.01$) was found between d-excess and RH. This may be possible since, both type of d-excess (high and low) is seen from Fig. 5a, b especially, in non-monsoon period. Therefore, the occurrence and mixture of precipitation from different moisture sources in non-monsoon months can be evidenced from here.

3.5.4 Vapour Pressure Effect

Vapour pressure (Vp) is a meteorological parameter which is connected to both RH and air temperature. In most of the cases the nature of influence (positive or negative r value) is synchronized with temperature. Effect of Vp is more prominent in non-monsoon period rather than the monsoon period. 46% of isotopic variation in CN station was found due to changes in vapour pressure during January to May. DN station also had a remarkable influence from Vp (28%), while unlike CN station, DN has negative correlation. During the monsoon period, most of the stations have positive or negligible correlation.

4 Conclusion

A total of 297 monthly data from GNIP stations in Bangladesh have been used in this study to observe the stable isotopic scenario of precipitation. Samples were very close to GMWL, when $\delta^{18}O$ values were plotted against $\delta^{2}H$. The overall LMWL for Bangladesh was found to be: $\delta^{2}H = 7.97 \times \delta^{18}O + 10.15$, which shows strong inclination to GMWL and equilibrium evaporation. Furthermore, it can be helpful to detect the recharge period and mechanisms related to groundwater for this region. For most instances, successive depletion of heavy isotopes have been observed from January to December. High humid conditions in monsoon season was found to generate greater amount of rainfall with slow rates of evaporation. Pre or post-monsoon rain results from different vapour sources which also includes fast evaporated moisture in low humid conditions. Amount effect is more prominent during January–May, while it occurs reversely during June–October in some stations. No single parameter could be established as a dominating one throughout the year in any of the stations.

Acknowledgement The authors acknowledge IAEA and Dr. Stefan Terzer, Isotope Hydrology section, IAEA for the help in providing stable isotopic data. Meteorological data was collected from BMD. The authors are also thankful to BMD for their support in GNIP sample collection work.

References

Aggarwal, P. K., Froehlich, K., Kulkarni, K. M., & Gourcy, L. L. (2004). Stable isotope evidence for moisture sources in the Asian summer monsoon under present and past climate regimes. *Geophysical Research Letters, L08203*(31), 2–5. https://doi.org/10.1029/2004GL019911.

Ahasan, M. N., Chowdhary, A. M., & Quadir, D. A. (2010). Variability and trends of summer monsoon rainfall over Bangladesh. *Journal of Hydrology and Meteorology, 7*(1), 1–17.

Araguas-Araguas, L., Froehlich, K., & Rozanski, K. (1998). Stable isotope composition of precipitation over southeast Asia. *Journal of Geophysical Research, 103*, 28721–28742.

Bowen, G. J., & Revenaugh, J. (2003). Interpolating the isotopic composition of modern meteoric precipitation. *Water Resources Research, 39*(10), 1–13. https://doi.org/10.1029/2003WR002086.

Chen, F., Zhang, M., Wang, S., Qiu, X., & Du, M. (2017). Environmental controls on stable isotopes of precipitation in Lanzhou, China: An enhanced network at city scale. *Science of the Total Environment, 609*, e1013–e1022. https://doi.org/10.1016/j.scitotenv.2017.07.216.

Clark, I. D., & Fritz, P. (1997). *Environmental isotopes in hydrogeology*. New York: CRC Press.

Climate Data. (2019). *Climate data for cities worldwide*. https://en.climate-data.org/

Craig, H. (1961). Isotopic variations in meteoric waters. *Science, 133*, 1702–1703.

Dansgaard, B. W. (1964). Stable isotopes in precipitation. *Tellus, 16*, 436–468.

Datta, P. S., Tyagi, S. K., & Chandrasekharan, H. (1991). Factors controlling stable isotope composition of rainfall in New Delhi, India. *Journal of Hydrology, 128*, 223–236.

Dublyansky, Y. V., Klimchouk, A. B., Tokarev, S. V., Amelichev, G. N., Langhamer, L., & Spötl, C. (2018). Stable isotopic composition of atmospheric precipitation on the Crimean peninsula and its controlling factors. *Journal of Hydrology, 565*(July), 61–73. https://doi.org/10.1016/j.jhydrol.2018.08.006.

Duy, N. L., Heidbüchel, I., Meyer, H., Merz, B., & Apel, H. (2018). What controls the stable isotope composition of precipitation in the Mekong Delta? A model-based statistical approach. *Hydrology and Earth System Sciences, 22*, 1239–1262.

Gat, J. R. (1996). Oxygen and hydrogen isotopes in the hydrologic cycle. *Annual Review of Earth and Planetary Sciences, 24*, 225–262.

Gonfiantini, R., Fröhlich, K., Araguas-Araguas, L., & Rozanski, K. (1998). Isotopes in groundwater hydrology. In *Isotope tracers in catchment hydrology* (pp. 203–246). Amsterdam: Elsevier Science.

Hughes, C. E., & Crawford, J. (2012). A new precipitation weighted method for determining the meteoric water line for hydrological applications demonstrated using Australian and global GNIP data. *Journal of Hydrology, 464–465*, 344–351. https://doi.org/10.1016/j.jhydrol.2012.07.029.

IAEA. (2014). *IAEA/GNIP precipitation sampling guide*. Vienna: International Atomic Energy Agency.

IAEA. (2019). *International Atomic Energy Agency (IAEA): Isotope hydrology information system*. The ISOHIS Database. http://isohis.iaea.org

Ichiyanagi, K. (2007). Review: Studies and applications of stable isotopes in precipitation. *Journal of Japanese Association of Hydrological Sciences, 37*(24), 165–185.

Khatun, M., Rashid, M., & Hygen, H. (2016). *Climate of Bangladesh* (pp. 17–60). MET Report No. 08/2016. Norwegian Meteorological Institute and Bangladesh Meteorological Department: Climate of Bangladesh.

Lachniet, M. S. (2009). Sea surface temperature control on the stable isotopic composition of rainfall in Panama. *Geophysical Research Letters, 36*, L03701. https://doi.org/10.1029/2008GL036625.

Liu, J., Song, X., Yuan, G., Sun, X., & Yang, L. (2014). Chemical and physical meteorology stable isotopic compositions of precipitation in China stable isotopic compositions of precipitation in China. *Tellus B, 66*, 22567. https://doi.org/10.3402/tellusb.v66.22567.

Managave, S. R., Jani, R. A., Rao, T. N., Sunilkumar, K., Satheeshkumar, S., & Ramesh, R. (2015). Intra-event isotope and raindrop size data of tropical rain reveal effects concealed by event averaged data. *Climate Dynamics, 47*(3–4), 981–987.

Merlivat, L., & Jouzel, J. (1979). Global climatic interpretation of the deuterium–oxygen 18 relationship for precipitation. *Journal of Geophysical Research, Oceans, 84*, 5029–5033.

Mullick, R. A., Nur, R. M., Alam, J., & Islam, K. M. A. (2019). Observed trends in temperature and rainfall in Bangladesh using pre-whitening approach. *Global and Planetary Change, 172*, 104–113.

Pfahl, S., & Sodemann, H. (2014). What controls deuterium excess in global precipitation? *Climate of the Past, 10*, 771–781. https://doi.org/10.5194/cp-10-771-2014.

Rozanski, K., Araguás-Araguás, L., & Gonfiantini, R. (1993). Isotopic patterns in modern global precipitation. *Climate Change in Continental Isotopic Records, Geophysical Monograph, American Geophysical Union, 78*, 1–36. https://doi.org/10.1029/GM078p0001.

RP5. (2018). *Reliable prognosis*. https://rp5.ru/Weather in_the_world

Shahid, S. (2010). Recent trends in the climate of Bangladesh. *Climate Research, 42*, 185–193.

Tian, C., Wang, L., Kaseke, K. F., & Bird, B. W. (2018). Stable isotope compositions (δ^2H, δ^{18}O and δ^{17}O) of rainfall and snowfall in the Central United States. *Scientific Reports, 8*(6712), 1–15. https://doi.org/10.1038/s41598-018-25102-7.

Vuille, M., Werner, M., Bradley, R. S., & Keimig, F. (2005). Stable isotopes in precipitation in the Asian monsoon region. *Journal of Geophysical Research-Atmospheres, 110*(23), 1–15.

WISER. (2018). *Water isotope system for data analysis, visualization and electronic retrieval*. https://nucleus.iaea.org/wiser

Impact of Internal Road Network on Water-Logging Inside Polders

Anika Tahsin, Sadmina Razzaque, Anisul Haque, Imran Hossain Newton, Abul Fazal M. Saleh, Rowshan Mamtaz, Md Ibnul Hasan, Md. Aminul Islam Khan, Flavia Simona Cosoveanu, and Cecilia Borgia

1 Introduction

During 1960s, polders (139 in total) were built in the coastal region of Bangladesh to protect the land from tides, storm surges, and salinity. Almost after 50 years of its construction, it is now popularly believed that polders are responsible for wide spreading water-logging inside the protected land (Tutu 2005; Noor 2018). The most accepted theory behind this belief is – the vast network of tide-dominated morphologically active estuaries brings a large volume of sediments from the sea into the estuarine systems (Rahman et al. 2015). A major part of these sediments is supposed to be deposited in the floodplain which has been hindered after polders were built and this leads to large scale sedimentation in the peripheral rivers of the polders, reducing the drainage capacity. As such, the area is now facing severe problems of water logging (Gain et al. 2017). Based on this theory, most of the studies mainly focused on the dredging (natural such as TRM or manual such as mechanized dredging) of peripheral rivers as a possible remedy of water-logging (Shampa and Paramanik 2012). None of these studies have considered the impact of the internal road network inside polders that obstruct the natural drainage route and ultimately causes water-logging. This study, for the first time, focused on this issue along with the impact of conveyance of peripheral rivers on water-logging inside polders. A numerical model (Delft3D) is applied in polder-26 (Fig. 1) where water-

A. Tahsin (✉) · S. Razzaque · A. Haque · I. H. Newton · A. F. M. Saleh · M. A. I. Khan
Institute of Water and Flood Management (IWFM), Bangladesh University of Engineering and Technology (BUET), Dhaka, Bangladesh

R. Mamtaz · M. I. Hasan
Department of Civil Engineering, BUET, Dhaka, Bangladesh

F. S. Cosoveanu · C. Borgia
MetaMeta Research, Wageningen, The Netherlands

A. Haque, A. I. A. Chowdhury (eds.), *Water, Flood Management and Water Security Under a Changing Climate*, https://doi.org/10.1007/978-3-030-47786-8_2

logging is identified as a problem (BWDB 2012). Several scenarios of drainage routes are sketched by considering basic hydrology, hydraulics and topography of the system where appropriate openings are allowed through culverts when the drainage route crosses a road. Impact of excavation of peripheral rivers are included with the culvert opening on the overall drainage pattern that reduces water-logging to an acceptable level. The culverts identified with this combination is termed as 'smart culverts'.

1.1 Study Area

Polder 26 (Fig. 1) is situated in the south-west coastal region of Bangladesh. It is located within latitudes 22^{0}-44′-40″ N and 22^{0}-48′-20″ N and longitudes 89^{0}-20′-14″ E and 89^{0}-25′-11″ E with an area of 2668 hectare. The polder was constructed by Bangladesh Water Development Board (BWDB) in 1967–1968 (CEGIS 2016). The polder is surrounded by Teligati river covering the north- west corner to the south-west part and Mora Bhadra river that used to run from the north-west corner to the south- east part of the polder. Several canals run within the poldered area which are very important in relation to drainage inside the polder. A number of internal roads are constructed inside the polder to facilitate a communication system.

The peripheral length of the polder is 28 km with an average top width of 3.8 m and crest level of 3.5 m above MSL. The side slope is maintained to ensure a 2 m hypotenuse length on both the riverside and the countryside. The setback distance is kept between 60 m and 80 m (CEGIS 2016).

Polder 26 experiences a tropical climate with an average temperature ranging from 19 °C (January) to around 29 °C (April). The maximum rainfall ever recorded in the area is 343 mm in the month of July and the lowest is 7 mm in the month of December. The area has a total population of 15,175 among which 85% are engaged in agricultural activities (BBS 2011). From the field visit, it was found that the scope

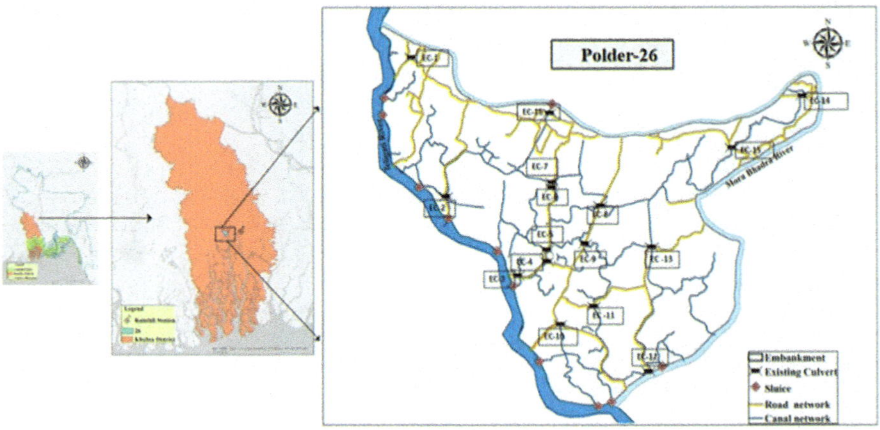

Fig. 1 Study area

of employment in agricultural sectors is gradually decreasing due to conversion of agricultural land into wetland for shrimp farming by opening the sluices and welcoming the saline water to intrude.

1.2 Existing State of Roads, Culverts and Rivers

From the DEM map (Fig. 2), it is seen that the land level of the north-western and the west-southern part is relatively lower than the north-eastern and east-southern part. Road networks and canal networks criss-cross in a complicated manner inside the polder. In the existing condition, roads and culverts are constructed within the polder without considering the land level and the drainage route. Most of the existing culverts are found to be wrongly located. Moreover, due to natural siltation and other man-made activities, many canals have lost conveyance and many culverts have lost adequate openings. Teligati is a flowing river but Mora Bhadra river lost its conveyance due to sedimentation (IWM 2015). From the field observations made by the study team it was found that due to this lost conveyance, the outfalls of the canals to Mora Bhadra river are virtually inactive. But this is not the case for the canal outfalls to the Teligati river.

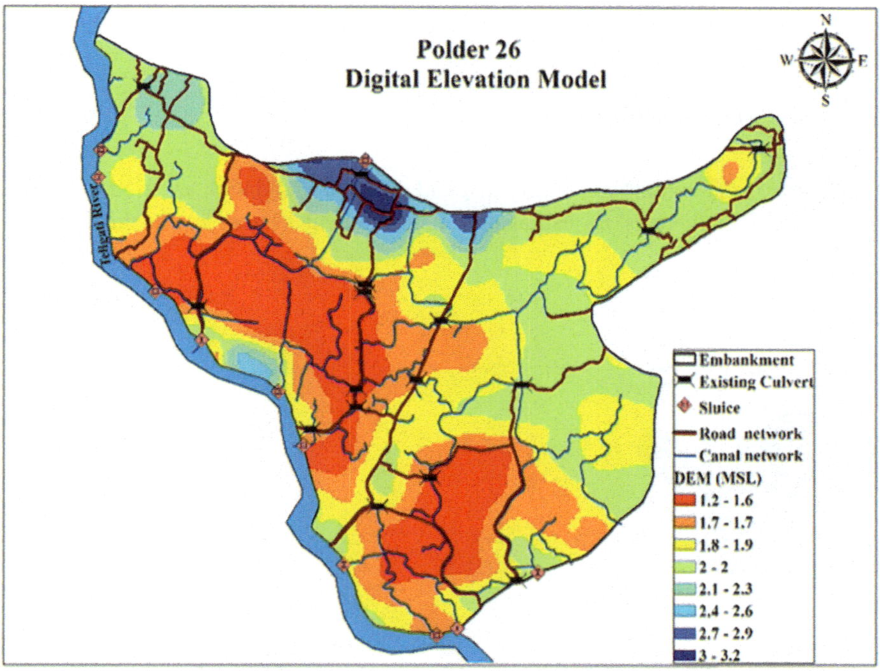

Fig. 2 DEM

2 Methodology

2.1 Hydrodynamic Modelling

An open source two-dimensional hydrodynamic model (Delft 3D) is applied to simulate the impacts of the road network, culverts and peripheral rivers on drainage inside the polder and to determine the optimum drainage route that removes water-logging from the polder. The model applied is the flow module of Delft3D modelling suite (Delft3D-FLOW 2014). The hydrodynamic model solves the two-dimensional depth-averaged equations for mass and momentum conservation and solves the system of equations by a finite difference numerical method. The model domain is bounded by the two peripheral rivers. In order to define land topography, DEM (Digital Elevation Model) constructed by using globally available SRTM (Shuttle Radar Topographic Mission) data was used. River and canal bathymetries are defined with data from Bangladesh Water Development Board, Google Earth and field surveys. Primary data on road network, canal network and locations of sluice gates are collected from mobile GPS field survey methods. Boundary conditions of the model (discharge in upstream and water level in downstream) are provided from the measured data of Bangladesh Water Development Board. To capture all the details within the model domain, grid resolutions varying from 1.5 m × 1.5 m to 15 m × 13.5 m are used (Fig. 3). This makes the total number of computational grid of the model equal to 518,700 (Fig. 3).

2.2 Model Validation

Field survey data on the inundated area and depth of inundation are used to validate the model. As there is no established measurement system in the field, data is

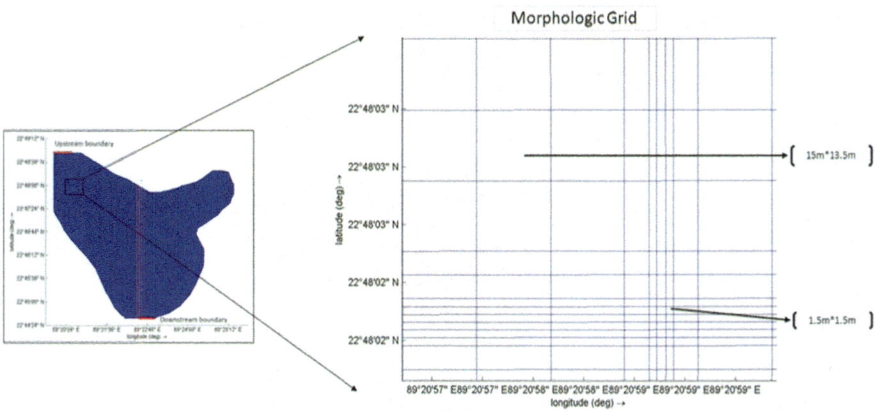

Fig. 3 Model grid

collected from visual inspection by taking the GPS coordinates, local people's perception and some in-situ judgement by the survey team. The field data collected in this way is used to validate the model by comparing the field and model inundated location, inundated area and inundation depth (Fig. 4). Inundation data collected from the field represents average inundation conditions where inundation depth varies from 60 cm to 150 cm. Model simulation accurately simulates inundation location and inundation extent. For inundation depth, a reasonable agreement is found between the field and the model by accepting the fact that field data represents an average condition (Fig. 4).

2.3 *Generation of Scenarios*

Scenarios are generated by varying the culvert numbers, culvert locations and conveyance of the peripheral river from the existing condition. In the existing conditions the following characteristics, that impact the drainage route of the system and cause water-logging, have been identified:

(a) Openings between road and canal crossings are inadequate.
(b) Some of the existing culverts on the road network are wrongly located which have no impact on the drainage route.
(c) Close to the canal outfall which may play a vital role in drainage; number of culverts are inadequate and/or wrongly placed.
(d) The conveyance of Mora Bhadra river (Fig. 1) is too low to convey the discharge coming from the internal canal systems.

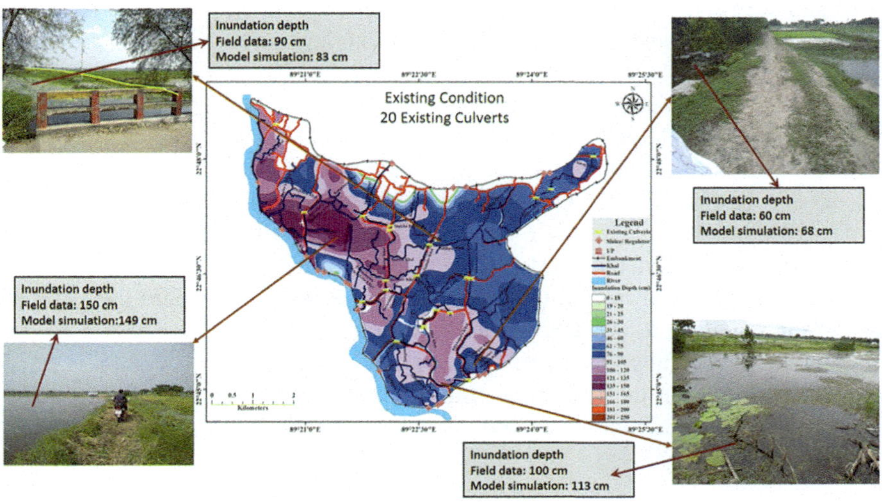

Fig. 4 Model validation results

Using different combination of (a), (b), (c) and (d), a total of six scenarios were generated (Table 1) which have been used to determine the most effective drainage route in the system to remove water-logging from the polder. In all these scenarios, the internal canal system is kept in its existing condition.

2.4 Generation of Inundation Scenarios Due to Rainfall

Water-logging in polder-26 is mainly caused by rainfall. Being a hydrodynamic model (not a hydrologic model), Delft3D cannot directly simulate the inundation due to rainfall for the entire domain. Application of available hydrologic models (for example open source HEC-HMS) can simulate the rainfall-runoff process, but those models give output in terms of discharge generated by rainfall at drainage outlets of the catchment. It is not possible to translate this output into inundation (depth and areal extent) in the area. In this study, a new modelling strategy is adopted where the total amount of rainfall volume (depending on the Intensity-Duration-Frequency or IDF curve) is distributed inside the polder based on inundation patterns formed due to an arbitrary volume of water entering the system. This creates the maximum possible inundation in the area based on the land topography, which is dynamically simulated by Delft3D. The actual inundation volume due to any rainfall is re-distributed in the system based on the ration of arbitrary volume (that causes the maximum inundation) and actual volume of rainfall for a particular IDF curve. This inundation can be equal to or less than the maximum inundation depending on the rainfall magnitude. In this way, it is possible to generate any inundation scenario accurately for any amount of rainfall. Drainage starts after a rainfall event when sluices are opened. The sluices which are operational are identified during the field visit and incorporated in the model. This drainage process is dynamically simulated by Delft3D. The validated (Fig. 4) Delft3D model is applied to simulate this drainage process.

Rainfall to the system is provided by using Intensity-Duration-Frequency (IDF) curve. To generate the IDF curve, 14 years daily rainfall data at Khulna station (see Fig. 1 for location of Khulna station. This is the nearest available rainfall station to polder-26) has been used. From the observed rainfall data, consecutive 1-day, 2-days, 3-days, 4-days, 5-days, 6-days and 7-days rainfall depth is calculated. For each of the consecutive rainfall depth, frequency analysis is performed to determine 10-years return period rainfall (10-years return period is the standard used in most designs in Bangladesh). To generate IDF curves for a 10-years return period, rainfall intensity corresponding to 10-years return period for each of the duration is calculated (Fig. 5). Using the IDF curve (Fig. 5), rainfall intensity that corresponds to 7-days continuous rainfall for a return period of 10-years is used in this study.

Table 1 Description of the scenarios

Scenario name	Number of internal culverts	Number of culverts close to canal outfall	Excavation of peripheral river	Scenario description
Existing condition	15	01	No	Inadequate openings between road and canal crossings inside the polder along with a number of wrongly placed culverts which have little or no impact on drainage. Moreover, close to canal outfall, inadequacy of number of culverts is observed. The peripheral river is silted up which makes it incapable of conveying discharge coming from inside the polder.
Scenario-1	20	01	No	In the existing condition inside the polder where the drainage rate is low the number of culverts is increased from 15 to 20. Here, the number of culverts close to canal outfall and conveyance of the peripheral river is kept same as existing condition. Objective to generate this scenario is to study the impact of increased number of internal culverts on water-logging compared to existing condition.
Scenario-2	15	03	No	Number of culverts close to canal outfall is increased from 1 to 3 in order to increase the drainage rate close to canal outfall. Number of culverts inside the polder and the conveyance of peripheral river is kept the same as existing condition. Objective of generating this scenario is to study increased number of culverts close to canal outfall on water-logging compared to existing condition.
Scenario-3	15	03	Yes	Number of culverts (both inside the culvert and close to canal outfall) is kept same as scenario-2, but the peripheral river is excavated to increase its conveyance capacity. Objective of generating this scenario is to study increased number of culverts close to canal outfall and increased conveyance of peripheral river on water-logging compare to existing condition.

(continued)

Table 1 (continued)

Scenario name	Number of internal culverts	Number of culverts close to canal outfall	Excavation of peripheral river	Scenario description
Scenario-4	20	03	No	Number of internal culverts is increased from 15 to 20, and the number of culverts close to the canal outfall increased from 1 to 3. Conveyance capacity of peripheral river remains the same as existing condition. Objective of generating this scenario is to study the impact of increased number of internal culverts and increased number of culverts close to canal outfall on water-logging compared to existing condition.
Scenario-5	20	01	Yes	Number of internal culverts is increased from 15 to 20, but number of culverts close to the canal outfall remains same as existing condition. Conveyance of peripheral river is increased. Objective of generating this scenario is to study the impact of increased number of internal culverts and increased conveyance of peripheral river on water-logging compared to existing condition.
Scenario-6	20	03	Yes	Number of internal culverts is increased from 15 to 20, number of culverts close to the canal outfall is increased from 1 to 3 and peripheral river is excavated to increase its conveyance capacity. Objective of generating this scenario is to study impact of the most favourable condition on water-logging compared to existing condition.

3 Results and Discussions

Drainage patterns that determines magnitude and extent of water-logging is simulated for all the scenarios shown in Table 1. Simulation is started after the rainfall event has stopped and when all the operational sluices are opened (data from field information). Each of the scenarios simulate inundation conditions and the associated drainage routes that produce this inundation. It is assumed that if no measure is taken to increase the drainage rate, the inundation condition will ultimately transform into a water-logged situation. From simulation results of all the 6 scenarios, an

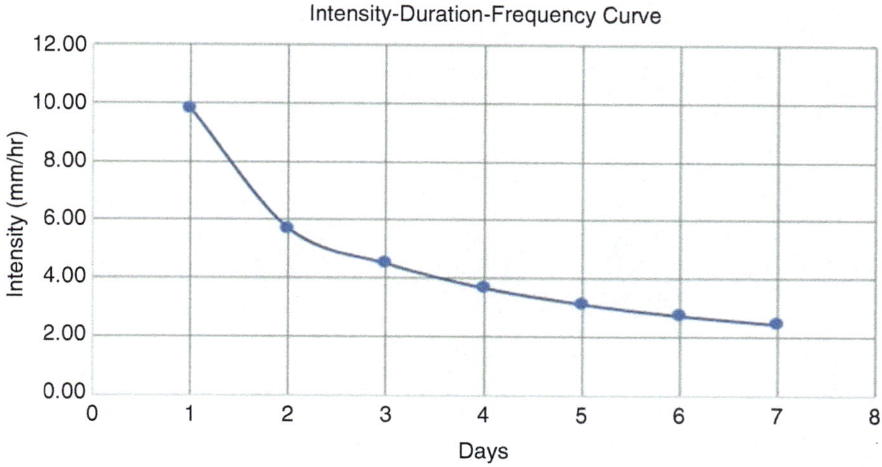

Fig. 5 IDF curve used to generate rainfall

optimum drainage route is found that ensures maximum rate of drainage and removes any possibility of water-logging inside the polder. The inundation that remains after the implementation of this optimum drainage route represents a perennial water body, not a water-logged situation (this is validated using people's perception from the field).

In all the scenarios, two types of culverts are considered – (1) Existing Culvert (EC) and (2) Smart Culvert (SC). Existing culverts are the culverts that already exist in the system. Smart culverts are the culverts which are placed in strategic locations by considering overall hydrology and hydraulics of the system and contribute to reducing water-logging to an acceptable level. Here SC-1, 2, 3,4,5,6 indicate internal smart culverts and SC-A and SC-B indicate smart culverts close to canal outfall.

Inundation depth inside the polder is divided into two land types: F0 (0–30 cm) and F1 (31–90 cm). Based on location and the water-logged condition, the polder is divided into 3 zones. Zones-1 & 2 are in the northern part of the polder whereas Zone-3 is in the southern part of the polder. Zone-1 represents the zone where water-logging is the maximum. A perennial water body also exists in this zone. Inundation is relatively less in Zone-3 whereas, Zone-2 represents the minimum inundated area.

3.1 *Existing Condition*

In the existing condition, a total of 16 exiting culverts (EC) are placed mainly at the canal-road intersection to allow drainage so that water can ultimately leave the polder to the peripheral rivers (Fig. 6). As mentioned before, all these culverts are placed without any understanding of the hydrology and hydraulics of the system. Due to this, it is found that EC-1 plays no role on the overall drainage of the system

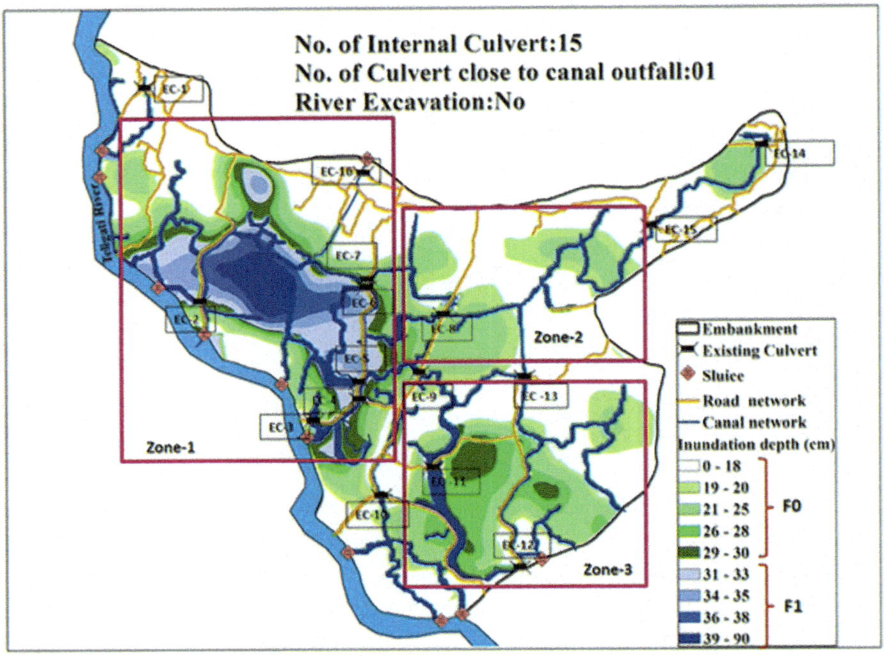

Fig. 6 Inundation map of existing condition

(Figs. 6 and 7). As Mora Bhadra river (Fig. 1) has no conveyance capacity, EC-12 and EC-16 are found to be non-functional (Figs. 6 and 7).

The zone adjacent to Teligati river is lower than the east side (data from both DEM and field visit). So, water flows towards the culverts from this region and the water flowing from inside of the polder drains out through EC-2, EC-3 and EC-14. Any obstacle of flow towards this direction creates water logging. This phenomenon is observed from the flow direction (Fig. 7) inside the polder in existing condition. Low drainage rate towards other existing culverts consequently lowers the drainage rate towards EC-2, EC-3 and EC-12 (Fig. 7). It was found that the slow drainage rate results in water-logging areas in Zone-1, Zone-2 and Zone-3. The extent of F0 and F1 inundated area is seen in Fig. 6.

3.2 *Scenario-1*

After identifying the water-logged areas and the low drainage rate in existing conditions, five locations inside the polders were selected for placement of 'smart culverts' (SC). Model simulation results with these five smart culverts shows less inundation compared to the existing condition in each of the three zones (Fig. 8).

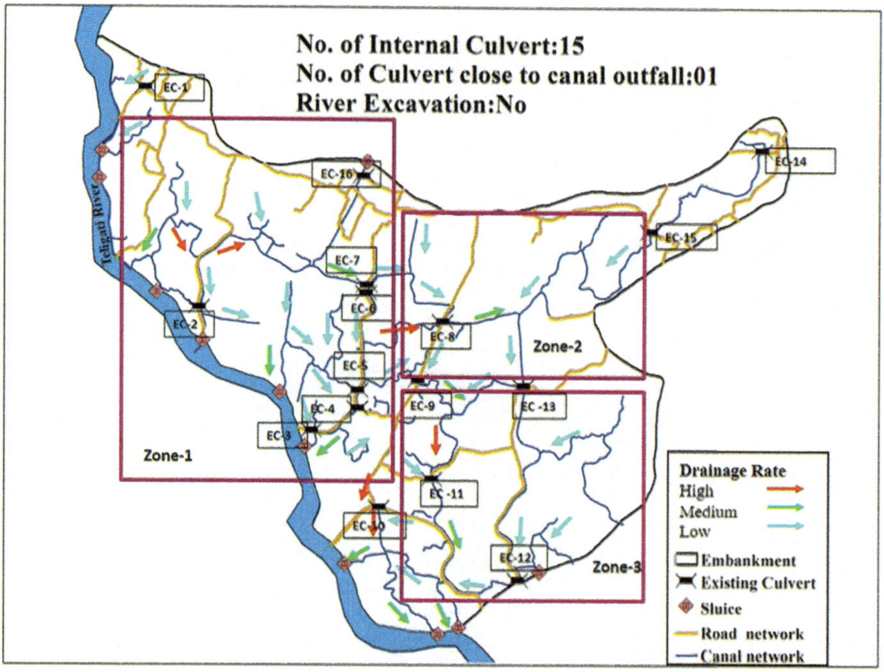

Fig. 7 Flow direction and drainage rate in existing condition

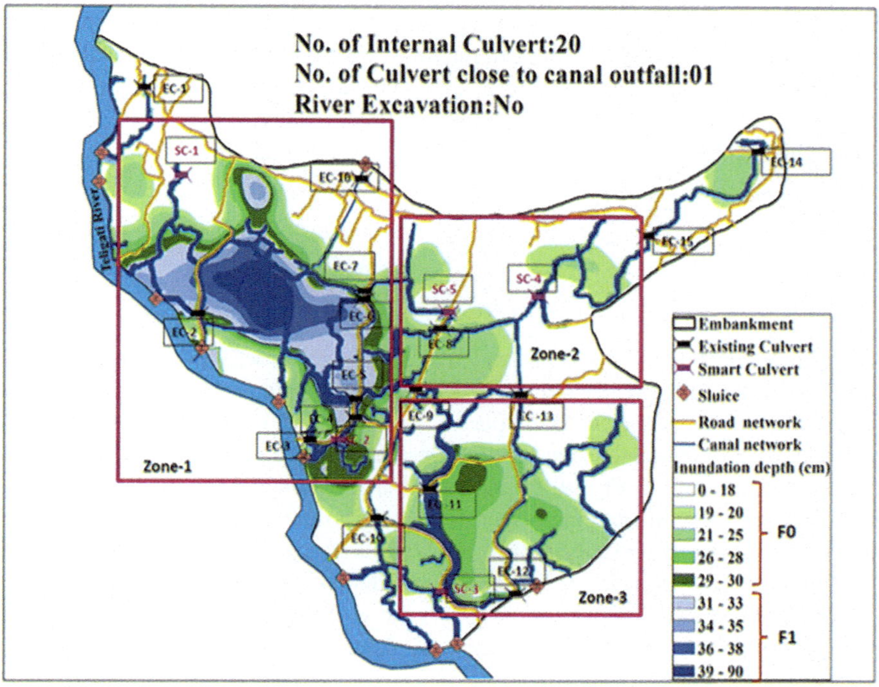

Fig. 8 Inundation map of scenario-1

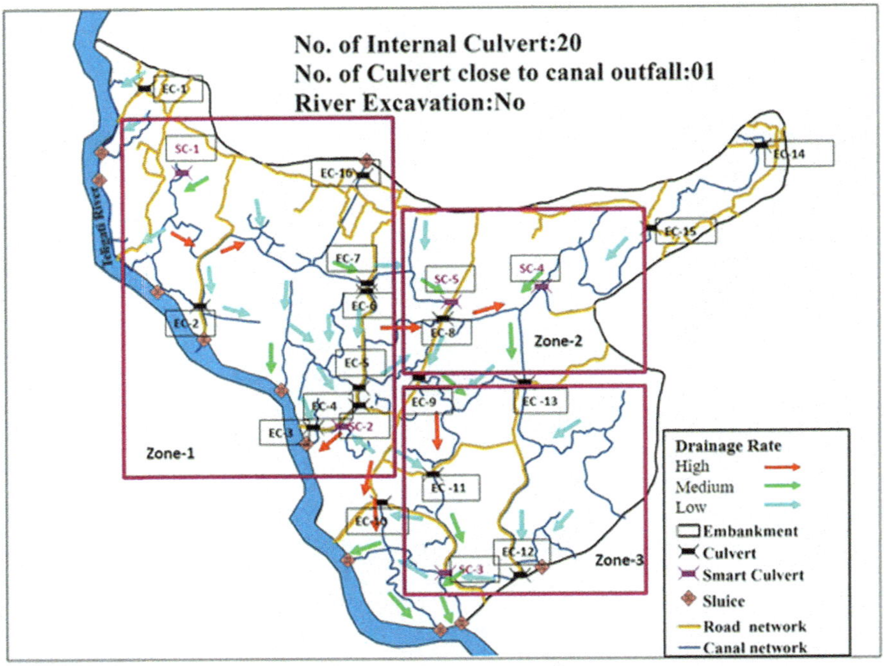

Fig. 9 Flow direction and drainage rate in scenario-1

Figure 9 shows the flow direction and drainage rate for scenario-1. It is observed that, for zone-1, when SC-1 was placed, the drainage rate increased from low to medium. Placement of SC-2 changed the flow direction from low land level towards the culvert. For Zone-2, placement of SC-5 and SC-4 results in increased drainage rate. In Zone-3, placement of SC-3 initiates flow of water to the nearby sluice. All these changes consequently reduce the extent of inundated area (Fig. 8).

3.3 *Scenario-2*

Even after placement of five smart culverts in scenario-1, low drainage rate is observed in EC-12 (Fig. 9). To increase the drainage rate at this location, two new smart culverts SC-A and SC-B are placed (Figs. 10 and 11). After placing these two culverts, overall inundated area inside the polder is found to be less compared to the situation in scenario-1 (Fig. 10). It indicates that location close to canal outfall plays an important role in draining water. This shows that increasing the number of culverts does not necessarily reduce the water-logging. Placement of the culverts in strategic locations has a major impact on reducing water logging.

The drainage rate in scenario-2 (Fig. 11) shows that after placing SC-A and SC-B along with EC-12, drainage rate towards these culverts increased, which then

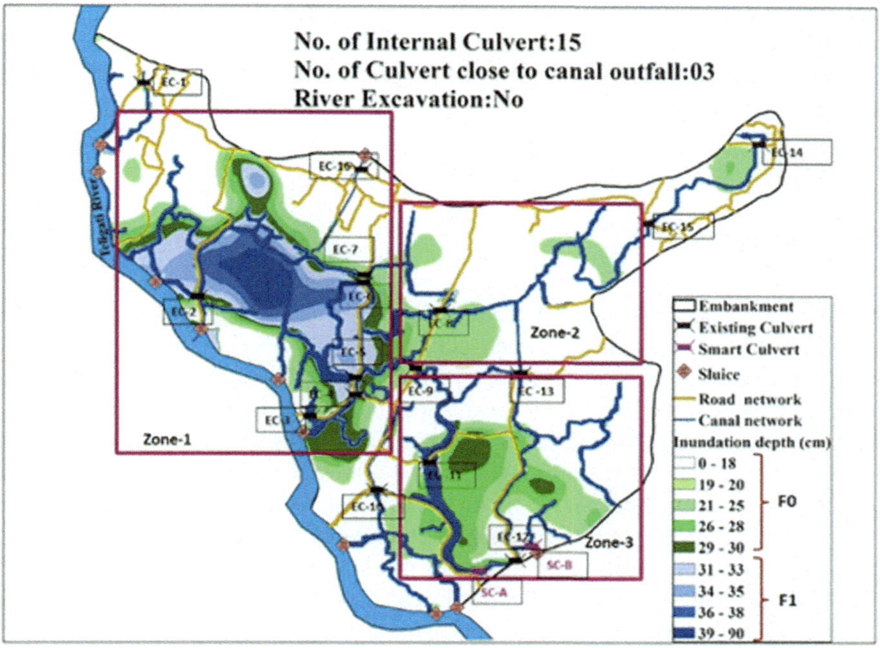

Fig. 10 Inundation map of scenario-2

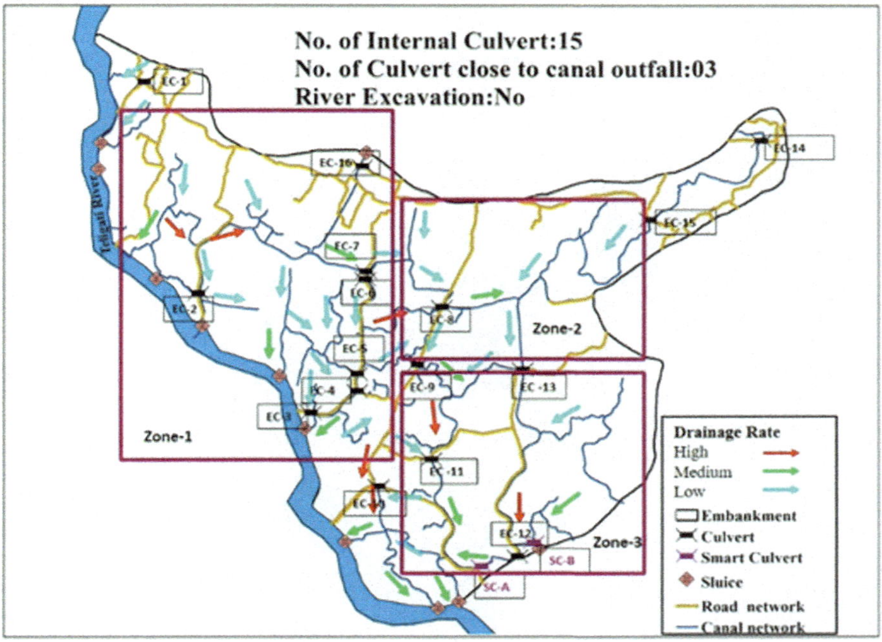

Fig. 11 Flow direction and drainage rate in scenario-2

reduces inundated area in Zone-1 and in Zone-2. But in Zone-3, extent of inundation along with the inundation depth increased. This shows that although drainage rate has increased but due to siltation in the adjacent river, these two culverts cannot perform to its full capacity.

3.4 *Scenario-3*

Scenario-3 is a combination of scenario-2 and dredging of the peripheral river (Mora Bhadra river). As mentioned earlier, culvert location close to canal outfall has a major role in reducing water-logging. To make these culvert locations more effective, the impact of dredging of the river adjacent to these culverts is observed and an improved water-logged condition is found compared to scenario-2 (Fig. 12).

The drainage rate in scenario-3 (Fig. 13) is almost identical to scenario-2 except for EC-16. As Mora Bhadra river is now functional, EC-16 seems to convey water with higher drainage rate. Reduction of inundated area in Zone-3 is observed (Fig. 12) as a result of proper drainage through SM-A and SM-B along with EC-12. This shows that culverts close to canal outfall can be effective to reduce water-logging if the conveyance of the adjacent river is improved.

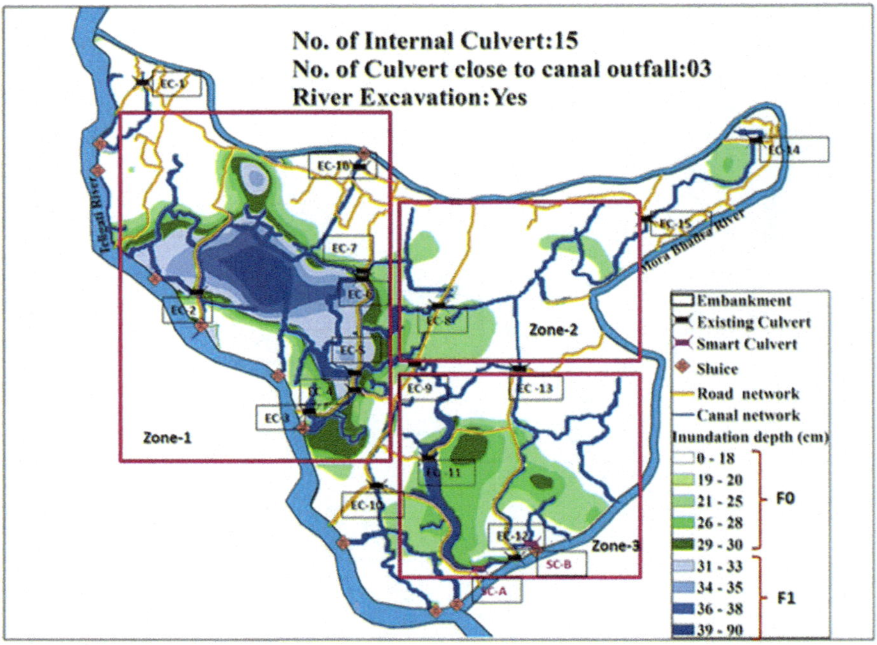

Fig. 12 Inundation map of scenario-3

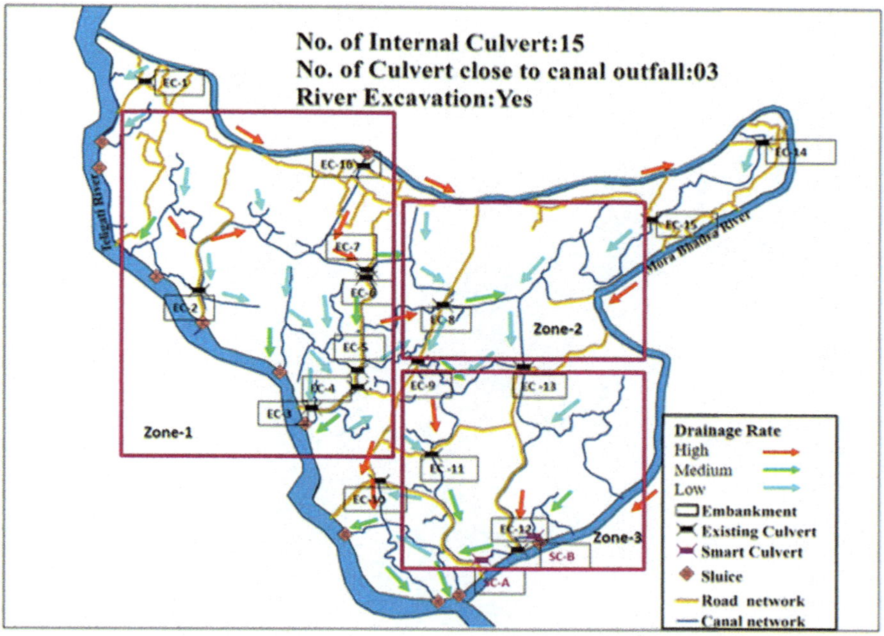

Fig. 13 Flow direction and drainage rate in scenario-3

3.5 Scenario-4

Without any river dredging, just by placing five smart culverts in five strategic locations (SM-1,2,3,4,5) and two culverts close to the canal outfall (SM-A, SM-B), more improvement in water-logged condition is observed in scenario-4 (Fig. 14). Reduction of F1 inundated area is seen in Zone-1 and in Zone-3.

From the flow direction in scenario-4 (Fig. 15), it is seen that the flows in canals of Zone-2 pass through EC-12. This culvert is seen to function effectively if conveyance of Mora Bhadra river is increased. In this entire domain, F1 inundated area reduces and drainage rate increases when compared to scenario-3. This indicates that river dredging is not the only solution to the water-logging problem. If culverts are placed in strategic locations, without any river dredging, the drainage condition inside polder can be improved.

3.6 Scenario-5

Figure 16 shows the inundation map in scenario-5 where the impact of Mora Bhadra river dredging along with five internal smart culverts is observed. In this scenario, F1

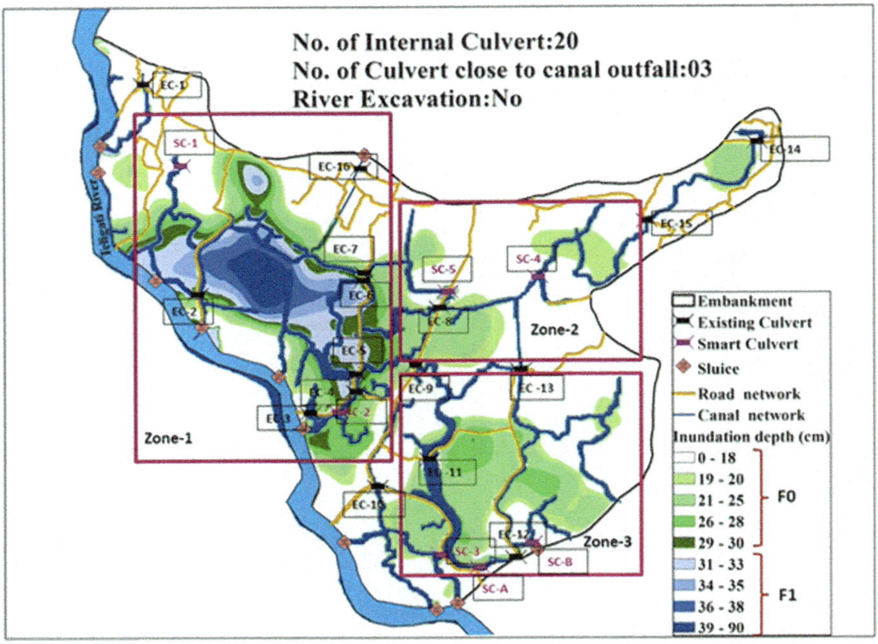

Fig. 14 Inundation map of scenario-4

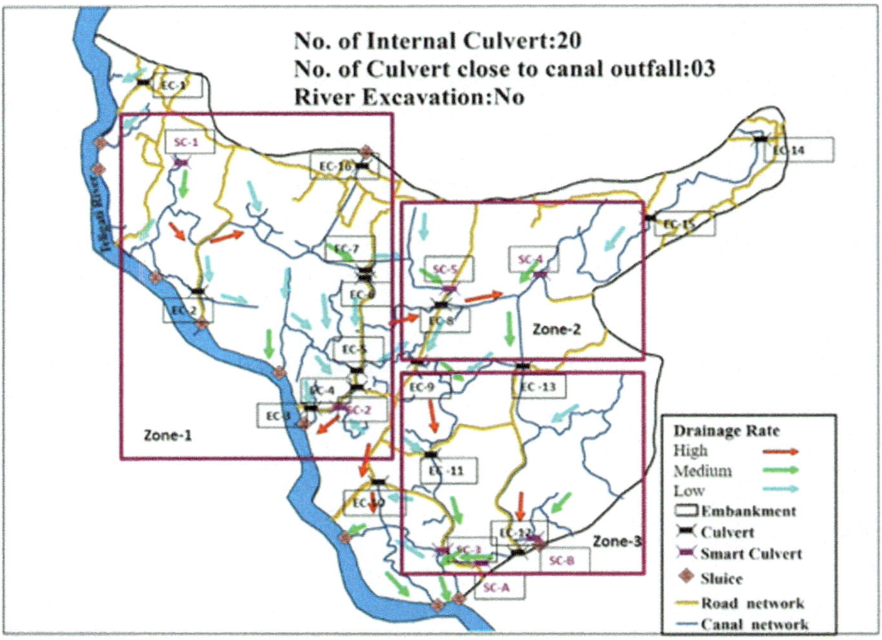

Fig. 15 Flow direction and drainage rate in scenario-4

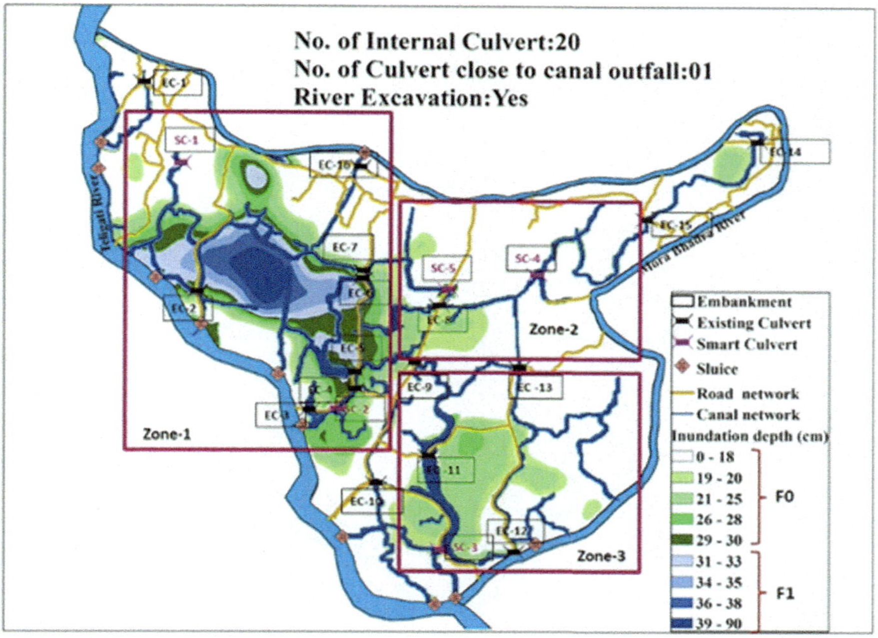

Fig. 16 Inundation map of scenario-5

inundated area reduces compared to previous scenarios. In all three zones, all the existing culverts along with five new smart culverts are found to be fully functional.

By observing the drainage rate (Fig. 17), it is found that due to the flow in the adjacent river, EC-16 started to convey water with a higher drainage rate. This water smoothly passes through SC-5 which ultimately drains out of the polder through EC-12 and SC-3. The dredging of Mora Bhadra river accelerates this process by making EC-12 functioning. This causes a higher drainage rate in this area and reduces the extent of the inundated area.

3.7 *Scenario-6*

The combined impacts of all 16 existing culverts, 5 smart culverts, 2 culverts close to canal outfall and dredging of Mora Bhadra river is observed in scenario-6. This scenario shows the lowest inundated area (Fig. 18) compared to all the previous scenarios. In this scenario it is found that Zone-2 has no inundation, no F1 inundated area in Zone-3 and the lowest F1 inundated area in Zone-1. The remaining inundation in Zone-1 is found in all the scenarios and it was found from the field visit that this inundated area is a perennial water body and is not categorized as a water-logged area.

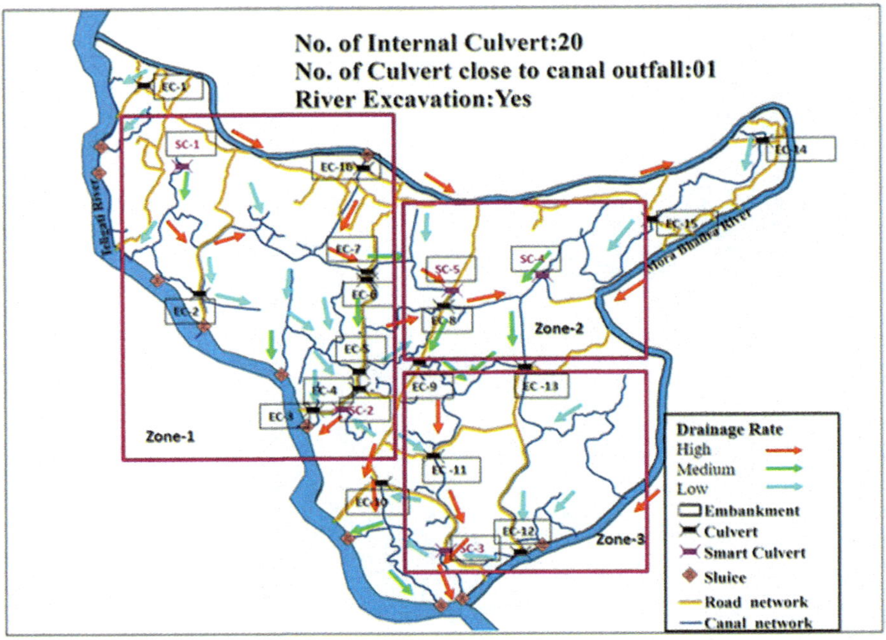

Fig. 17 Flow direction and drainage rate in scenario-5

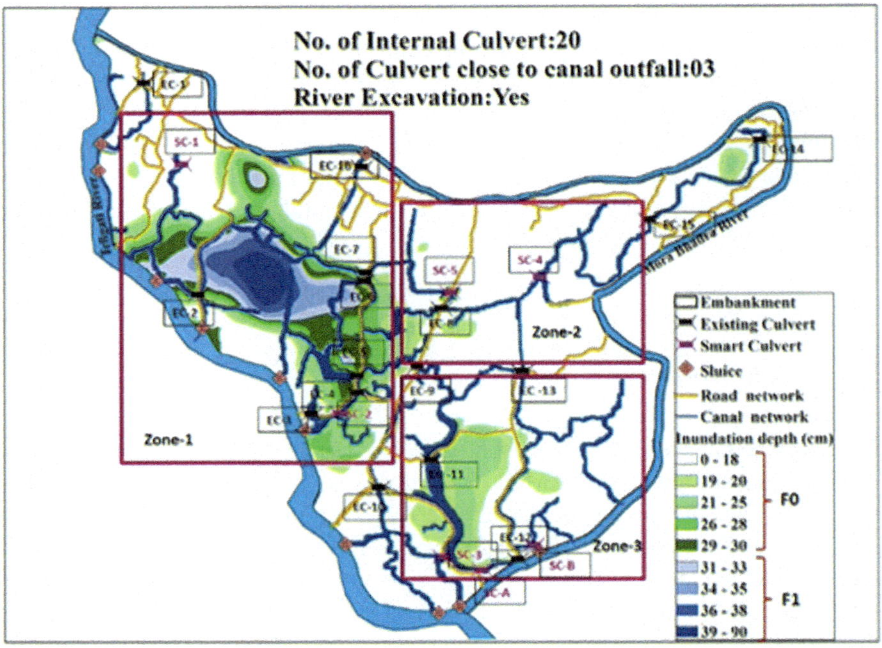

Fig. 18 Inundation map of scenario-6

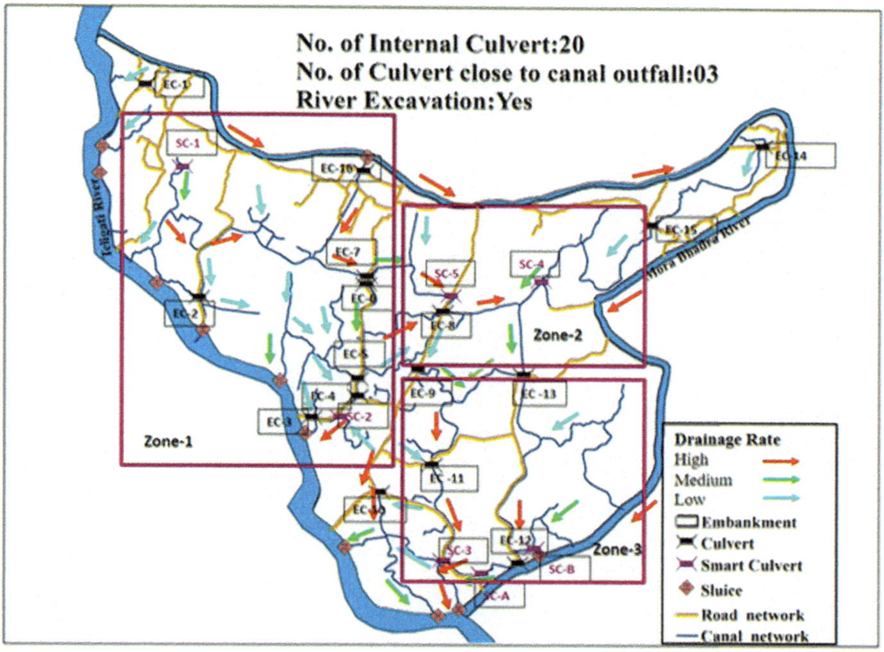

Fig. 19 Flow direction and drainage rate in scenario-6

Scenario-6 shows the highest drainage rate (Fig. 19). Here, all the culverts are fully functional with minimum obstruction of flow, which eventually increases the drainage rate. The dredging of Mora Bhadra river accelerates the overall drainage process through proper functioning of SC-3, SC-A, SC-B and EC-12.

3.8 *Optimum Drainage Route*

Based on flow direction and drainage rates in different scenarios, a schematic diagram of optimum drainage route inside the polder that effectively removes water-logging is shown in Fig. 20.

The model simulation results show gradual improvement of water-logging condition from scenario-1 to scenario-6. The decrease of F1 inundated area is seen in Fig. 21 where it is observed that the existing 1528 acres of inundated area is reduced to 1250 acres when optimum drainage route is effective. This reduction of inundated area indicates that along with river dredging, placement of culverts in strategic locations to accelerate drainage of water plays an important role in reducing water-logging inside the polder.

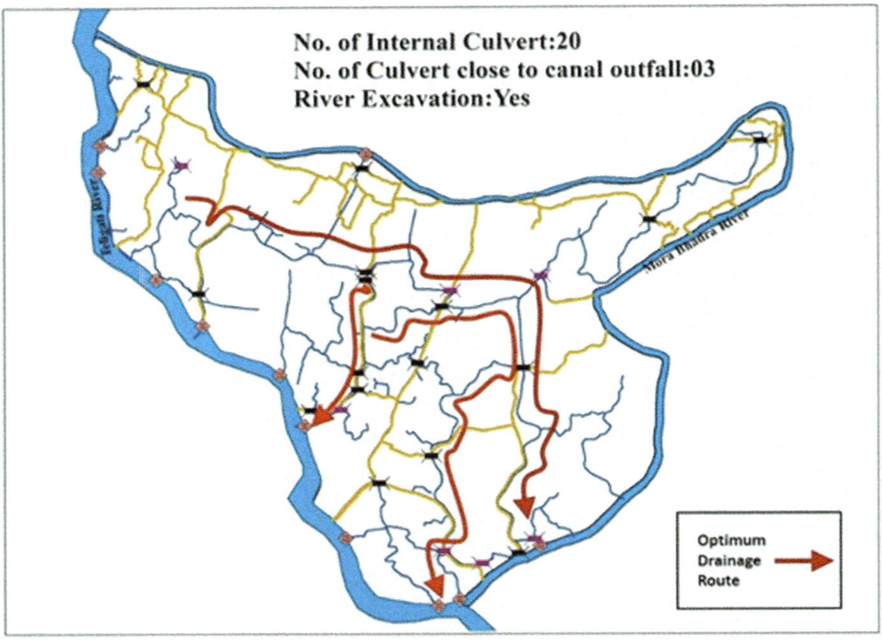

Fig. 20 Optimum drainage route

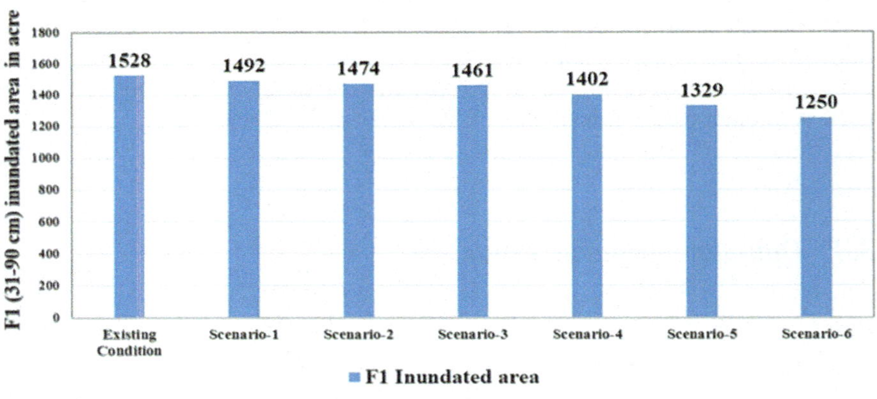

Fig. 21 Reduction of F1 inundated area

4 Conclusion

Road network plays an important role in creating water-logging inside polders by obstructing free flow of water along the drainage route. Introducing culverts in strategic locations of the road systems remove these obstacles and accelerate the drainage rate. Dredging of peripheral rivers outside the polder increases the conveyance of the rivers. This increased river conveyance is effective to reduce water-logging inside polders when accelerated drainage rate due to culverts convey the water from water-logged area to the dredged river systems. In this way river dredging also facilitates in reducing water-logging. This indicates that river dredging is only effective when the internal drainage condition is also improved.

Acknowledgement The authors are grateful to MetaMeta Research, The Netherlands for funding the project and Bangladesh Water Development Board (BWDB) for providing data without which this study could not have been carried out.

References

Bangladesh Water Development Board. (2012). *Water management review in polder 26 may 2012*. Dhaka: Ministry of Water Resources.

BBS. (2011). *Population and housing census 2011*. Final result. Bangladesh Bureau of Statistics: Statistics and Informatics Division, Ministry of Planning, Government of the People's Republic of Bangladesh.

CEGIS. (2016). *Environmental Impact Assessment (EIA) on rehabilitation of polder 26*. Final report. Government of People's Republic of Bangladesh, Ministry of Water Resources, Bangladesh Water Development Board.

Delft3D-FLOW, D. (2014). *Simulation of multi-dimensional hydrodynamic flows and transport phenomena, including sediments*. User Manual Delft3D-FLOW, the Netherlands, p. 690.

Gain, A. K., Benson, D., Rahman, R., Datta, D. K., & Rouillard, J. J. (2017). Tidal river management in the south West Ganges-Brahmaputra delta in Bangladesh: Moving towards a transdisciplinary approach? *Environmental Science & Policy, 75*, 111–120.

IWM. (2015). *Preparation of dem and delineation of hydrological boundaries for polder-2, 26 and 31(Part)*. Final report.

Noor, S. (2018). *Investigation on polderization induced water logging and feasible adaptation measures in Dumuria upazila under Khulna district*. Unpublished masters thesis. Institute of Water and Flood Management, BUET, Dhaka.

Rahman, M. Z., Islam, M. S., & Khan, Z. H. (2015). Tidal River Management (TRM) – An innovative scientific approach for sustainable sediment management. In *International conference on Recent Innovation in Civil Engineering for Sustainable Development (IICSD-2015)*, pp. 954–959.

Shampa, M. I. M. P., & Paramanik, M. (2012). Tidal River Management (TRM) for selected coastal area of Bangladesh to mitigate drainage congestion. *International Journal of Scientific & Technology Research, 1*(5), 1–6.

Tutu, A. A. (2005). River management in Bangladesh: A people's initiative to solve water-logging. *Participatory Learning and Action, 51*, 117–123.

Application of Machine Learning Algorithms for Local Level Flood Prediction: A Simplest Way of Likelihood Predictive Model of Monsoon River Flood

Arif Hasan Khan, Md. Abdulla Hel Kafi, Shah Mostafa Khaled, and Mollah Md. Awlad Hossain

1 Introduction

Flood is a globally occurring, often prolonged, recurrent hazard which frequently incur damage to agriculture, households, infrastructures etc. (WMO 2011, 2017). Consequences of river floods are often made more severe as a result of torrential rainfall, land use change in upstream catchment as well as demographic and socio-economic changes in flood vulnerable areas (WMO 2017). Generally, damages caused by floods significantly affect the national GDP of less developed countries (WMO 2017).

Bangladesh is a renowned deltaic plain, formed around the confluence of Ganges, Brahmaputra and Meghna rivers and their tributaries (FAO 2012). Originating from the Himalayas, Ganges and Brahmaputra rivers join and then flow through Bangladesh and finally discharge into the Bay of Bengal (Rahman et al. 2014). The topography of Bangladesh is mainly flat, except some hilly areas to the east of the country, and nearly 90 percent of its ground surface is about 10 m above mean sea level (Rahman et al. 2014). Being located in between the Himalayan mountains (to its north) and the Bay of Bengal (to its south), Bangladesh is characterized by a tropical monsoon climate, which results in high volumes of rainfall during the monsoon season (Rahman et al. 2014). In addition, geographical location of Bangladesh is in the delta of a massive river basin (Rahman et al. 2014). This special combination of geo-hydrological location and climatic conditions, makes Bangladesh highly vulnerable to recurring floods (Rahman et al. 2014). About 7%

A. H. Khan (✉) · M. A. H. Kafi · M. M. A. Hossain
Institute of Water Modelling, Dhaka, Bangladesh

S. M. Khaled
Institute of Information Technology, University of Dhaka, Dhaka, Bangladesh

A. Haque, A. I. A. Chowdhury (eds.), *Water, Flood Management and Water Security Under a Changing Climate*, https://doi.org/10.1007/978-3-030-47786-8_3

of the total area of Bangladesh contains rivers and inland water bodies (FAO 2012), and over the surrounding areas and riverines, floods with varying intensity are anticipated as annual occurrences (Rahman et al. 2014). Being a flood prone area, Bangladesh typically faces four different types of floods which are flash floods, monsoon river floods, local rain-fed floods, and cyclone and storm surge floods (WMO/GWP 2003).

Initiatives for implementing appropriate structural and non-structural countermeasures at both the community and national level are required (WMO 2017). The Bangladesh Water Development Board (BWDB) has undertaken several flood control, drainage and irrigation (FCDI) projects, entailing both structural and non-structural measures, all over the country to minimize the risk of floods (BWDB 2009). Of the non-structural measures, flood forecasting through early warning system (EWS) is commonly used to warn communities prior to a flood and disseminate information on flood preparedness actions (Kundzewicz 2002).Several initiatives on flood forecasting are already in place. A 5-day lead time forecast system is operational at the national level under FFWC of BWDB, which is based on the simulation of deterministic 1-dimensional mathematical model with real-time data (BWDB 2015). However, such a deterministic model requires complex simulation, cross section topographic data, geo-special data, various hydro-meteorological data, as well as boundary and river alignment data. These operations require substantial time and involvement of expert professionals, which might not be available at the local level. In contrast, the target model of our study is free from such complexities.

The major purpose of this study is to develop a local level flood prediction model with likelihood occurrence of monsoon river flood by using machine learning algorithms and regression techniques for extended lead time. The model has been built to anticipate floods in surrounding areas of three gauge stations of BWDB, situated in different locations within the Manikganj district of Bangladesh. Previously, a study titled 'Comparative Presentation of Machine Learning Algorithms in Flood Prediction Using Spatio-Temporal Data' was conducted to predict sudden floods in North Texas, USA (Jangyodsuk et al. 2017), in which influential features have been considered for flood warning at local finer scale with maximum 6 h of lead time. However, some machine learning algorithms such as Naïve Bayes or Artificial Neural Networks (ANN) had not been attempted, which are well known in the context of predictive models. In addition, the earlier study used some attributes of data which require special setup for collection. Such setup might not be available at the local level in Bangladesh.

In contrast, this study focuses on the application of the simplest way of predictive model in local level flood prediction for extended lead time which requires relatively less data, and could be applicable for local agencies and non-government organizations (NGO). A number of conventional machine learning algorithms for a predictive model, such as Decision Tree, Naïve Bayes, Support Vector Machine (SVM), Logistic Regression and Artificial Neural Networks (ANN), and regression technique of Least Absolute Shrinkage and Selection Operator (LASSO) have been used in this study. Features of data sets in our study includes temporal data of water level and rainfall reading. In addition, with floods being a seasonal event, the nth day of year for prediction has also been considered (Fig. 1).

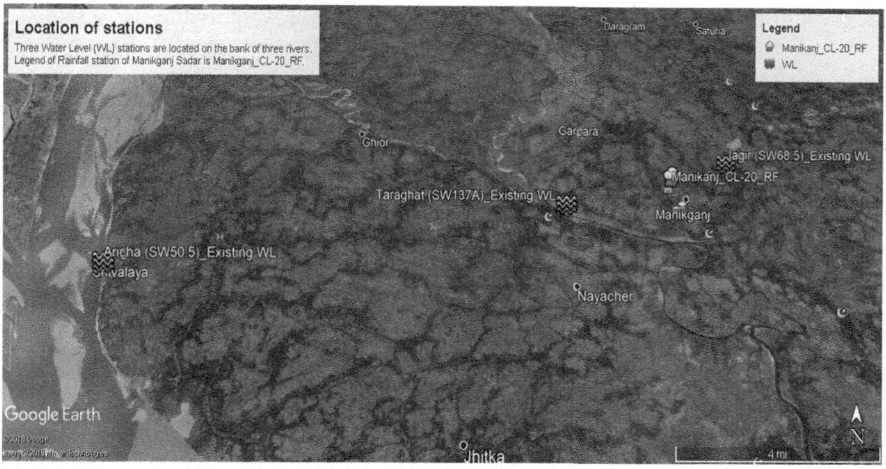

Fig. 1 Study area

The study has been conducted on flood conditions over three locations within the Manikganj district. The locations are on the bank of three rivers:

- Taraghat: Located on the bank of Kaliganga river
- Jagir: Located on the bank of Dhaleshwari river
- Arichaghat: Located at the bank of giant Jamuna river

2 Methodology

The study set out to develop models to foresee the occurrence of a flood after 10 days from a day, for specific stations. Models were developed to predict the likelihood of flood occurrence around these stations and hence they are referred to as predictive models. Development of the model has been instructed by Machine Learning algorithms and Regression techniques, which are implemented with different computer programmes. Machine Learning concerns with the construction of such a computer programme that automatically improves itself from experience (BWDB 2016). Besides, by using a regression technique, the relationship between input attributes and response variables can be elicited (Ciaburro 2018).

Here, we used six algorithms for programming the models and one model is based only on a specific algorithm. Thus, for one specific station, six different models have been built, based on different algorithms. Subsequently, results found from models based on different algorithms have been compared to identify the most suitable algorithm. A brief introduction to the algorithms used for building the models is provided in the following section.

2.1 Compared Algorithms

As mentioned earlier, models have been programmed with algorithms of Decision Tree, Naïve Bayes, SVM, Logistic Regression and ANN, and LASSO.

2.1.1 Decision Tree Learning

Decision Tree is one of the most widely used learning methods for inductive hypothesis. This method approximates discrete-valued target functions which is capable of learning as disjunctive expressions (Mitchell 1997). Learned trees can be represented as human recognizable if-then, else sets. In this study, we used C4.5 algorithm for building decision tree.

2.1.2 Naïve Bayes Classifier

Naïve Bayes is a probabilistic approach of assumption. This is vastly practical Bayesian learning which is based on simplifying assumption by considering attributes which are conditionally independent. Naïve Bayes classifier is applied to learn to classify instances where each instance is a combination of certain attribute values, and the target function can take any value from a predefined set. The probability of observing the instance with attribute value set (a_1, a_2, $a_3 \ldots a_n$) is the arithmetic product of the probabilities of a certain hypothesis for individual attributes:$\prod_i P(a_i|v_j)$. Target value output by Naïve Bayes classifier can be expressed with the following equation: v_{NB}

$$v_{BN} = \operatorname{argmax}_{v_j \in V} P(v_j) \prod_i P(a_i|v_j) \tag{1}$$

Here, v_j is the observed class at training dataset and P(x|c) is the probability of the instance to be classify as *c* when value of a certain attribute be $a_i = x$, see (Mitchell 1997).

Consequently, data set from a station is fed into each of the models, which are being developed to predict flood occurrence for the associated station. This data feeding process is sometimes referred to as training the model. Then, each of the six models become independently capable of predicting the likelihood of flood occurrence for the station. We trained the models with one-to-one sets of reading of observed features (water level, rainfall etc.) with associated observed inundation condition (true/false for flood) in later.

2.1.3 Artificial Neural Networks

Artificial Neural Networks (ANN) is a robust and powerful approach for approximating target function. At the very beginning, ANN was inspired by the observation on how the biological learning mechanism is built of interconnected neurons. In our study, ANN is based on a unit, referred to as perceptron. Perceptron takes the value of attributes as input and for different attributes, different weights are initially assigned. Perceptron multiplies attributes' value with associated weights and sums all the products. If the result exceeds some threshold value, then the output is 1 and otherwise −1 ((Mitchell 1997).

However, a single perceptron is only able to represent linear decision surface. In contrast, occurrence of flood is not linearly correlated with the attributes which we have considered. In our study, we used the network of multilayer perceptron, which can express various nonlinear decision surfaces. The network may contain one or more internal or hidden units having nodes with different weights. Weights are tuned by employing 'BACPRPOPAGATION algorithm' (Mitchell 1997) with training data (Mitchell 1997).

2.1.4 Support Vector Machine

Support Vector Machines (SVM) is a learning method which is mostly used for classification. In SVM, classes are bounded by planes and another hyperplane separates classes, where distance from the separator hyperplane to class boundary is referred to as margin. Optimal class separator plane is drawn with the aim so that margins become maximum. In many real world problems, such as flood prediction in this study, instead of being linear, separator hyperplane is complicated and spread over a higher dimension space. One solution provided by the SVM is that, from the input space data will be mapped into a high dimensional space by a nonlinear transformation (Fig. 2).

This transformation is performed by Kernel function $K(x_i, x_j)$between pairs of data points (Byun and Lee 2002).

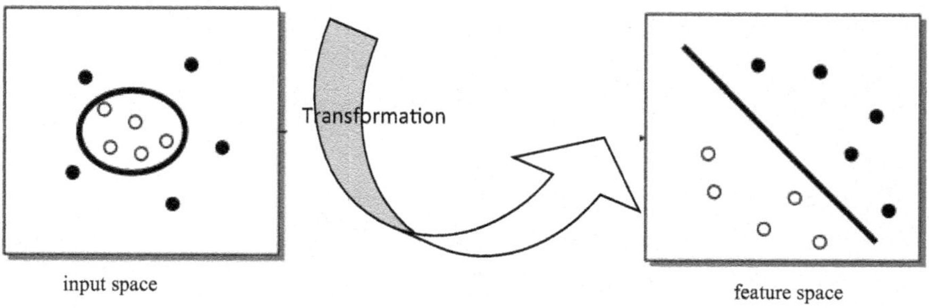

Fig. 2 Transformation toward high dimensional space

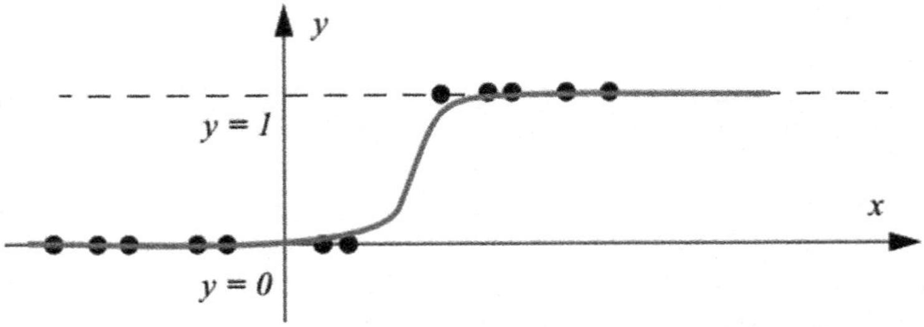

Fig. 3 Fitting input points with logistic curve

2.1.5 Logistic Regression

Logistic Regression is a statistical method often used in machine learning for binary classification. Model based on Logistic Regression maps relationship from a set of independent variables to dichotomous dependent variables, which are the element of classes (Ahmed 2017). Real-life events however may not always be able to classify such a straightforward way with respect to its attribute set values.

Logistic Regression does this synchronization by fitting input training data points with a logistic curve (grey curve in the 'Fig. 3'). This logistic function is referred to as Sigmoid function, which is a monotonic and continuous function and changes between 1 and 0 (Deng).

2.1.6 LASSO

Least absolute Shrinkage and Selection Operator' (LASSO) is a statistical technique for building models (Tibshirani 1996; Zou et al. 2007). LASSO has been explored in this study because, a previous study found it to produce the best results compared to other algorithms (Jangyodsuk et al. 2017).

LASSO attempts to minimise the residual sum of squares abiding by the constraint that, the sum of the absolute value of the coefficients should not exceed a constant. Considering this constraint, it tends to shrink some coefficients and makes some coefficients 0. As a result, the final model may use fewer features and sometimes prevents overfitting. This way LASSO builds interpretable models and retains good properties of both ridge regression and subset selection.

2.2 *Performance Measurement Method*

Classification result generated during testing a model is represented by Confusion Matrix off our different validation of predictions which are True Positive (TP), True Negative (TN), False Positive (FP) and False Negative (FN) (Davis and Goadrich 2006) (Table 1):

Table 1 Confusion matrix

		Predicted class	
		Yes	No
Actual class	Yes	Number of observations with **True Positive(TP)**	Number of observations with **False Negative(FN)**
	No	Number of observations with **False Positive(FP)**	Number of observations with **True Negative(TN)**

Here, count of True Positive indicates number of actually observed floods, perfectly classified by model as flood positive. Similarly, count of True Negative represents number of samples having no real flood observed and truly classified by model as flood negative. These two results situated diagonally denote the model's success. True positives and true negatives are the observations that are correctly predicted and therefore shown in green.

Conversely, count of False Positive indicates number of samples having no real flood observed but falsely classified by model as flood positive. Again, count of False Negative indicates number of actually observed floods but falsely classified by the model as flood negative, which would cause damage. Later, two counts indicate the model's failure. The aim is to minimize false positives and false negatives and hence they are highlighted in red.

Performance of the predictive model is measured with a set of indicators such as Precision, Recall etc. (Goutte and Gaussier 2005; Davis and Goadrich 2006):

Precision – Precision is the ratio of correctly predicted positive observations to the total predicted positive observations. The question that this indicator answers is: Of all the days that have been labeled as flooded, how many have actually been observed as floods? High precision relates to the low false positive rate.

$$\text{Precision} = \text{TP}/(\text{TP} + \text{FP})$$

Recall (Sensitivity) – Recall is the ratio of correctly predicted positive observations to all the observations in actual class – yes. The question this indicator answers is: Of all the days that in which flood was actually observed, how many were classified as flood positive?

$$\text{Recall} = \text{TP}/(\text{TP} + \text{FN})$$

2.3 Experiments

The mentioned algorithms and regression techniques are applied to train flood prediction models in three locations within the Manikganj district of Bangladesh, with associated data collected from nearby stations. Based on different algorithms, six different models have been trialled for three different stations. Each of the three data sets from different stations has been applied on models based on different algorithms and their performances have been compared in the study.

2.3.1 Data Sets

In this study, data set includes time series data for the period of 1983 to 2016 of following attributes or features:

- Date of observation
- Water level (WL)observed from nearby stations of Taraghat, Jagir and Aricha
- Rainfall (RF)observed from Manikganj Sadar station

Data have been collected from the Bangladesh Water Development Board (BWDB). Models for 3 different locations have been trained with WL data of the nearby associated station only. However, as Taraghat, Jagir and Aricha are themselves nearby in the context of rainfall observation, all the models take same RF station's (Manikganj Sadar) data for training and execution. Metadata is shown in following Table 2:

Occurrence of flood is identified by exceeding water level from danger level (DL) specified by BWDB (BWDB 2015). Water level higher than this limit indicates the surrounding area being flooded.

2.3.2 Data Processing

Following steps of data processing activities have been done with SQL (Structured Query Language) on time series data of each station which had been stored in the database (Fig. 4):

1. Calculating day difference between the day of observation and 1st January of the same year of observation day
2. Finding maximum water level of each day
3. Calculating accumulated rainfall recorded for 3 consecutive days starting from 3 days prior to the day of observation
4. Looking forward on maximum WL of 10 days lead from day of observation

Table 2 Metadata of BWDB gauge-stations used in the experiments

Station name	River name	Total records	Danger water level	Observed flood (no of days)
Tara ghat	Kaliganga	11,700	8.38	496
Aricha ghat	Jamuna	11,974	9.40	362
Jagir	Dhaleshwari	4495	8.23	294

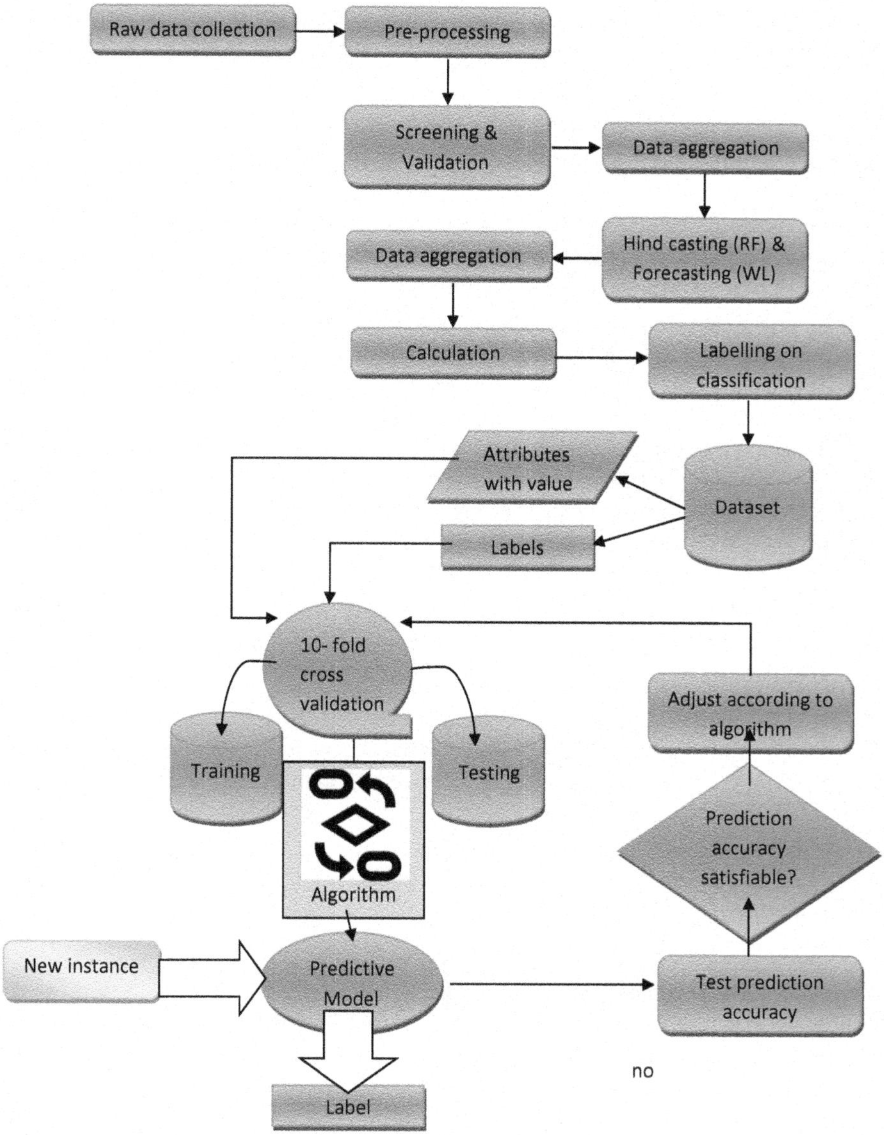

Fig. 4 Schematic overview

5. Finding whether flood was observed or not on lead day (found in step 4) by comparing WL with danger level (DL) of associated station; thus classification ('Positive as Flood' or 'Negative as Flood') has been found
6. Labeling instances found in step 5 based on classification
7. Combining the all the features and observation together by date

The consistency of data sets has been checked using traditional methods such as visual screening, double mass curve, etc.

2.3.3 Training and Validation

After consistency checking, the temporal data has been processed in a compatible format to feed into the models. After data preparation, the model has been trained and tested with tenfold cross validation method (Stone 1974). In deployment of tenfold cross validation, the data set is divided into 10 subsets subsequently iterated by the holdout method for 10 times. Chronologically in each time, one sub-set out of the 10 is used as the test set while remaining 9 sub-sets are put together to form a training set. Finally, the average error across all of the 10 trials is computed.

The following Table 3 shows distribution of samples for training and testing on each fold, for the three different stations:

Every data point gets to be in a test set exactly once, and gets to be in a training set 9 times.

3 Results

In this study, we have trained and tested different models for flood prediction with data from three different stations and generated validation results during testing.

3.1 Validation Result

Validation of experiment results is represented as a confusion matrix. Then, from the confusion matrix, performance indicators Precision and Recall have been calculated for each trial. The performance of different models for the different stations has been sequentially presented below.

Table 3 Distribution of samples for cross validation on each fold

Station name	Total no. of samples (n)	Training samples (approximately) (n/10)∗9	Test samples (approximately) n/10
Taraghat	11,700	10,530	1170
Arichaghat	11,974	10,773	1197
Jagir	4495	441	449

3.1.1 Aricha Station

Data set of Aricha station has been applied on models based on different algorithms. Validations on classification produced by different models are represented by confusion matrix followed by name of algorithm on which the model was built.

Decision Tree

Confusion matrix:

		Prediction	
		Positive as flood	Negative as flood
Observation	Positive as flood	47	315
	Negative as flood	35	11,577

Therefore, Precision = TP/ (TP + FP) = 47 / (47 + 35) = 0.57
and Recall = TP/ (TP + FN) = 47 / (47 + 315) = 0.13.

Naïve Bayes

Confusion matrix:

		Prediction	
		Positive as flood	Negative as flood
Observation	Positive as flood	174	188
	Negative as flood	619	10,993

Therefore, Precision = TP/ (TP + FP) = 174 / (174 + 619) = 0.22
and Recall = TP/ (TP + FN) = 174 / (174 + 188) = 0.48.

SVM

Confusion matrix:

		Prediction	
		Positive as flood	Negative as flood
Observation	Positive as flood	0	362
	Negative as flood	0	11,612

As the model did not classify any event as 'Positive as Flood', hence calculation of Precision and Recall score is not required for this trial.

ANN

Confusion matrix:

		Prediction	
		Positive as flood	Negative as flood
Observation	Positive as flood	45	317
	Negative as flood	32	11,580

Therefore, Precision = TP/ (TP + FP) = 45 / (45 + 32) = 0.58
and Recall = TP/ (TP + FN) = 45 / (45 + 317) = 0.12.

Logistic Regression

Confusion matrix:

		Prediction	
		Positive as flood	Negative as flood
Observation	Positive as flood	56	306
	Negative as flood	38	11,574

Therefore, Precision = TP/ (TP + FP) = 56/ (56 + 38) = 0.6
and Recall = TP/ (TP + FN) = 56 / (56 + 306) = 0.15.

LASSO

Confusion matrix:

		Prediction	
		Positive as flood	Negative as flood
Observation	Positive as flood	0	2
	Negative as flood	63	2347

As the model could not classify any observed flood event as 'Positive as Flood', hence calculation of Precision and Recall score is not required for this trial.

3.1.2 Taraghat Station

Data set of Taraghat station has been applied on models based on different algorithms. Validations on classification produced by different models are represented by confusion matrix followed by name of algorithm on which the model was built.

Decision Tree

Confusion matrix:

		Prediction	
		Positive as flood	Negative as flood
Observation	Positive as flood	134	362
	Negative as flood	67	11,137

Therefore, Precision = TP/ (TP + FP) = 134 / (134 + 67) = 0.67
and Recall = TP/ (TP + FN) = 134 / (134 + 362) = 0.27.

Naïve Bayes

Confusion matrix:

		Prediction	
		Positive as flood	Negative as flood
Observation	Positive as flood	346	150
	Negative as flood	681	10,523

Therefore, Precision = TP/ (TP + FP) = 346 / (346 + 681) = 0.34
and Recall = TP/ (TP + FN) = 346 / (346 + 150) = 0.7.

SVM

Confusion matrix:

		Prediction	
		Positive as flood	Negative as flood
Observation	Positive as flood	0	496
	Negative as flood	0	11,204

As the model did not classify any event as 'Positive as Flood', hence calculation of Precision and Recall score is not required for this trial.

ANN

Confusion matrix:

		Prediction	
		Positive as flood	Negative as flood
Observation	Positive as flood	166	330
	Negative as flood	129	11,075

Therefore, Precision = TP/ (TP + FP) = 166 / (166 + 129) = 0.56
and Recall = TP/ (TP + FN) = 166 / (166+ 330) = 0.33

Logistic Regression

Confusion matrix:

		Prediction	
		Positive as flood	Negative as flood
Observation	Positive as flood	148	348
	Negative as flood	118	11,086

Therefore, Precision = TP/ (TP + FP) = 148 / (148 + 118) = 0.56
and Recall = TP/ (TP + FN) = 148 / (148 + 348) = 0.3.

LASSO

Confusion matrix:

		Prediction	
		Positive as flood	Negative as flood
Observation	Positive as flood	12	4
	Negative as flood	64	2273

Therefore, Precision = TP/ (TP + FP) = 12 / (12 + 64) = 0.16
and Recall = TP/ (TP + FN) = 12 / (12 + 4) = 0.75.

3.1.3 Jagir Station

Data set of Jagir station has been applied on models based on different algorithms. Validations on classification produced by different models are represented by confusion matrix followed by name of algorithm on which the model was built.

Decision Tree

Confusion matrix:

		Prediction	
		Positive as flood	Negative as flood
Observation	Positive as flood	78	216
	Negative as flood	33	4168

Therefore, Precision = TP/ (TP + FP) = 78 / (78 + 33) = 0.7
and Recall = TP/ (TP + FN) = 78 / (78 + 216) = 0.26.

Naïve Bayes

Confusion matrix:

		Prediction	
		Positive as flood	Negative as flood
Observation	Positive as flood	74	220
	Negative as flood	98	4103

Therefore, Precision = TP/ (TP + FP) = 74 / (74 + 98) = 0.43
and Recall = TP/ (TP + FN) = 74 / (74 + 220) = 0.25.

SVM

Confusion matrix:

		Prediction	
		Positive as flood	Negative as flood
Observation	Positive as flood	0	294
	Negative as flood	0	4201

As the model did not classify any event as 'Positive as Flood', hence calculation of Precision and Recall score is not required for this trial.

ANN

Confusion matrix:

		Prediction	
		Positive as flood	Negative as flood
Observation	Positive as flood	79	215
	Negative as flood	46	4155

Therefore, Precision = TP/ (TP + FP) = 79 / (79 + 46) = 0.63
and Recall = TP/ (TP + FN) = 79 / (79 + 215) = 0.27.

Logistic Regression

Confusion matrix:

		Prediction	
		Positive as flood	Negative as flood
Observation	Positive as flood	80	214
	Negative as flood	53	4148

Therefore, Precision = TP/ (TP + FP) = 80 / (80 + 53) = 0.6
and Recall = TP/ (TP + FN) = 80 / (80 + 214) = 0.27.

LASSO

Confusion matrix:

		Prediction	
		Positive as flood	Negative as flood
Observation	Positive as flood	5	1
	Negative as flood	52	876

Therefore, Precision = TP/ (TP + FP) = 5 / (5 + 52) = 0.08
and Recall = TP/ (TP + FN) = 47 / (47 + 35) = .083.

3.2 *Practical Testing of Models*

To determine the accuracy of prediction of these models in practice, they were tested with new data, which are entirely unknown to them. As mentioned earlier, the models have been trained with data from 1983 to 2016. Some random test cases

Table 4 Test cases and output of testing models

Model	Observation day				Forecast day			
For the station at	Date	nth day	Observed WL	3 days ∑Rf	Date	nth day	Model's prediction	Real observation
Arichaghat	14-July-17	195	9.21	10	24-July-17	205	Negative	Negative
	13-Aug-17	225	8.98	60	23-Aug-17	235	Positive	Positive
	19-Aug-17	231	10.12	0	29-Aug-17	241	Negative	Positive
Jagir	23-July-17	204	7.67	84	2-Aug-17	214	Negative	Negative
	21-Aug-17	233	9.3	0	31-Aug-17	243	Negative	Positive
	12-Sept-17	255	7.67	132	22-Sept-17	265	Negative	Negative
Taraghat	18-Aug-17	230	8.93	0	28-Aug-17	240	Positive	Positive
	21-Aug-17	233	9.28	37	1-Sept-17	243	Positive	Positive
	5-Sept-17	248	7.68	0	15-Sept-17	258	Negative	Negative

from the monsoon season in the year 2017 were selected, to test the three models (for three stations) based on Naïve Bayes algorithm. These cases were new to the models. The test cases and outputs from the models are as follows:

The test was done for three models, each associated with our WL station. For each stations, we selected tree instances (days) during monsoon in a random choice manner, as displayed in Table 4. Result of the test experiment is demonstrated with Fig. 5. In this figure, the symbol at left side contains a number which represents the nth day of year when the attributes (WL, Rf) have been read from the station. For example, the symbol 

represents the observation (WL, Rf reading) on 248th day of 2017 (5 September, 2017) at Taraghat station. After prediction on 10 days lead (258th day of 2017 or 15 September, 2017) the result is represented with symbol 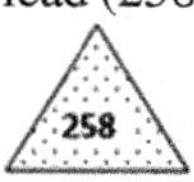

on the right side of the same figure. The dotted symbol represents negative prediction for flood and waved symbol represents positive prediction for flood. The bottom section of the right side (dashed pattern) of the figure contains the instances which really fetched flood. Upper section of right side contains instances which did not actually fetch flood.

From the figure above, it can be seen that the model associated with Arichaghat predicts two instances as negative and one as positive for flood. In contrast, the observation suggests that there was a flood observed on the day of 29 August, 2017 (241st day), which has been misclassified by the model, because prediction results

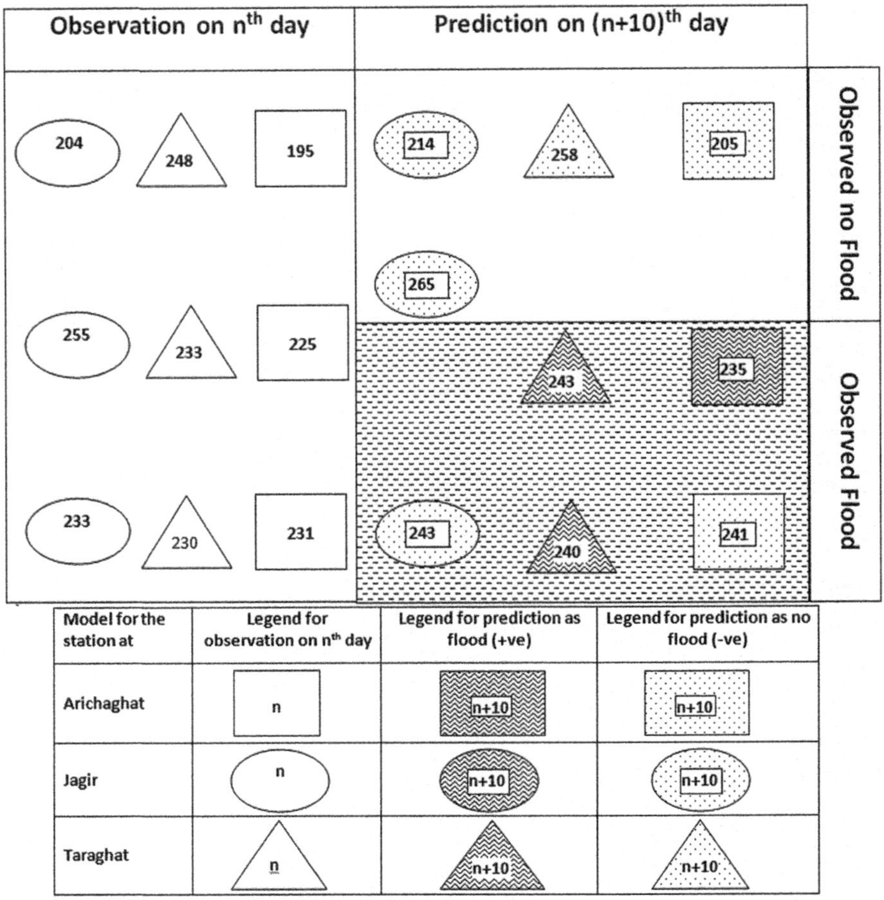

Fig. 5 Result of the test experiment on models of Naïve Bayes algorithm

based on data of 19 August, 2017 (231st day) indicates no flood occurrence on 29August2017. However, predictions on first two instances are exactly the same as the observation. Therefore, in case of the model for Arichaghat, the model passed for classifying 2 instances and failed for 1 instance out of three in real but unfamiliar instances.

Again, the model associated with Jagir predicts all the three instances as negative for flood. In contrast, the observation says there was flood observed on the day of 31 August, 2017 (243rd day), which has been misclassified by the model, because prediction result using the data of 21 August, 2017 (233rd day) indicates no flood occurrence on 31 August 2017. However, observation and prediction results of the model are same for the other two instances. Hence, in case of the model for Jagir, the model passed for classifying 2 instances and failed for 1 instance out of three, in actually observed but unfamiliar instances.

Finally, the model associated with Taraghat predicts two instances as positive and one as negative for flood. On the other hand, the observation also found the same results as prediction for associated instances. Therefore, in case of the model for Taraghat, the model passed for all three out of three predictions in real but unfamiliar instances.

4 Discussion

In this study, six different prediction models have been compared for predicting monsoon river floods in three locations within the Manikganj district, namely Aricha, Taraghat and Jagir situated on the banks of Jamuna, Kaliganga and Dhaleswari rivers respectively. Data set from each of the three stations has been individually fed into different models. The scores of performance indicators were calculated from the confusion matrix found from each trial.

Precision is the ratio of number of observed flood days correctly predicted by the model to total number of days predicted as flood by the model. In contrast, Recall is the ratio of the number of observed flood days correctly predicted by the model to total number days actually flood observed. An ideal predictive model should not predict for false classes. Therefore, 'False Positive' and 'False Negative' both should be 0 in terms of an ideal predictive model. Thus, for such a model, Precision and Recall scores should be as follows:

$$\begin{aligned}\text{Precision} &= \text{True Positive}/(\text{True Positive} + \text{False Positive})\\ &= \text{True Positive}/(\text{True Positive} + 0)\\ &= 1\end{aligned}$$

$$\begin{aligned}\text{Again, Recall} &= \text{True Positive}/(\text{True Positive} + \text{False Negative})\\ &= \text{True Positive}/(\text{True Positive} + 0)\\ &= 1\end{aligned}$$

However, such an absolute probabilistic model with 0 false classifications barely exists. Generally, predictive models tend to wrongly classify some actual events, in which case:

If, False Positive >0 then Precision <1.
and, False Negative >0 then Recall <1.

Both Precision and Recall increase with the number of actually observed floods correctly classified by the model. However, Precision decreases with incorrect prediction of the number of real safe days, i.e. classifying as 'Positive as Flood' for days where no actual flood has been observed. In contrast, Recall decreases with the number of incorrect predictions on actually observed flood days by classifying them as 'Negative as Flood' days, when the actual observation should classify them 'Positive as Flood'.

Precision gives us an idea on the probability of classifications being correct or incorrect. However, Precision does ignore one validation criterion which is highly important in the context of our study, which is the 'False Negative'. This implies that Precision does not concern with the extent to which the number of observed flood days has been wrongly classified by the model as 'Negative as Flood'. In practice, such a model might misguide its users by suggesting no flood preparedness is required prior to an actual occurrence of a flood. In contrast, Recall concerns with 'False Negative', because Recall score of a model decreases with the increasing count of 'False Negative'. However, Recall neglects, the extent to which days with no actual observed floods are wrongly classified as 'Positive as Flood', which in turn ignores the liability of unnecessary effort. Therefore, to determine a suitable model, the trade-off between Recall and Precision needs to be accounted for. The comparison of Recall and Precision scores produced by models for different stations are presented in a set of charts and tables below (Table 5).

To visualise the comparison lets plot the data into a chart (Fig. 6):

Table 5 Comparison of performance indicators scores found from Aricha ghat station

	Performance indicators	
Method/algorithm	Precision	Recall
Logistic regression	0.6	0.15
Decision tree	0.57	0.13
SVM	–	–
Naïve Bayes	0.22	0.48
LASSO	–	–
ANN	0.58	0.12

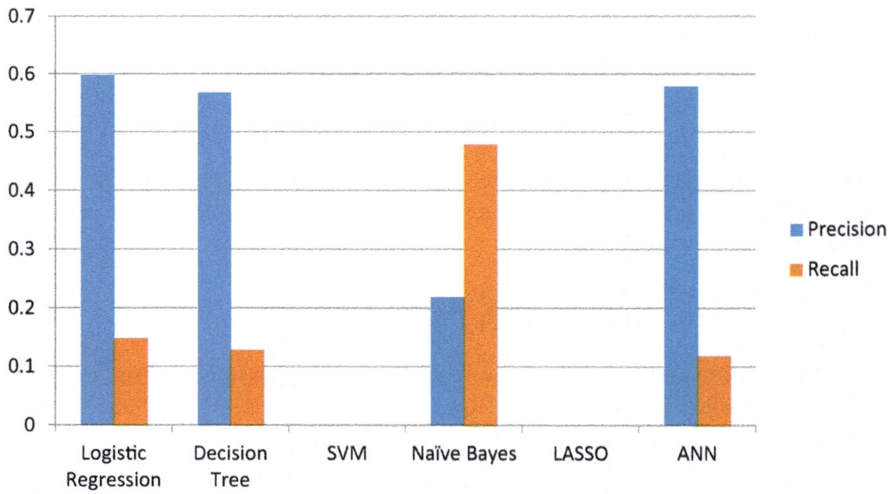

Fig. 6 Comparison of performance indicator scores found from Aricha ghat station

From the trials executed on models based on different algorithms with data from Aricha ghat station, we find that the models based on LASSO and SVM have failed. Although, the model based on Logistic Regression, ANN and Decision Tree score good at Precision, the Recall score is low for all three of them. However, the model based on Naïve Bayes appears to be highest scorer in terms of Recall for this station (Table 6).

To visualise the comparison lets plot the data into a chart (Fig. 7):

From the trials executed on models based on different algorithms with data from Tara ghat station, we find that the model based on SVM has failed. Although, the model based on Logistic Regression, ANN and Decision Tree score good at Precision, the Recall score is low for all three of them. In contrast, the model based on LASSO scores highest at Recall among all, but the Precision score is significantly low. The model based on Naïve Bayes produces a very good score for Recall and an acceptable score for Precision (Table 7).

To visualise the comparison lets plot the data into a chart (Fig. 8):

From the trials executed on models based on different algorithms with data from Jagir station, we find that the model based on SVM has failed. Although, the model based on LASSO produces excellent Recall score, the Precision score is considerably low. On the other hand, Recall scores for models based on Logistic Regression,

Table 6 Comparison of performance indicators scores found from Tara ghat station

	Performance indicators	
Method/algorithm	Precision	Recall
Logistic regression	0.56	0.3
Decision tree	0.67	0.27
SVM	–	–
Naïve Bayes	0.34	0.7
LASSO	0.16	0.75
ANN	0.56	0.33

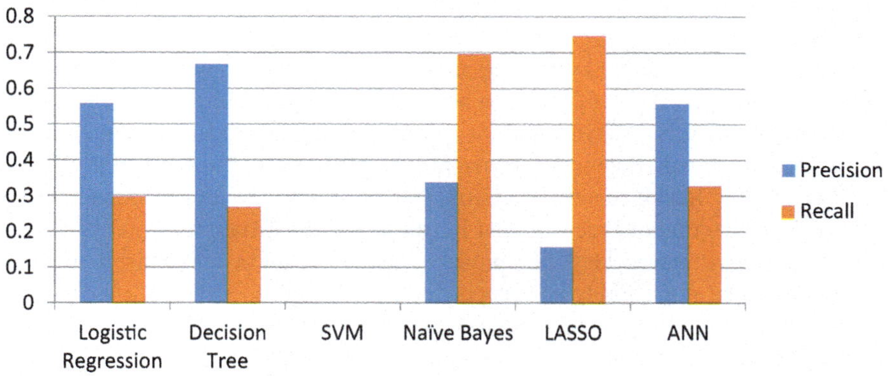

Fig. 7 Comparison of performance indicators scores found from Tara ghat station

Table 7 Comparison of performance indicators scores found from Jagir station

	Performance indicators	
Method/algorithm	Precision	Recall
Logistic regression	0.6	0.27
Decision tree	0.7	0.26
SVM	–	–
Naïve Bayes	0.43	0.25
LASSO	0.08	0.83
ANN	0.63	0.27

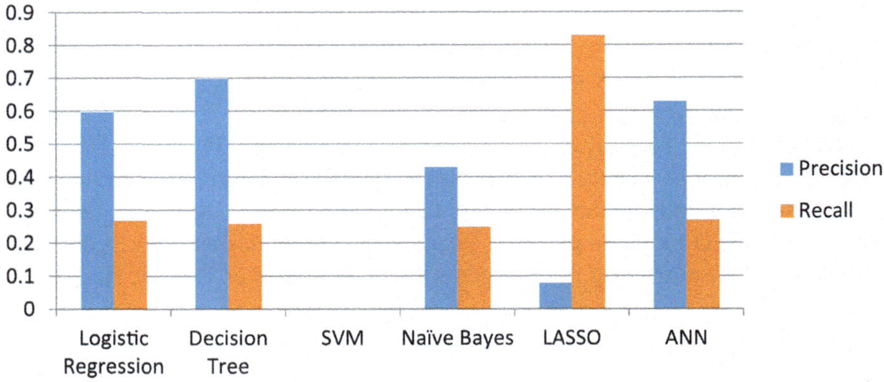

Fig. 8 Comparison of performance indicators scores found from Jagir station

ANN, Naïve Bayes and Decision Tree are close to each other. Considering the trade-off between Precision and Recall, the model based on ANN is said to be a relatively better performer for this station.

The comparison of Precision scores for models based on different algorithms applied in all three stations together, is presented in the following chart (Fig. 9):

Similarly, comparison of Recall scores for models based on different algorithms applied in all three stations together is presented in the following chart (Fig. 10):

From the figures (Figs. 7 and 8), it is visible that, in spite of having a history of successful implementation in many applications, SVM fails in our experiment. In a previous study, LASSO was found to be most appropriate. However, in our experiment, LASSO seems to have failed for Aricha station. In contrast, the models based on ANN and Logistic Regression generates good Precision scores, but both of them suffer from low Recall scores. 'Decision Tree' however provides moderate performance both in terms of Precision and Recall for all three stations. Finally, models based on Naïve Bayes classifier produce higher recall in comparison to other algorithms, after maintaining an acceptable Precision score.

As mentioned earlier, models with higher Precision scores denote that there are fewer 'Negative as Flood' days wrongly classified as 'Positive as Flood'. Precautionary measures for flood preparedness require significant effort and costs.

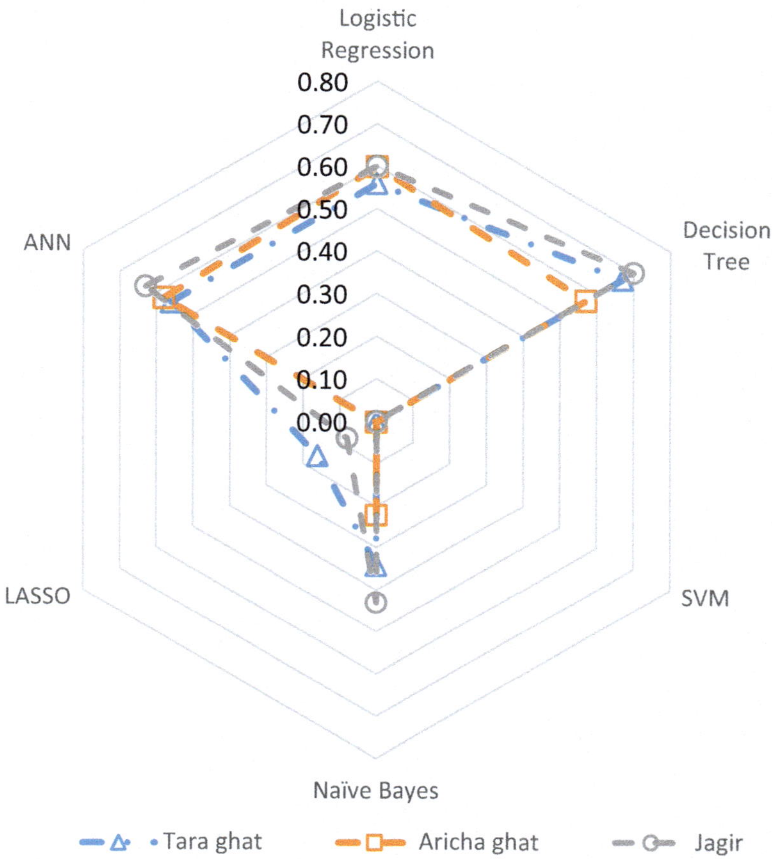

Fig. 9 Comparison of precision scores of all models found from all three stations

Signalling a false alert by classifying 'Negative as Flood' as 'Positive as Flood' when there will be 'Negative as Flood', would lead to inefficient use of resources. Thus, an ideal predictive model should strive to maintain Precision score at an acceptable level.

In contrast, Recall score is a more important indicator in terms of predicting a hazardous event like flood. Because Recall refers to the sensitivity, a model that scores low on Recall would misclassify a potential 'Positive as Flood' day as 'Negative as Flood'. Floods often incur substantial economic losses and damages and therefore such misclassification is unacceptable in the context of flood prediction. Therefore, in terms of selecting an algorithm for a flood prediction model, the importance of Recall score should not be disregarded by any means and a model with high Recall score would be most appropriate.

Considering the above context, after comparing results from our experiment, we can establish that the model based on Naïve Bayes algorithm is likely to perform comparatively better than the other algorithms in terms of food prediction.

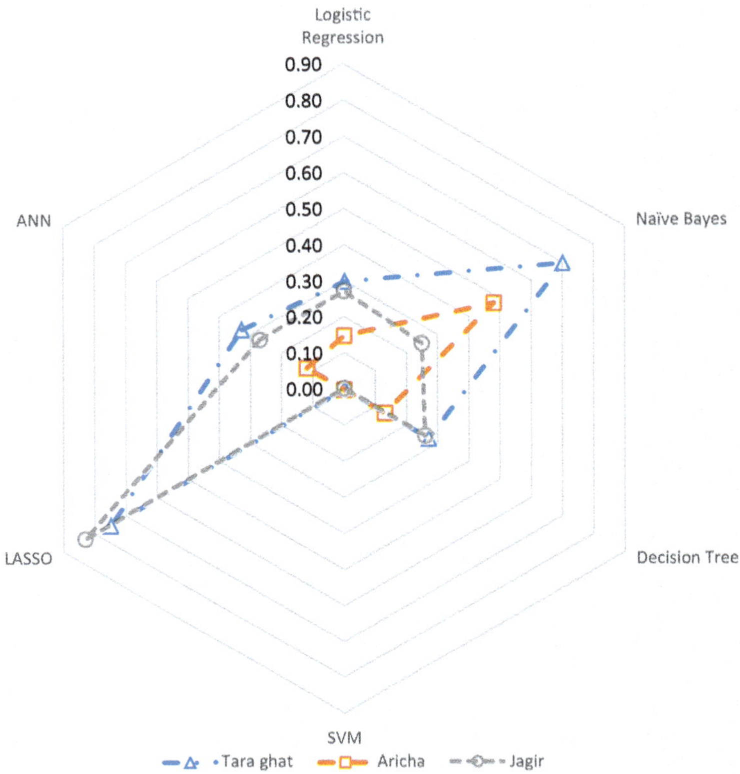

Fig. 10 Comparison of recall scores of all models found from all three stations

5 Conclusion

Flood forecasting or prediction is a very important non-structural measure in flood risk management (Kundzewicz 2002; Menzel and Kundzewicz 2003). This study has been conducted with the aim of developing a set of predictive models using different algorithms and regression techniques, and the suitability of these algorithms in local level flood prediction was tested. Models have been developed with machine learning algorithms such as Decision Tree, Logistic Regression, Naïve Bayes, SVM, ANN and the regression technique of LASSO. The models were then trained and subsequently tested with standard classification model testing technique. Following training and testing, validation on classification was figured out by employing the confusion matrix. Finally, performance of predictive model was measured against the indicators of Precision and Recall.

Precision has significance in minimizing unnecessary effort towards flood preparedness, thereby saving time, manpower and expenses. Whereas, the Recall value is important for indicating the probability of wrongly classifying real flood events

(Jangyodsuk et al. 2017). Since floods result in significant socioeconomic losses and damages, the model which demonstrates lower instances of failing to classify actual flood events, should be nominated as most suitable.

The models based on SVM and LASSO has lost their candidateship for being selected by scoring 0 for Precision and Recall in one or more stations. In spite of scoring well at Precision, the model based on ANN and Logistic Regression, could not be selected for their low score at Recall, which is a key concern under the study considering the possibility of damage. In contrast, the model based on Decision Tree has been found as a moderate performer both in terms of both precision and recall. However, the model based on Naïve Bayes classifier produces a relatively higher Recall score and maintains an acceptable Precision score. In order to minimize potential damages from floods, the model with a comparatively higher recall value is preferable. Therefore, the Naïve Bayes classifier can be considered as a stronger candidate than the Decision Tree. In conclusion, comparing all six algorithms, Naïve Bayes classifier has been found as better performer than the other algorithms employed in this study, in the context of local level prediction of monsoon river floods.

Acknowledgement I would like to express my deep gratitude to Professor Dr. Mohammad Shoyaib, IIT, University of Dhaka and Mostafa Khaled, Associate Professor, IIT, University of Dhaka, my research supervisors, for their patient guidance, enthusiastic encouragement and useful critiques of this research work. I would also like to thank Abdulla Hel Kafi, Head of Research and Development Unit, Institute of Water Modelling(IWM) for his help in doing the meteorological and hydrological data analysis.

References

Ahmed, L. A. (2017). Using logistic regression in determining the effective variables in traffic accidents. *Applied Mathematical Sciences, 11*(42), 2047–2058.

BWDB. (2009). *Five Year Strategic Plan of BWDB, Roadmap for realizing Organizational Goals (2009–2014)*. Published by the Task Force on BWDB Reform Process with the financial support from the Integrated Planning for Sustainable Water Management (IPSWAM) Project funded by GoN grant.

BWDB. (2015). *Annual Flood Report 2015*. Flood Forecasting & Warning Center, Bangladesh Water Development Board.

BWDB. (2016). *Annual Flood Report 2016*. Flood Forecasting & Warning Center, Bangladesh Water Development Board.

Byun, H., & Lee, S. W. (2002). Applications of support vector machines for pattern recognition: A survey. In *International workshop on support vector machines* (pp. 213–236). Berlin/Heidelberg: Springer.

Ciaburro, G. (2018). *Regression analysis with R: Design and develop statistical nodes to identify unique relationships within data at scale*. Birmingham: Packt Publishing Ltd.

Davis, J., & Goadrich, M. (2006). The relationship between precision-recall and ROC curves. In *Proceedings of the 23rd international conference on machine learning* (pp. 233–240). New York: ACM.

Deng, K. *OMEGA: On-line memory-based general purpose system classifier*. Dissertation.

FAO. (2012). *Irrigation in Southern and Eastern Asia in figures*. AQUASTAT Survey – 2011.

Goutte, C., & Gaussier, E. (2005). A probabilistic interpretation of precision, recall and F-score, with implication for evaluation. In *European conference on information retrieval* (pp. 345–359). Berlin/Heidelberg: Springer.

Jangyodsuk, P., Seo, D. J., Elmasri, R., & Gao, J. (2017). *Comparative presentation of machine learning algorithms in flood prediction using spatio-temporal data.* In Proceedings of the 2015 international conference on communications, signal processing, and systems.

Kundzewicz, Z. W. (2002). Non-structural flood protection and sustainability. *Water International, 27*(1), 3–13.

Menzel, L., & Kundzewicz, Z. W. (2003). Non-structural flood protection-a challenge. In *International conference 'Towards national flood reduction strategies'*, Warsaw, pp. 6–13.

Mitchell, T. M. (1997). *Machine learning*. New York: McGraw-Hill. ISBN:978-0-07-042807-2.

Rahman, D. M. M., Hossain, M. A., & Bhattacharya, D. A. K. (2014). An analytical study of flood Management in Bangladesh. *IOSR Journal of Engineering (IOSR-JEN), 4.*, (1, 01–06.

Stone, M. (1974). Cross-validatory choice and assessment of statistical predictions. *Journal of the Royal Statistical Society. Series B (Methodological)*, 111–147.

Tibshirani, R. (1996). Regression shrinkage and selection via the lasso. *Journal of the Royal Statistical Society: Series B (Methodological), 58*(1), 267–288.

WMO. (2011). *Manual on flood forecasting and warning*. World Meteorological Organization. Retrieved from http://www.wmo.int/pages/prog/hwrp/publications/flood_forecasting_warning/WMO%201072_en.pdf

WMO. (2017). *Hydrology and Water Resources Programme (HWRP): Flood forecasting initiative.* World Meteorological Organization. Retrieved from http://www.wmo.int/pages/prog/hwrp/FloodForecastingInitiative.php

WMO/GWP. (2003). *The Associated Programme on Flood Management, Integrated Flood Management Case Study Bangladesh: Flood Management.* Technical Support Unit, World Meteorological Organization. Retrieved from https://cleancookstoves.org/binary-data/RESOURCE/file/000/000/78-1.pdf

Zou, H., Hastie, T., & Tibshirani, R. (2007). On the "degrees of freedom" of the lasso. *The Annals of Statistics, 35*(5), 2173–2192.

Developing a Composite Map of Vulnerability to Rainfall Extremes in Sri Lanka

E. M. G. P. Hemachandra, N. D. K. Dayawansa, and R. P. De Silva

1 Introduction

Climate change is now widely recognized as a global phenomenon. It can be defined as "a change in the state of the climate that can be identified (e.g., by using statistical tests) by changes in its properties and that persists for an extended period, typically decades or longer" (IPCC 2014). Even though climate change is a natural phenomenon, it can also be triggered by anthropogenic factors. Climate change manifests in the form of climatic extremes such as prolonged wet spells, lengthy dry spells and heavy rainfalls that can wreak substantial damage to human settlements and infrastructure as well as agricultural production and ecosystems.

Extreme weather events are typically a result of changes in overall temperature and precipitation. "An extreme (weather or climate) event is generally defined as the occurrence of a value of a weather or climate variable above (or below) a threshold value near the upper (or lower) ends ('tails') of the range of observed values of the variable. Some climate extremes (e.g., droughts, floods) may be the result of an accumulation of weather or climate events that are, individually, not extreme themselves (though their accumulation is extreme)" (WMO 2016). Additionally, these climatic extreme events can be defined in terms of impacts of climate change on people, society or the environment.

Impacts of climate change manifest in the form of sea level rise, melting of glaciers, cyclones etc. which have recently become prevalent throughout the world (Parry et al. 2007). The region of South Asia, attributing to its socio-economic context characterized by high poverty incidence and weak governance structures, remains highly vulnerable to extreme climatic events. According to recent studies,

E. M. G. P. Hemachandra (✉) · N. D. K. Dayawansa · R. P. De Silva
Department of Agricultural Engineering, Faculty of Agriculture, University of Peradeniya, Peradeniya, Sri Lanka

A. Haque, A. I. A. Chowdhury (eds.), *Water, Flood Management and Water Security Under a Changing Climate*, https://doi.org/10.1007/978-3-030-47786-8_4

majority of the countries in South Asia demonstrate a trend of increasing average annual temperatures. However, both increasing and decreasing trends can be observed for rainfall volumes across these countries depending on their spatial variability (Sivakumar and Stefanski 2010). Severe floods and landslides had occurred due to high intense rainfall events in many parts of Asia. Similarly, the frequency of drought incidence has also increased in the South Asian region (Hijioka et al. 2014).

Between 1961 and 1990, an increase in air temperature has been observed across all meteorological stations in Sri Lanka (Fernando and Chandrapala 1992). High annual variability with alternative dry and wet periods has been observed from 1880 until 1970, while significant reduction has occurred thereafter (Eriyagama et al. 2010). Hence, it can be inferred that the change in rainfall regime in the country can be attributed to changes in climate.

With a significant number of people engaged in agriculture, rainfall and temperature are crucial factors for Sri Lanka particularly for the dry zone in the country due to the unpredictability of rainfall variability there. Rainfall patterns play a significant role in driving decision-making in agricultural practices such as land use and management, selecting plants and selecting breeds for livestock, crop production practices namely irrigation, pest and disease control, etc. (Hutchinson 1995). Hence, traditional livelihoods are at high risk to the impacts of climate change.

Even though, climate change models indicate future predictions towards both increasing and decreasing rainfall (Karl et al. 1996; Easterling et al. 2000b), it is not always reliable due to the driving rainfall dynamics (The World Bank Group 2011). Within this background, further studies need to be conducted to estimate the changes in rainfall pattern and variability of the extremes (Easterling et al. 2000a).

Analysis of historical rainfall data provides proper guidance for future predictions on rainfall variability. Accordingly, annual, monthly and daily rainfall records can be used. Quantitative analysis of mean monthly and annual rainfall does not accurately reflect the patterns in which extreme weather events have occurred (Jayawardene et al. 2005). However, daily rainfall data can be used to deduce the change of rainfall regime in a quantitative and qualitative manner. Accordingly, daily rainfall data can be analyzed to derive several indicative parameters of rainfall extremes, which can be used for decision-making in different sectors. Daily rainfall data are therefore important to derive some facts which cannot be derived using either monthly aggregate or annual aggregate of rainfall volumes, such as consecutive dry and wet periods in a year, frequency of heavy rainfall events, general dryness and wetness of a year, etc. Hence, spatial database of rainfall related indicative parameters of rainfall extremes can be a comprehensive means to guide decision-making in crop planning and management, soil conservation practices, flood risk management, drought management and environmental conservation practices.

In the context of this concern, this study has been conducted to develop maps of extreme rainfalls in Sri Lanka by deriving six different parameters using daily rainfall data for a recent thirty year period from 1986 to 2015. The final composite map would indicate the vulnerability to rainfall extremes influenced by global climate change.

2 Methodology

2.1 *Study Area*

Sri Lanka is situated in the Northern Hemisphere in between North latitudes 5°–10° and East Longitudes 79°–82°. The average annual temperature of the country is approximately 27 °C and the average annual rainfall in the country varies between 900 and 5000 mm (Punyawardana 2002). Accordingly, Sri Lanka is divided into three major climatic zones; namely dry, wet and intermediate predominantly based on mean annual rainfall. There are four major seasons -; First Inter-Monsoon, Southwest Monsoon, Second Inter-Monsoon and Northeast Monsoon considering temporal variability of rainfall in the country (Ministry of Mahaweli Development and Environment 2016).

2.2 *Data Collection*

Non recording type daily rainfall data from 103 rain gauging stations (Fig. 1) which represent all three major climatic zones in Sri Lanka were collected from the national government's Irrigation Department, Meteorological Department and Natural Resources Management Centre for the period of 1986–2015. In this study, the distribution of rain gauging locations is representative of the rainfall variability in the country.

The density of rain gauging locations in the wet zone is high due to high variability of rainfall in the region. Comparatively, rainfall variability is less in the dry zone and thereby a lesser number of rain gauge locations can be observed. Furthermore, locations were selected according to the availability of data for the Thirty year period with only a few cases of missing rainfall data. Consistency of the daily rainfall data was verified. Missing rainfall data were estimated using data from the nearby representative stations.

2.3 *Data Processing and Analysis*

2.3.1 Derivation of Inferential Statistical Parameters

Using daily rainfall data, the following six different indicative parameters were derived to assess rainfall extremes. The parameters were developed to emphasize the changes of rainfall pattern using multiple viewpoints which are more important in decision making. The calculations were done using Microsoft Excel.

Fig. 1 Distribution of the rain gauging locations

Thirty Year Maximum of Highest Number of Consecutive Wet Days in a Year

This parameter looks for all the continuous wet days recorded during the year. It takes the highest number of continuous wet days with daily rainfall equal or higher than 1 mm in each year. Calculation was done for all 30 years. Thirty year maximum of highest number of consecutive wet days in each location was used for mapping.

Thirty Year Maximum of Highest Number of Consecutive Dry Days in a Year

This parameter looks for all the continuous dry days recorded during the year. It takes the highest number of continuous dry days with daily rainfall less than 1 mm in each year. Calculation was done for all 30 years. Thirty year maximum of the highest number of consecutive dry days in each location was used for mapping.

Thirty Year Maximum of the Number of Wet Days in a Month

Initially, number of days where daily rainfall is equal or higher than 1 mm in a month was calculated for each month within the thirty year period. Then the maximum number of wet days in a month for each of the years was obtained for each location. Thirty year maximum of the wet days in a month was used for mapping.

Thirty Year Maximum of the Number of Dry Days in a Month

Initially, number of days where daily rainfall is less than 1 mm in a month was calculated for each month for the thirty year period for all locations. Then the maximum number of dry days in a month for the 30 years was obtained for each location. Thirty year maximum of dry days in a month was used for mapping.

Thirty Year Maximum of the Ratio of Wet Days to Dry Days in a Month

The ratio of wet days to dry days was calculated for each month within the thirty year period and the maximum ratio was used in mapping.

Thirty Year Minimum of the Ratio of Wet Days to Dry Days in a Month

The ratio of wet days to dry days was calculated in each month within the thirty year period and the minimum ratio was used in mapping.

2.3.2 Development of Descriptive Statistical Parameters

Descriptive statistical parameters such as thirty year maximum, minimum, average, and standard deviation were calculated for each indicative parameter.

2.4 Development of a Composite Map

The composite map of extreme positive rainfalls during the Thirty year period in Sri Lanka was developed using the derived parameters. Z distribution of each parameter was derived. Those values were rearranged to indicate wet extremes as positive and dry extremes as negative. Cumulative Z values of each parameter were obtained for each location. Multiparameter map was developed using IDW interpolation in Geographical Information Systems (GIS). IDW determines the cell values using a linear weighted combination set of sample points. In this interpolation method a function of a distance of the input point from the output cell location is taken to assign the weights. The map was classified to present the vulnerability of rainfall extremes. Figure 2 presents the flow diagram of the methodology.

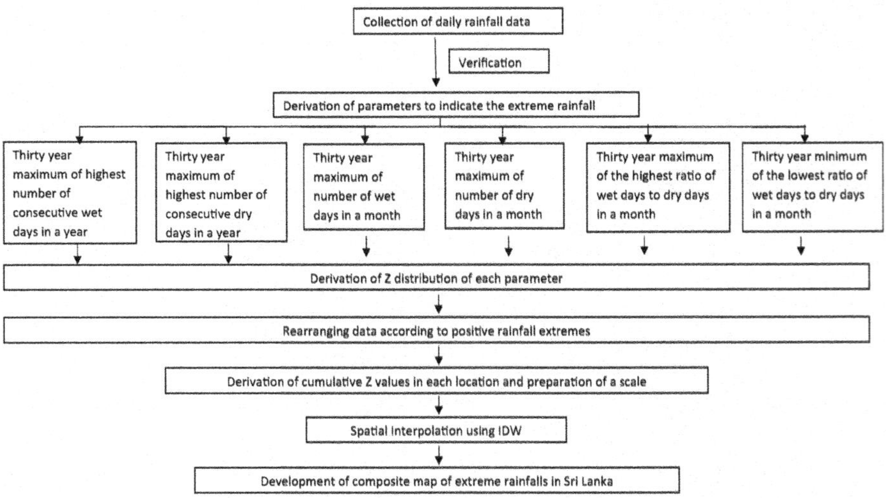

Fig. 2 Flow diagram of the methodology

3 Results

Climate change and associated variations in climatic factors are increasingly rendering the livelihoods of people as highly vulnerable. People are subject to a range of adverse consequences due to unpredictable rainfall patterns (Ministry of Mahaweli Development and Environment 2016). Various parameters have been developed in different studies to indicate climate change and its variables including climatic extremes (Karl et al. 1996; Nastos and Zerefos 2009; Wang et al. 2014). Some studies have developed the Climatic Extreme Indices indicating drought, extreme moisture surplus, deviation of maximum and minimum temperatures from the mean, wet and dry days and heavy precipitations (Karl et al. 1996). Under the study, six parameters have been developed which include the total number of wet and dry days as well as the number of consecutive wet and dry days. Additionally, the ratio of wet and dry days is an important indicator to express the average dryness or wetness. Hence, the lowest ratio indicates the maximum dryness of a particular area while the highest ratio indicates the maximum wetness of that particular area. According to this study, the developed parameters are representing different aspects of rainfall variability.

3.1 Highest Number of Consecutive Wet and Dry Days in a Year

Consecutive dry and wet periods are critical factors for agricultural production. For example, rice is one of the major crops which requires a waterlogging condition in a particular stage of its growth. In rainfed rice cultivation, it is necessary to have consecutive rainy days within a particular period. Figure 3 is related to thirty year maximum of highest number of consecutive wet days in a year. Accordingly, the highest number of consecutive wet days in the wettest year in almost all the areas in dry and intermediate zones is around 14–30 (less than 1 month). There are few areas (Hambantota and Ampara) which are experiencing consecutive wet periods of less than 2 weeks. If irrigation water is not adequate, agricultural practices in these areas are likely to suffer from severe water scarcity due to lack of rainfall. However, most of the areas in the wet zone have experienced prolonged consecutive wet periods of 1–2 months during this thirty year period. In some areas in Nuwara Eliya, the period extends up to 2–3 months even.

Consecutive dry periods are important for both agricultural and non-agricultural practices. Some crops like rice and legumes require a dry period during the harvesting stage. However, long dry spells create problems in most of the other field crop cultivations. According to Fig. 4, few areas in Ampara district is having extremely high consecutive dry periods. Approximately the dry zone of Sri Lanka has experienced the highest consecutive dry period of 3–6 months during the thirty year period. Most areas in the wet zone have shown the highest consecutive dry

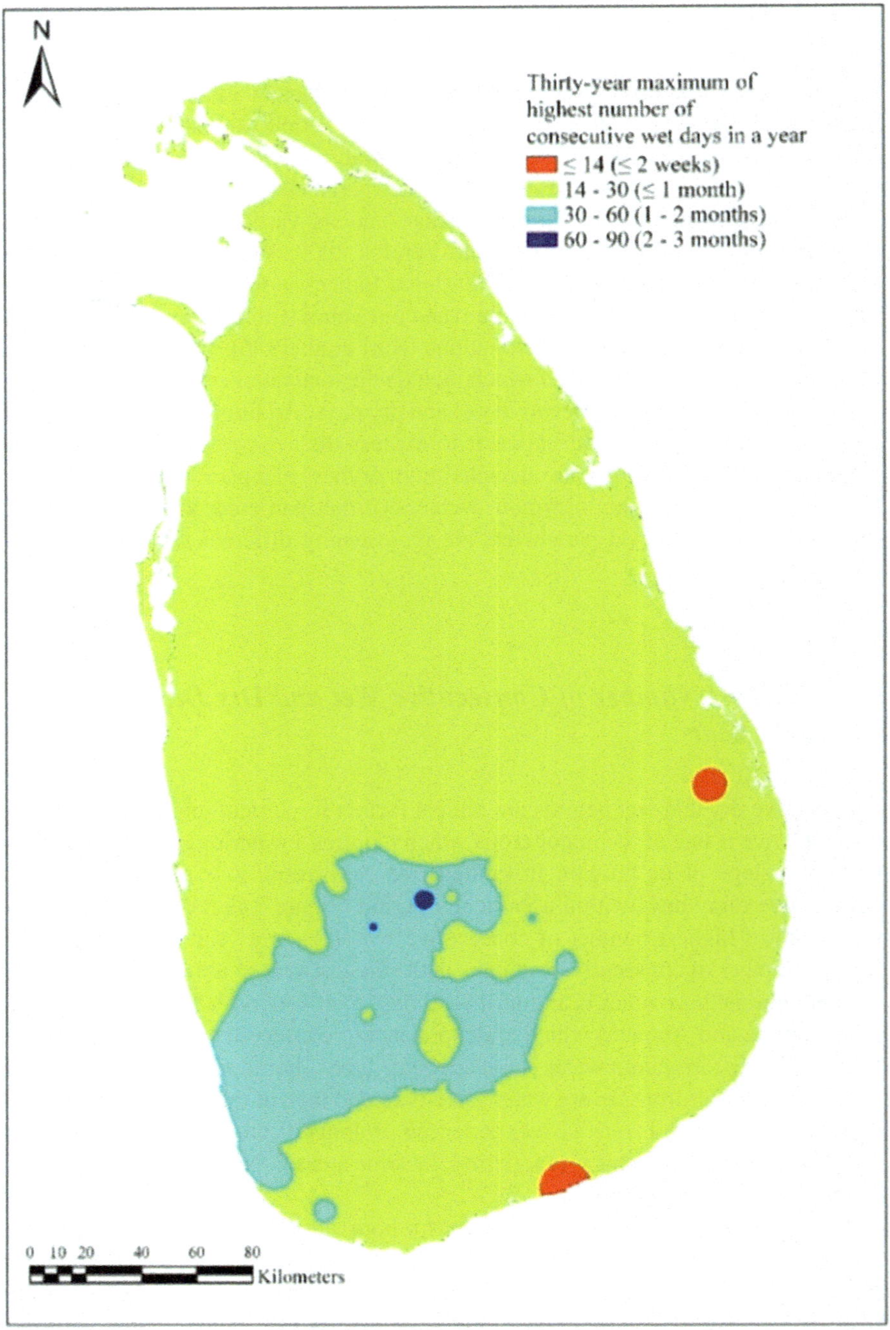

Fig. 3 Thirty year maximum of highest number of consecutive wet days in a year

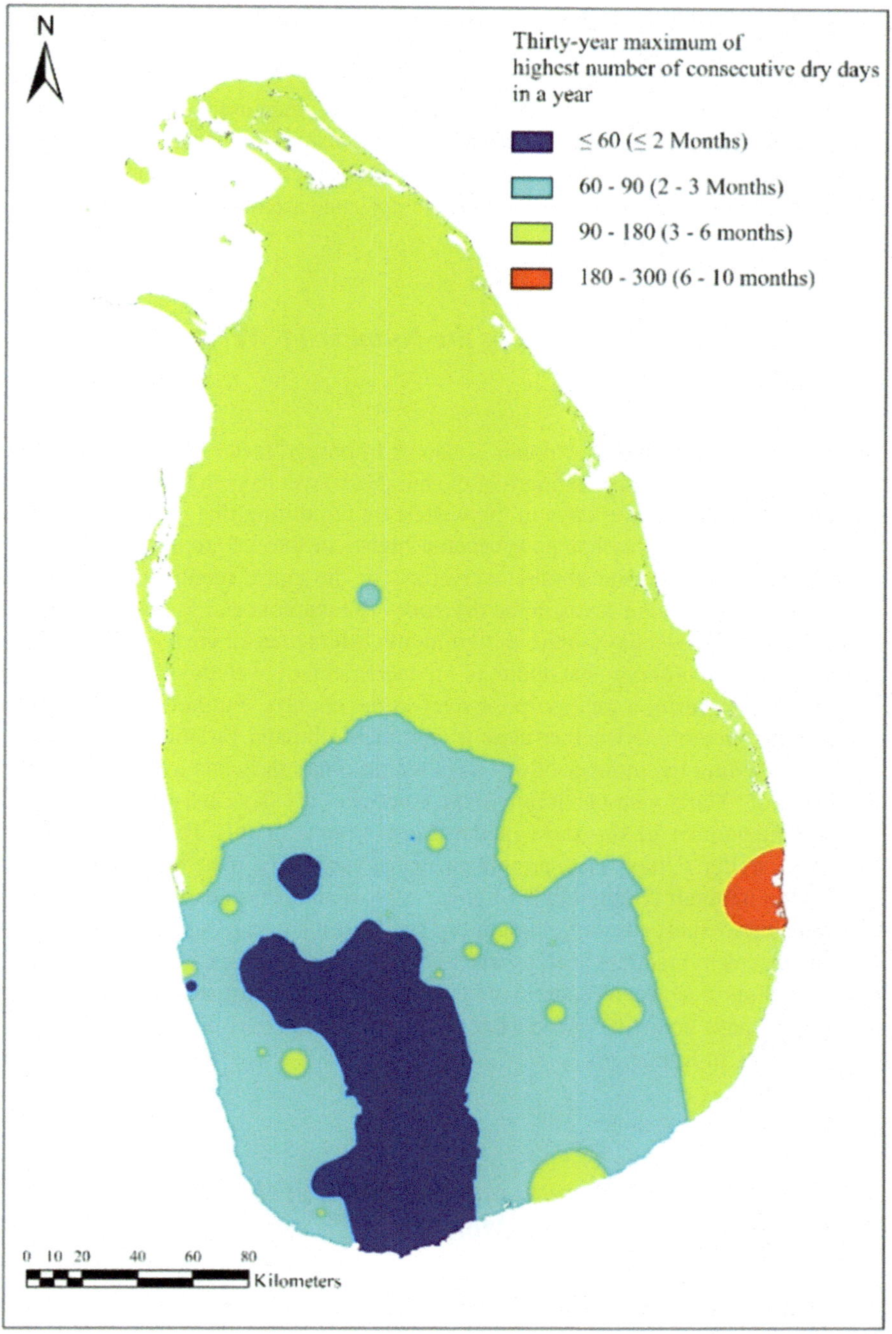

Fig. 4 Thirty year maximum of highest number of consecutive dry days in a year

period of 2–3 months during this thirty year period. In some areas in the wet zone such as Ratnapura, Nuwara Eliya, Kaluthara and Matara, this has been less than 2 months.

Long dry spells can result in low water tables, increased temperature, high evaporation and a reduction of water levels in reservoirs. Hence, it can create challenges not only for agriculture but also for domestic water consumption, hydropower generation, industrial activities etc. in dry zone areas of the country.

3.2 Thirty Year Maximum of the Number of Wet and Dry Days in a Month

Number of wet days during a month is also an important factor for agriculture. The parameter of "Thirty year maximum of the number of wet days in a month" looks for the highest number of wet days in the wettest month during this thirty year period. According to Fig. 5, this has now become higher in the wet zone. In some rainy periods, rain has occurred almost every day for the entire month in those areas. However, there are some areas in the dry zone (Hambantota and Puttlam) where the highest number of wet days in the wettest month is less than 22 days. Additionally, it is interesting to observe that there is an increment of wet days in the area of Mahaweli river basin, a multipurpose river in the country. Further analysis can be conducted to determine whether there is any micro-climatic variation in that area.

Understanding the number of dry days within a month is just as important. The parameter of "Thirty year maximum of the number of dry days in a month" identifies the highest number of dry days in the driest month for this Thirty year period. According to Fig. 6, there is no prominent spatial variability in the highest number of dry days in a month throughout Sri Lanka. Almost all areas except few locations in wet zone have experienced one-month dry spell during a year. Agriculturists should plan management practices such as water conservation practices, shading, irrigation, etc. according to seasonal patterns of the rainfall and during such dry spells. The areas which have not experienced a one-month dry period over this thirty year period is found to be in Ratnapura.

3.3 Thirty Year Maximum and Minimum of the Ratio of Wet Days to Dry Days in a Month

The parameter of "Thirty year maximum of the ratio of wet days to dry days in a month" identifies the number of wet days per one dry day in wettest month within thirty year period. It shows the maximum potential of wetness in an area. Spatial distribution of the parameter is presented in Fig. 7.

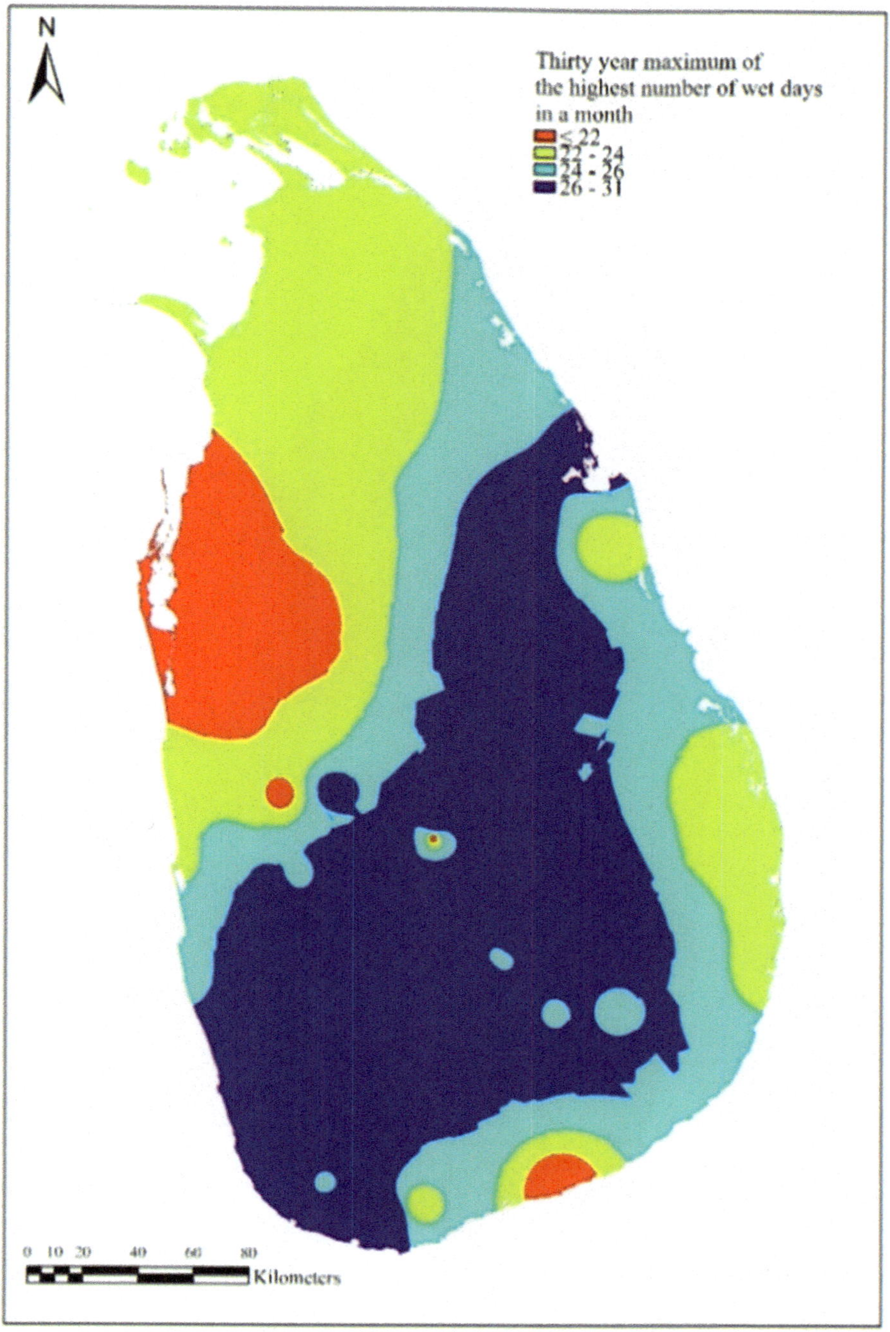

Fig. 5 Thirty year maximum of the number of wet days in a month

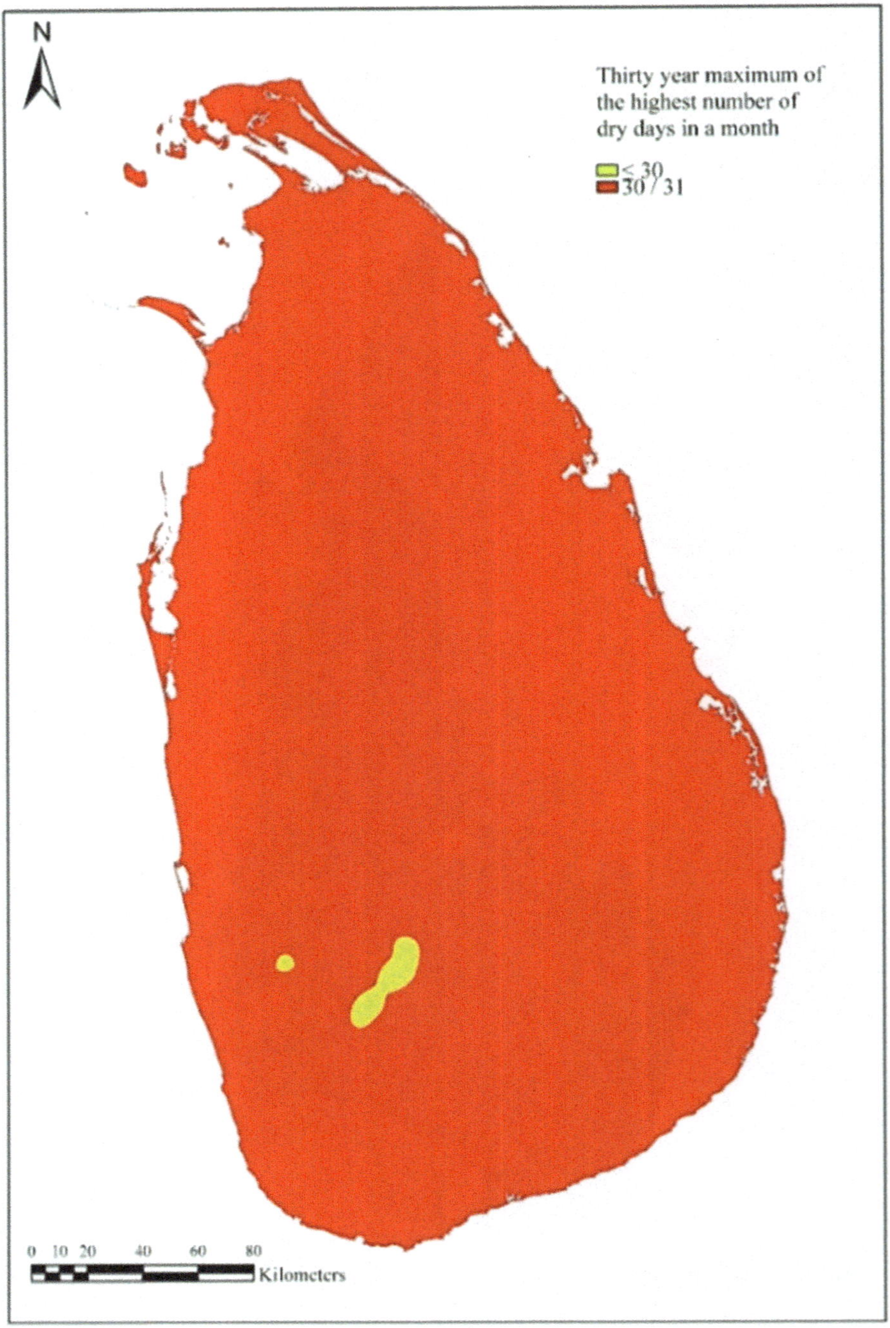

Fig. 6 Thirty year maximum of the number of dry days in a month

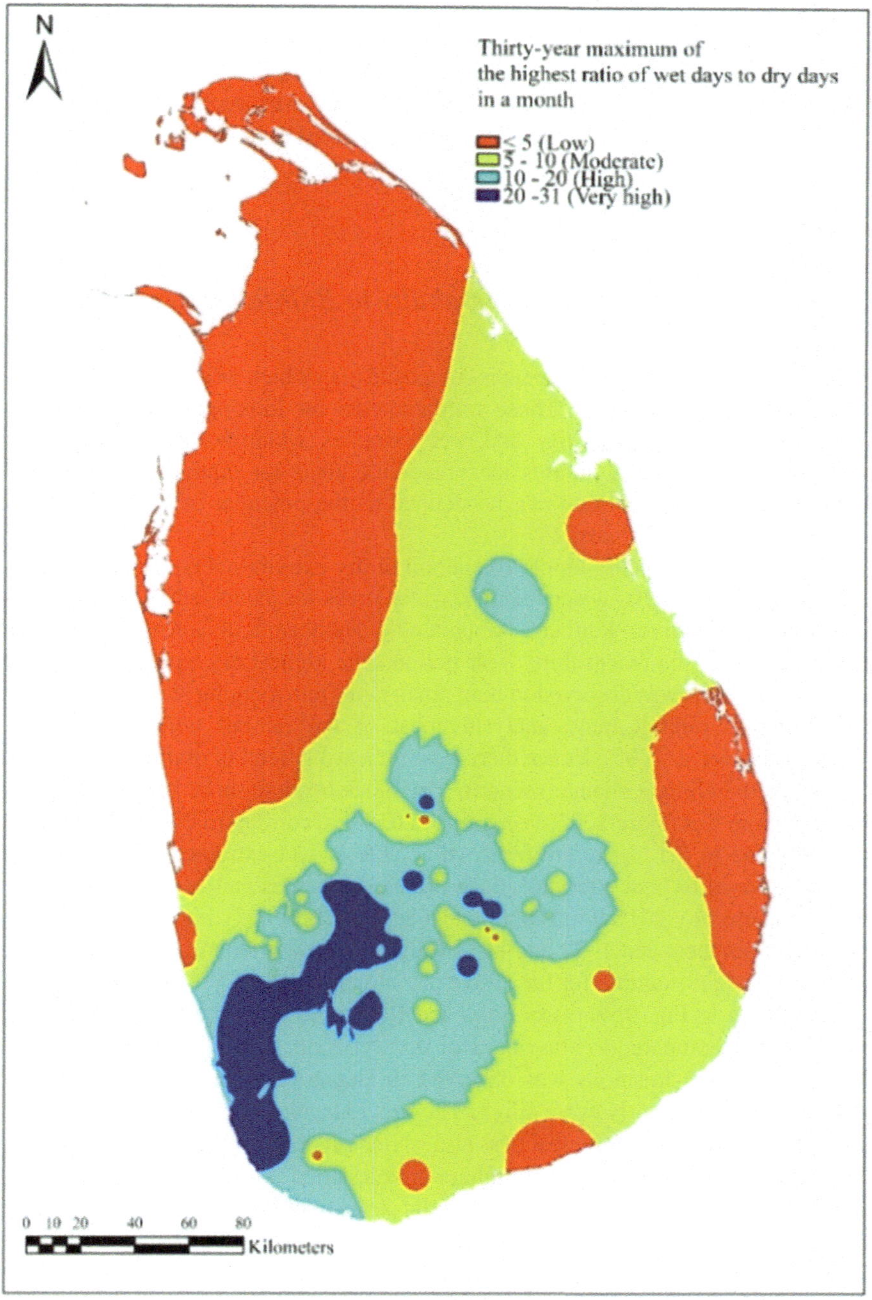

Fig. 7 Thirty year maximum of the ratio of wet days to dry days in a month

The parameter of "Thirty year minimum of the ratio of wet days to dry days in a month" identifies the number of wet days per one dry day in driest month within Thirty year period. According to Fig. 8, there is no spatial variability of this parameter. Accordingly, almost all the areas of Sri Lanka are having a potential of one-month dry spell within a year. It results the lowest ratio of wet days to dry days to be zero.

3.4 Composite Map of Vulnerability to Rainfall Extremes

Figure 9 presents the multi-parameter composite map which was developed using six of the indicative parameters. Those six parameters are looking for extreme rainfall conditions including both dry and wet extremes which have occurred during 1986–2015. The map highlights the vulnerable areas for rainfall extremes in Sri Lanka, wherein vulnerability can be defined as the potential of either dry or wet extremes in a particular area.

The areas experiencing higher likelihood of dry extremes (Fig. 9) are vulnerable to prolonged dry spells, water scarcity, water stress for plants, etc. The map shows that Jaffna, Mulativu, Puttlam, Ampara and Hambanthota areas had most dry extremes during the recent thirty year period. The highest dry extreme of 103 rain gauging locations was observed in Sangaman tank (rain gauging station) in Ampara. According to rainfall trend analysis, most of the rainfall trends are negative (Jayawardene et al. 2005). Hence there is an increasing trend of dryness in Sri Lanka.

As per the climate change scenario, dry areas become more dry and wet areas become more wet (Dore 2005; Wang et al. 2014). According to Fig. 9 some areas in the wet zone in Sri Lanka are more vulnerable to wet extremes. Those areas are highly exposed to heavy rainfall, flash floods, landslides and long wet spells. The damage caused by 2016 floods and landslides was extremely higher than the worst flood disasters between 1992 and 2011 (https://www.worldbank.org) which can be considered as an example for the above phenomenon.

According to Fig. 9, Nuwara Eliya and Ratnapura areas are experiencing more wet extremes compared to other parts of the wet zone. The highest wet extreme of 103 rain gauging locations was observed in Hapugasthenna estate (rain gauging station) in Maskeliya, Nuwara Eliya. The previous studies have proven that those areas are vulnerable to landslides also. Nine districts in Sri Lanka including Ratnapura, Nuwara Eliya and Kaluthara are classified under landslide prone areas (Ajith 2018). The occurrences as well as the severity of the landslides has increased during recent times (The World Bank Group 2011).

The moderate areas in Fig. 9 are moderate in terms of vulnerability. Both wet and dry extremes are low in that area. According to Fig. 9, the moderate area has shrunken. The Intermediate zone which is a major climatic zone in Sri Lanka falls on the category of "dry extremes". Hence, Intermediate Zone in the country is displaying a trend of drying and there is a shift of dry zone towards the wet zone

Fig. 8 Thirty year minimum of the ratio of wet days to dry days in a month

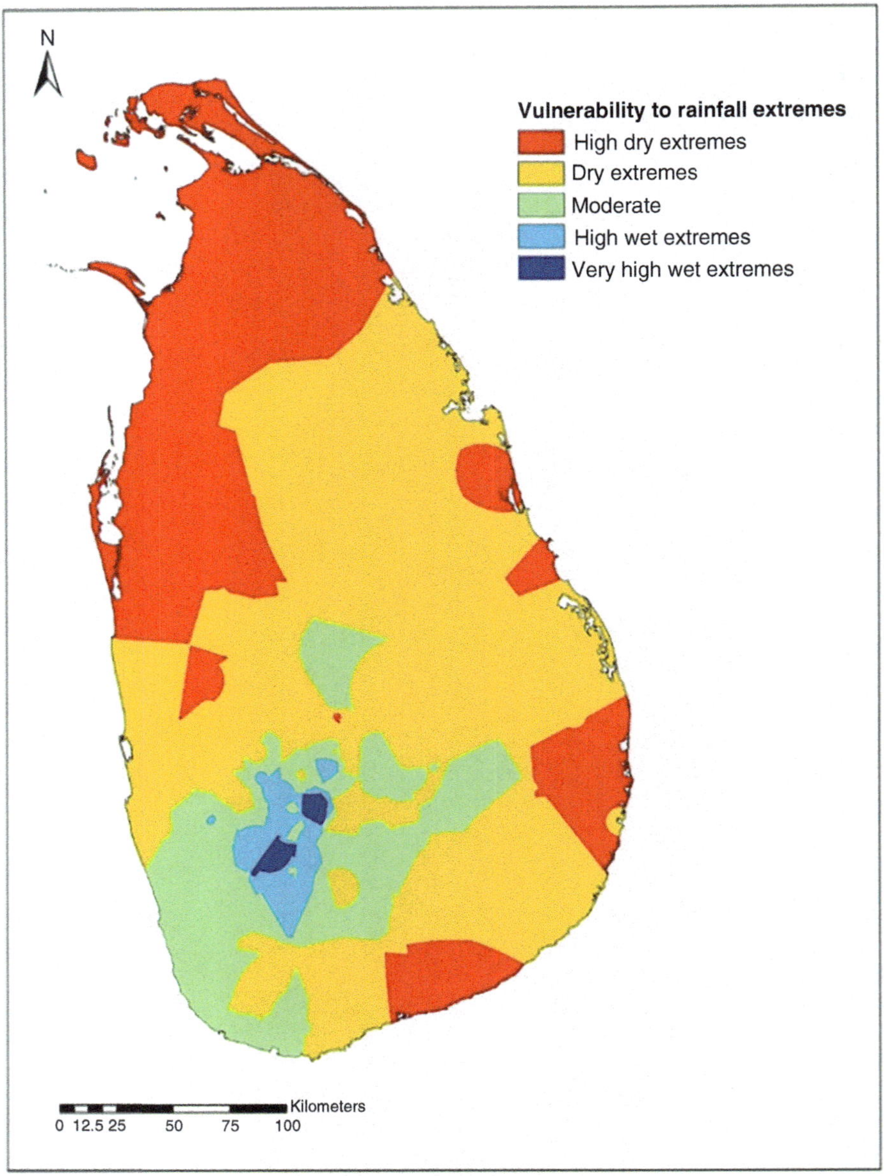

Fig. 9 Vulnerability to rainfall extremes

of the country. Accordingly, there is a reduction of the wet zone in the country (Ministry of Environment, Sri Lanka 2011).

Some evidences of increasing intensive rainfalls together with increased prolonged dry spells were reported by previous studies. Thereby, occurrences of extreme climatic events will be high in the future (De Costa 2008; Manawadu and

Fernando 2008; USAID 2015). Additionally, Colombo district has been categorized as the most vulnerable district where, the vulnerability has been defined as the degree to which a system is susceptible or unable to cope with the adverse effects of climate change (Niranjan et al. 2015)

4 Discussion

The maps developed in this study can be used to guide agricultural decision-making and practices, urban planning, disaster preparedness etc. From an agricultural perspective, the crops which are cultivated should be changed with the changing climate. Therefore, these maps can be used to identify the emerging trends of climatic factors and it is easy to perform agricultural planning and management practices such as crop suitability analysis, crop selection, irrigation planning, soil and water conservation practices and so on. It has been reported that the dry zone in Sri Lanka is at high risk of losing agricultural production due to increasing dryness (Seo et al. 2005).

These developed maps will also provide useful information in the planning and design of infrastructure. The areas which have a potential of high intensive rainfall should have more water collection methods and proper drainage systems to remove excess water.

Disaster preparedness with the changing climate is important to save lives and assets by identifying vulnerable areas before an extreme climatic event. Figure 9 indicates the potential areas for wet and dry extremes. Hence, the map can be used in flood control, drought management, designing soil conservation practices. Additionally, these maps are useful in environmental conservation and management.

In this analysis, only 103 rain gauging locations of Sri Lanka was selected to analyze daily rainfall data due to limitations in accessing data from a large number of stations for a consecutive Thirty year period. However, use of high number of rain gauging stations is expected to improve the accuracy of analysis and mapping. Additionally, it is better to analyze the daily rainfall data for a very long period. In such a situation, accessing the continuous daily rainfall data for more than thirty years from the same rain gauging location is another limitation.

This analysis can be extended to increase the number of parameters to develop maps which could visualize annual, seasonal and monthly variations of rainfall.

5 Conclusion

The extreme rainfall related parameters developed under the study provide an insight into the characteristic behavior of rainfall conditions in Sri Lanka. It will not be possible to identify and recognize these changes if long-term averages of rainfall are not analyzed. These parameters highlight the existence of high spatial variability of rainfall in the country. The analysis identifies that the wet zone of Sri Lanka is experiencing increasing wet extremes making certain areas in the wet zone highly vulnerable to floods, landslides, etc. Whereas, the dry zone of the country is increasingly prone to dry extremes, exacerbating the risk of droughts, water scarcity, etc. in the area. The spatial diversity of these parameters provides information for the farmers and decision makers to identify suitable crop management practices according to climatic conditions of their localities. Additionally, this information will be useful in disaster preparedness and management aspects.

The study proves the usefulness of analyzing climate extremes to derive an understanding regarding the anticipated changes in climatic conditions and their possible impacts on lives and livelihoods.

Appendix

No	Station	District	Longitudes (km)	Latitudes (km)
1	Trincomalee	Trincomalee	252,000	371,734
2	Puttlam	Puttlam	96,596	312,344
3	Katugastota – Kandy	Kandy	183,750	233,000
4	Katunayake	Gampaha	102,086	217,916
5	Hambantota	Hambantota	239,000	105,000
6	Dyraaba Estate-Badulla	Badulla	218,500	187,500
7	Ledgerwatta	Badulla	227,258	202,544
8	Mapakadawewa	Badulla	229,500	231,750
9	Beause (Lower)	Galle	151,496	105,920
10	Vincit Estate – Waharaka	Gampaha	134,000	211,000
11	Norton	Nuwara Eliya	172,358	190,466
12	Vannikulam	Mulattivu	151,496	429,830
13	Holmwood Estate – NuwaraEliya	Nuwara Eliya	193,500	183,500
14	Kenilworth – Nuwara Eliya	Nuwara Eliya	168,500	199,500
15	Alupolla group-Ratnapura	Ratnapura	177,500	167,500
16	Ambewela – Nuwara Eliya	Nuwara Eliya	202,500	185,750
17	Amparai tank	Ampara	296,738	231,123
18	Anamaduwa Dispensary	Puttlam	115,000	297,750
19	Anuradhapura	Anuradhapura	156,986	345,284
20	Baddegama	Galle	134,250	108,600
21	Badulla	Badulla	231,500	199,000
22	Balangoda Post Office	Ratnapura	192,122	160,820
23	Bandaraeliya	Badulla	227,258	176,192
24	Bandarawela – Badulla	Badulla	222,500	180,000
25	Campion Estate	Nuwara Eliya	192,500	175,500
26	Canawarella Grp.	Badulla	238,238	188,270

(continued)

No	Station	District	Longitudes (km)	Latitudes (km)
27	Colombo	Colombo	99,500	190,250
28	Debedda	Badulla	238,238	195,956
29	Dehiwala Zoo	Colombo	100,500	184,500
30	Denagama – Matara	Matara	186,500	98,500
31	Detanagala(Balangoda Grp.)-Ratnapur	Ratnapura	192,000	171,000
32	Digalla – Dehiowita	Kegalle	148,202	193,760
33	Diyatalawa survey camp	Badulla	220,250	178,000
34	Duckwari Estate	Kandy	201,500	239,500
35	Dunedine	Kegalle	146,006	202,544
36	Geekiyanakanda Estate	Kalutara	128,438	155,330
37	Goluwawatta	Matara	167,966	100,430
38	Hakgala Bot. Gard.	Nuwara Eliya	205,500	191,500
39	Halwatura	Kalutara	137,222	168,506
40	Hanwella Group	Colombo	128,438	186,074
41	Hapugastenna Est. – Maskeliya	Nuwara Eliya	179,000	190,500
42	Hellbodde Estate – Pussellawa	Nuwara Eliya	188,000	209,750
43	Henarathgoda Bot. Grdns.-Gampaha	Gampaha	116,500	208,000
44	Hiyare	Galle	150,398	97,136
45	Horana	Kalutara	127,500	168,000
46	Kamalasram – Udubaddawa	Kurunagala	113,066	251,954
47	Kantali tank	Trincomalee	222,866	347,480
48	Sangamam Tank	Ampara	312,902	213,524
49	Keragala Estate – Kuruwita	Ratnapura	153,250	177,000
50	Kirklees Estate	Badulla	217,376	197,054
51	Kottukachchi	Puttlam	111,968	300,266
52	Koulwewa	Kurunagala	135,250	248,500
53	Kurundu Oya – Nuwara Eliya	Nuwara Eliya	206,500	208,000
54	Kurunagala	Kurunagala	153,692	250,856
55	Labuduwa – Galle	Galle	140,000	97,500
56	Labugama tank	Colombo	135,026	180,584
57	Labookelle Est.	Nuwara Eliya	194,250	202,250
58	Lellopitiya – Deltota	Ratnapura	170,500	163,500
59	Liddesdale	Nuwara Eliya	209,000	202,500
60	Lower Spring Valley	Badulla	236,042	190,466
61	Mapalana	Matara	177,000	95,000
62	Murukkan	Mannar	120,752	400,184
63	Nawalapitiya	Kandy	175,000	205,500
64	Navatikiri aru	Ampara	292,772	250,490
65	Negambo	Gampaha	96,596	223,406
66	New Forest – Kandy	Kandy	190,000	216,000
67	Nuwara Eliya	Nuwara Eliya	199,250	197,000

(continued)

No	Station	District	Longitudes (km)	Latitudes (km)
68	Peradeniya agmet.	Kandy	181,000	231,000
69	Ratmalana	Colombo	102,086	179,486
70	Ratnapura	Ratnapura	166,500	167,500
71	Rayigama	Kalutara	135,026	173,996
72	Tabbowa Irrigation	Puttlam	109,772	316,736
73	USK Valley	Kalutara	140,516	152,036
74	Vavuniya	Vavuniya	170,162	391,400
75	Wariyapola Expt. Station	Kurunagala	142,250	270,250
76	Welimada group – Badulla	Badulla	215,500	190,500
77	Pallegama group-Niyadurupola	Kegalle	141,797	218,465
78	Wellawaya	Monaragala	236,250	170,000
79	Udaveriya, West Haputale	Badulla	208,000	177,500
80	Wewalthalawa Est	Kegalle	157,500	205,250
81	Wewessa estate	Badulla	235,500	195,000
82	Canyon	Nuwara Eliya	173,456	186,074
83	Castlereigh	Nuwara Eliya	177,848	184,976
84	Diyabeduma	Polonnaruwa	210,788	301,364
85	Polontalawa	Kurunagala	115,262	278,306
86	Galoola Estate	Badulla	241,532	206,936
87	Galphele	Kandy	192,122	237,680
88	Giritale tank	Polonnaruwa	216,500	316,000
89	Katukitula – Helbodde	Nuwara Eliya	188,000	209,750
90	High Forest Estate	Nuwara Eliya	206,396	206,936
91	Hingurakgoda	Polonnaruwa	220,500	315,500
92	Hope Estate-Daraoya	Nuwara Eliya	197,000	210,250
93	Kalarr	Trincomalee	254,708	341,990
94	Kalatuwawa	Colombo	135,000	185,000
95	Kandekatiya	Badulla	227,258	217,916
96	Kandy Kings Pavillion	Kandy	185,000	233,000
97	Kundasale farm – Kandy	Kandy	191,500	232,250
98	Undugoda	Kegalle	155,888	213,524
99	Mahagalkadawala – Galgamuwa	Kurunagala	146,006	316,736
100	Maussakelle	Nuwara Eliya	175,250	183,800
101	Maha Illuppallama – Anuradhapura	Anuradhapura	167,000	322,250
102	Okkampitiya – Moneragala	Monaragala	259,000	171,750
103	Pottuvil	Ampara	316,196	186,074

References

Ajith, S. (2018). *Climate change and Sri Lanka*. Retrieved from https://unfccc.int/files/adaptation/application/pdf/nwp_integration_silva.pdf on 01 Apr 2018.

De Costa, W. A. J. M. (2008). Climate change in Sri Lanka: Myth or reality? Evidence from long-term meteorological data. *Journal of the National Science Foundation of Sri Lanka, 36*, 63–88.

Dore, M. H. (2005). Climate change and changes in global precipitation patterns: What do we know? *Environment International, 31*(8), 1167–1181.

Easterling, D. R., Evans, J. L., Groisman, P. Y., Karl, T. R., Kunkel, K. E., & Ambenje, P. (2000a). Observed variability and trends in extreme climate events: A brief review. *Bulletin of the American Meteorological Society, 81*(3), 417–425.

Easterling, D. R., Meehl, G. A., Parmesan, C., Changnon, S. A., Karl, T. R., & Mearns, L. O. (2000b). Climate extremes: Observations, modeling, and impacts. *Science, 289*(5487), 2068–2074.

Eriyagama, N., Smakhtin, V., Chandrapala, L., & Fernando, K. (2010). *Impacts of climate change on water resources and agriculture in Sri Lanka: A review and preliminary vulnerability mapping* (Vol. 135). Colombo: IWMI.

Fernando, T. K., & Chandrapala, L. (1992). Global warming and rainfall variability–Sri Lankan Situation. In *Proceedings of the 47th annual session of Sri Lanka Association for the Advancement of Science (SLASS)*, p. 138.

Hijioka, Y., Lin, E., Pereira, J. J., Corlett, R. T., Cui, X., Insarov, G. E., Lasco, R. D., Lindgren, E., & Surjan, A. (2014). Asia. In V. R. Barros, C. B. Field, D. J. Dokken, M. D. Mastrandrea, K. J. Mach, T. E. Bilir, M. Chatterjee, K. L. Ebi, Y. O. Estrada, R. C. Genova, B. Girma, E. S. Kissel, A. N. Levy, S. MacCracken, P. R. Mastrandrea, & L. L. White (Eds.), *Climate change 2014: Impacts, adaptation, and vulnerability. Part B: Regional aspects. Contribution of Working Group II to the fifth assessment report of the Intergovernmental Panel on climate change* (pp. 1327–1370). Cambridge/New York: Cambridge University Press.

https://www.worldbank.org/en/country/srilanka/publication/sri-lanka. Development-update retrieved on 06 Aug 2019.

Hutchinson, M. F. (1995). Interpolating mean rainfall using thin plate smoothing splines. *International Journal of Geographical Information Systems, 9*(4), 385–403.

IPCC. (2014). Annex II: Glossary [K. J. Mach, S. Planton, & C. von Stechow (Eds.)]. In Core Writing Team, R. K. Pachauri, & L. A. Meyer (Eds.), *Climate change 2014: Synthesis report. Contribution of Working Groups I, II and III to the fifth assessment report of the Intergovernmental Panel on climate change* (pp. 117–130). Geneva: IPCC.

Jayawardene, H. K. W. I., Sonnadara, D. U. J., & Jayewardene, D. R. (2005). Trends of rainfall in Sri Lanka over the last century. *Sri Lankan Journal of Physics, 6*, 7–17.

Karl, T. R., Knight, R. W., Easterling, D. R., & Quayle, R. G. (1996). Indices of climate change for the United States. *Bulletin of the American Meteorological Society, 77*(2), 279–292.

Manawadu, L., & Fernando, N. (2008). Climate changes in Sri Lanka. University of Colombo Review. *Journal of the University of Colombo, Sri Lanka, New Series, 1*, 2.

Ministry of Environment. (2011). *Sri Lanka's second national communication on climate change.*

Ministry of Mahaweli Development and Environment. (2016). *National adaptation plan for climate change impacts in Sri Lanka 2016 to 2025.*

Nastos, P. T., & Zerefos, C. S. (2009). Spatial and temporal variability of consecutive dry and wet days in Greece. *Atmospheric Research, 94*(4), 616–628.

Niranjan, F., Uddika, N. C. P., Bantilan, M. C. S., & Singh, N. P. (2015). Assessing agricultural vulnerability to climate change in Sri Lanka. *Sri Lanka Journal of Food and Agriculture, 1*(1), 85–92.

Parry, M., Parry, M. L., Canziani, O., Palutikof, J., Van der Linden, P., & Hanson, C. (Eds.). (2007). *Climate change 2007-impacts, adaptation and vulnerability: Working group II contribution to the fourth assessment report of the IPCC* (Vol. 4). Cambridge: Cambridge University Press.

Punyawardana, B. V. R. (2002). *Climate change: Challenges and opportunities in Sri Lanka.* Dhaka: Natural Resource Management Centre.

Seo, S. N. N., Mendelsohn, R., & Munasinghe, M. (2005). Climate change and agriculture in Sri Lanka: A Ricardian valuation. *Environment and Development Economics, 10*(5), 581.

Sivakumar, M. V., & Stefanski, R. (2010). Climate change in South Asia. In *Climate change and food security in South Asia* (pp. 13–30). Dordrecht: Springer.

The World Bank Group. (2011). *Vulnerability, risk reduction, and adaptation to climate change, Sri Lanka.* Climate risk and adaptation country profile April 2011. The World Bank Group, Washington.

USAID (United States Agency International Development). (2015). *Climate change information factsheet Sri Lanka.*

Wang, S., Huang, J., He, Y., & Guan, Y. (2014). Combined effects of the Pacific decadal oscillation and El Nino-southern oscillation on global land dry–wet changes. *Scientific Reports, 4*, 6651.

WMO (World Meteorological Organization). (2016). *Guidelines on the definition and monitoring of extreme weather and climate events.*

Deciphering of Groundwater Recharge Potential Zones in Dhaka City, Bangladesh by RS and GIS Techniques

Imran Hossain Newton, Rashed Uz Zaman, Sara Nowreen, A. K. M. Saiful Islam, Sadmina Razzaque, and G. M. Tarekul Islam

1 Introduction

Attributing to rapid urbanization and population growth, the demand for fresh water is fast increasing in Dhaka city of Bangladesh (Hoque et al. 2007). About 87% of the water supply in the city comes from groundwater sources (Uddin and Baten 2011). Overreliance on groundwater has resulted in an average annual declination of about 2 m of groundwater levels (Akther et al. 2009). This consequence has resulted in the drawing down of water levels up to 70 m (Hoque et al. 2007) which leads to drying out the wells (Wada et al. 2010). As a result, city dwellers are facing severe shortages of water. The problem is more severe during the dry and hot summer season because of high demand and reduction in abstraction quantity (Akther et al. 2009). In contrast, the obstruction in infiltration due to the expansion of infrastructures hinders the groundwater recharge process which further exacerbates the situation. Hence, there is an urgent need to seek out a plausible solution or technique to manage this resource. In view of this, this study explored plausible artificial recharge zones for restoring groundwater resources to a better state.

Field survey-based investigation for artificial recharge zones is highly time consuming, costly, and relatively difficult. Considering these limitations, an integrated approach of remote sensing (RS) and geographical information system (GIS) techniques were employed in this study, as this methodology has proven to be an effective method in investigation of artificial recharge sites (Srinivasa Rao and Jugran 2003). In recent years, many studies (Shaban et al. 2006; Kumar et al. 2007; Chowdhury et al. 2009; Yeh et al. 2009; Fashae et al. 2014; Bhuvaneswaran et al. 2015; Fenta et al. 2015; Deepa et al. 2016; Souissi et al. 2018; Thakur et al.

I. H. Newton (✉) · R. U. Zaman · S. Nowreen · A. K. M. S. Islam · S. Razzaque · G. M. T. Islam
Institute of Water and Flood Management (IWFM), Bangladesh University of Engineering and Technology (BUET), Dhaka, Bangladesh

A. Haque, A. I. A. Chowdhury (eds.), *Water, Flood Management and Water Security Under a Changing Climate*, https://doi.org/10.1007/978-3-030-47786-8_5

2018) have been undertaken, which used the RS and GIS techniques in the exploration of artificial recharge sites as well as the demarcation of groundwater potential zones.

However, these studies reported that many factors (i.e., drainage density, lineament density, land use/land cover (LULC), slope, soil texture, rainfall, lithology, etc.) are responsible for groundwater movement and occurrence and it is essential to understand their properties in artificial recharge sites investigation. In this study, some selected influencing factors (i.e., drainage density, lineament density, LULC, slope, and lithology) were investigated and thematic layers were prepared and then integrated to create a spatial mapping that demarcated different categories of (artificial) groundwater recharge potential zones for Dhaka city.

2 Study Area

Dhaka city (as shown in Fig. 1), the capital of Bangladesh, is situated in the centre of the country and lies between 23°40′ and 23°54′ North latitude and 90°20′ and 90°30′ East longitude, covering an area of 300 km^2 and having an alleviation of 6.5–9 m above mean sea level. The city experiences hot, humid and wet tropical climate with an annual average rainfall of 2500 mm (using weather data for the years of 2005–2018), where about 80% of rainfall occurs during July and August. The annual average temperature is 26 °C. The city is bounded by the Balu River in the east, the Turag River in the west, the Tongi Khal in the north and the Buriganga River in the south and southeast.

3 Methodology

The study considered five influential factors that are largely responsible for groundwater movement and occurrence, viz. drainage density, lineament density, LULC, slope, and lithology (Fenta et al. 2015). The thematic maps on drainage density and slope were generated using the Digital Elevation Models (DEM) data from Shuttle Radar Topography Mission (SRTM) images. Landsat-8 OLI satellite images were used to prepare the lineament density map. At the beginning, the lineament features were extracted from the false color composite images of Landsat-8 OLI using Geomatica 2018 software and then the features were converted to lineament density map using ArcGIS 10.3 software. The LULC pattern was assessed from the Landsat-8 OLI satellite images of January 2018. Supervised classification approach with Maximum Likelihood Classification (MLC) algorithm of ArcGIS 10.3 software was employed to assess the LULC. During supervised classification, about 40 training samples (considering at least 300 pixels) were considered to train a single LULC type. This process was repeated further to identify different LULC classes. Error analysis was performed by considering two several measures of accuracy

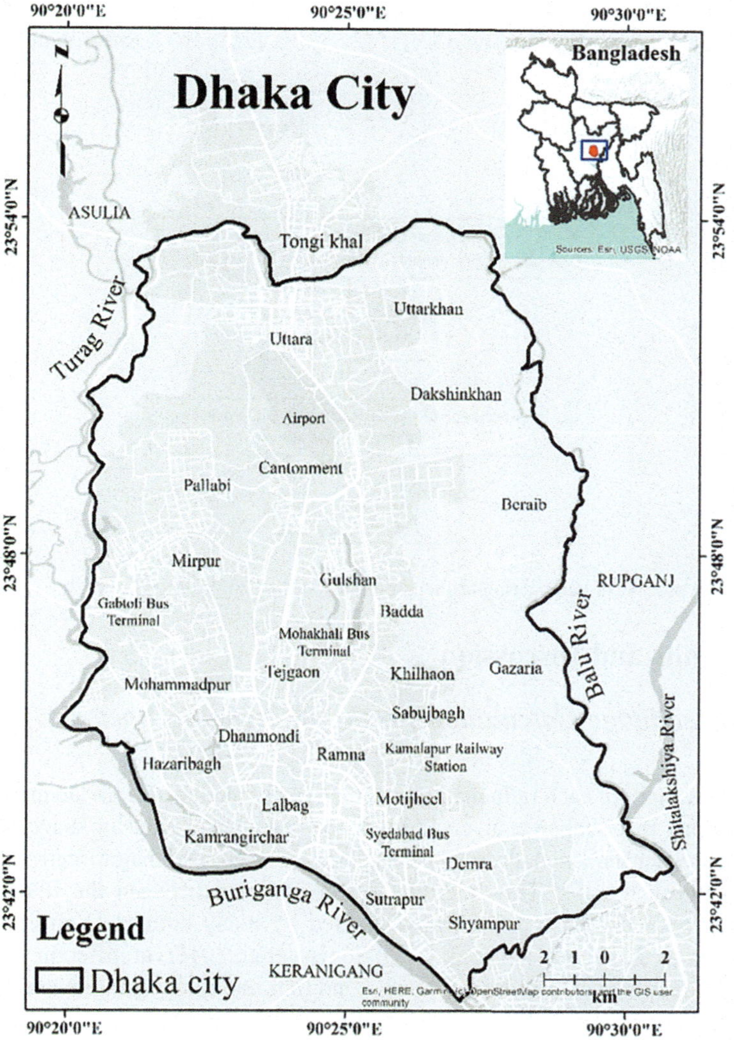

Fig. 1 Map of Dhaka city, Bangladesh

assessment – overall accuracy and Kappa coefficient (Cohen 1960; Fung 1990). The lithology map was prepared by digitizing the geological map provided by the United States Geological Survey (USGS) as shown by Alam et al. (1990). All the thematic layers were then transformed to raster data with 30 m resolution using ArcGIS 10.3 software. Subjective weights were assigned to the respective thematic layers and then integrated using weighted overlay method to create a single thematic map which demonstrates the artificial recharge sites. The total process of deciphering artificial recharge zones is shown in Fig. 2 by a schematic diagram.

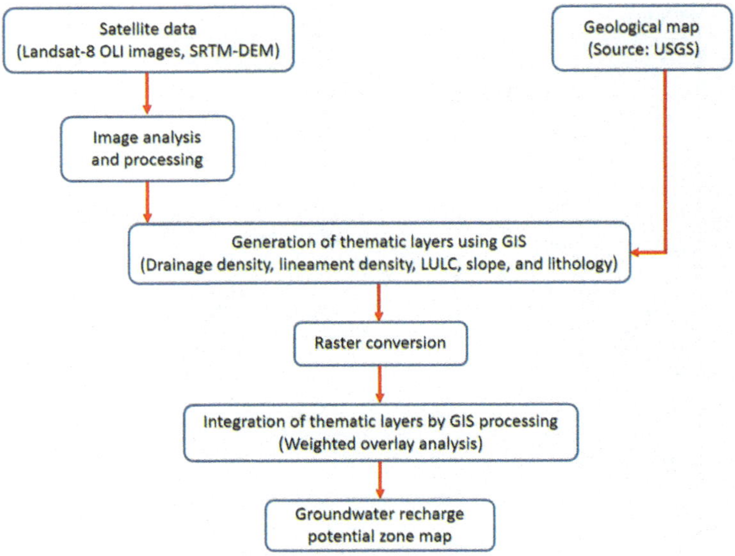

Fig. 2 The process of deciphering artificial groundwater recharge zones

4 Results and Discussion

4.1 Weightage Calculation for the Influencing Factors

Here, the weight of each influencing factor (IF) was calculated from the interrelationships among IFs. Shaban et al. (2006) proposed an interrelationship diagram (Fig. 3) of groundwater IFs which described that there are minor and major interrelationships exist between the IFs. This interrelationship diagram between the IFs related to groundwater recharge process was also used by many authors (Yeh et al. 2009; Magesh et al. 2012; Selvam et al. 2015; Souissi et al. 2018;) at different topographical, hydrogeological, and hydrological condition to explore groundwater potential zone or to decipher plausible artificial recharge sites. As shown in Fig. 3, each factor that has a major relationship with the other factors was assigned a weightage of 1.0 and those with a minor relationship are considered to have the weightage of 0.5. Here, an IF having the highest number of major and/or minor relationships with other factors gets the highest weights. The sum of weights from minor and major effects for each factor is the representing relative rate of the IF (Table 1). Further, the relative rate was used to calculate the score of each IF using the formula shown in Table 1. Afterward, the proposed score of each IF was reclassified and put for sub factors (Table 2). Subsequently, the scores were used to integrate all the thematic maps by superimposing one over another (named as weighted overlay method) to produce a final map of potential zones for recharge by using ArcGIS 10.3 software.

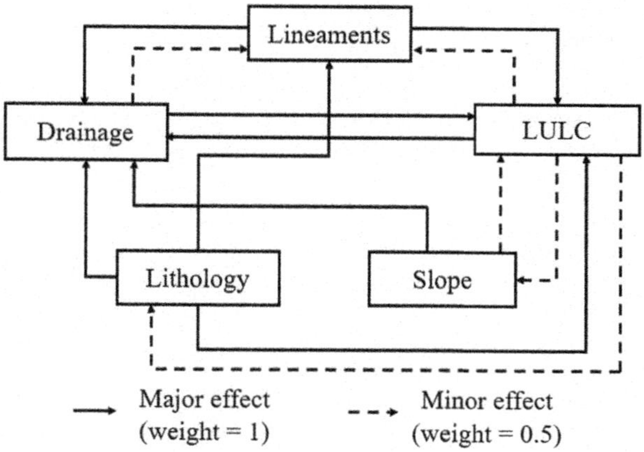

Fig. 3 Interrelationship between the IFs related to groundwater recharge process. (Modified from Shaban et al. 2006)

Table 1 Effect of influencing factors, relative rates and scores for each recharge potential factor

Factor	Major effect (A)	Minor effect (B)	Proposed relative rates (A + B)	Proposed score of each influencing factor $\frac{(A+B)*100}{\sum 10.5}$
Drainage	1	0.5	1.5	14
Lineaments	1 + 1	0	2	19
LULC	1	0.5 + 0.5 + 0.5	2.5	24
Slope	1	0.5	1.5	14
Lithology	1 + 1 + 1	0	3	29
			∑10.5	∑100

4.2 *Evaluation of the Influencing Factors Related to Groundwater Recharge*

Drainage Density

The drainage density indicates the measure of the total drainage-length in a unit area and is determined by the following equation proposed by Greenbaum (1985) (see Eq. 1).

$$D_d = \frac{\sum_{i=1}^{i=n} S_i}{A} \tag{1}$$

Where, D_d = drainage density, $\sum_{i=1}^{i=n} S_i$ = total length of the drainage network (L), and A is the unit area (L^2). The higher the drainage density, the lesser the recharge potentiality. The drainage density is an inverse function of permeability, which means the more the drainage density, the less possibility of groundwater recharge

Table 2 Weights assigned to sub-factors of each IF based on their influence to groundwater recharge

Factor	Sub-factors	Assigned weight
Drainage density	Very low (0–0.15 km/km^2)	14
	Low (0.16–0.30 km/km^2)	11
	Moderate (0.31–0.49 km/km^2)	7
	High (0.50–0.83 km/km^2)	3
Lineament density	Very low (0–0.058 km/km^2)	7
	Low (0.059–0.22 km/km^2)	12
	Moderate (0.23–0.43 km/km^2)	15
	High (0.44–0.67 km/km^2)	19
LULC	Water body	24
	Vegetation	18
	Bare soil	11
	Built-up area	5
Slope	0–1.68°	14
	1.69°–3.48°	11
	3.49°–6.45°	7
	6.46°–32.8°	3
Lithology	Alluvial silt with clay	29
	Marsh clay with peat	18
	Madhupur clay residuum	7

(Magesh et al. 2012). Considering the natural drainage system, a very low drainage density of 0–0.15 km/km^2 is noticed in the major part (34%) of the study area followed by low 0.16–0.30 km/km^2 (27%), moderate 0.31–0.49 km/km^2 (22%), and high 0.50–0.83 km/km^2 (17%) (Fig. 4).

Lineament Density

Lineaments are generally referred to linear features like faults, joints, and fractures which can be identified from satellite imagery. The area with high lineament density have high potential for groundwater recharge and the study of this parameter provides a good guide for groundwater exploration (Bhuvaneswaran et al. 2015). The lineament density indicates the total length of lineaments in a unit area (Greenbaum 1985). The lineament density is given by the following equation (see Eq. 2).

$$L_d = \frac{\sum_{i=1}^{i=n} L_i}{A} \tag{2}$$

Where, L_d = lineament density, $\sum_{i=1}^{i=n} L_i$ = the total length of the lineaments (L), and A denotes the unit area (L^2). The lineament density map (Fig. 5) shows that very low (0–0.058 km/km^2) lineament density occupied major part (85%) of the study area

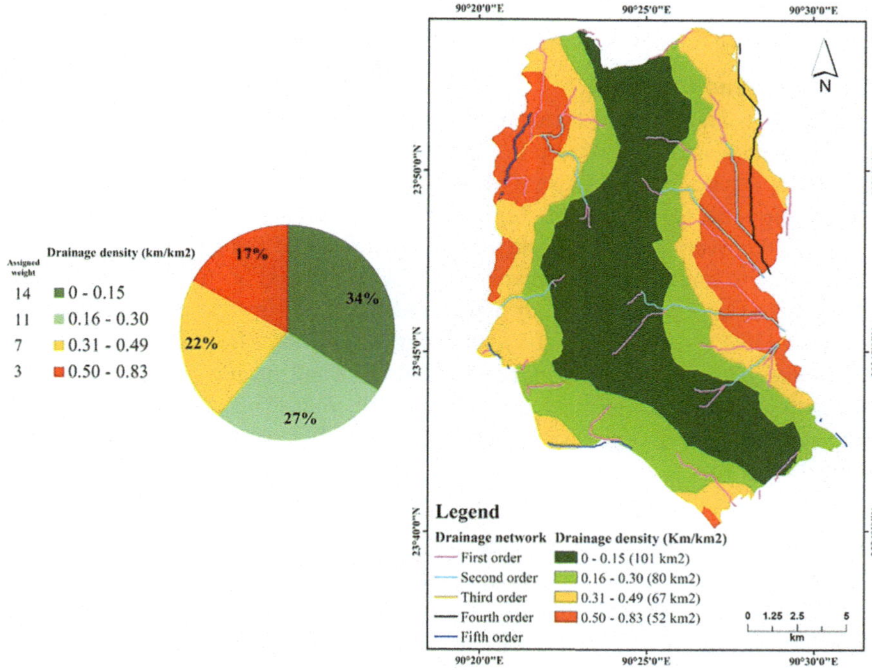

Fig. 4 Drainage density map of Dhaka city, Bangladesh

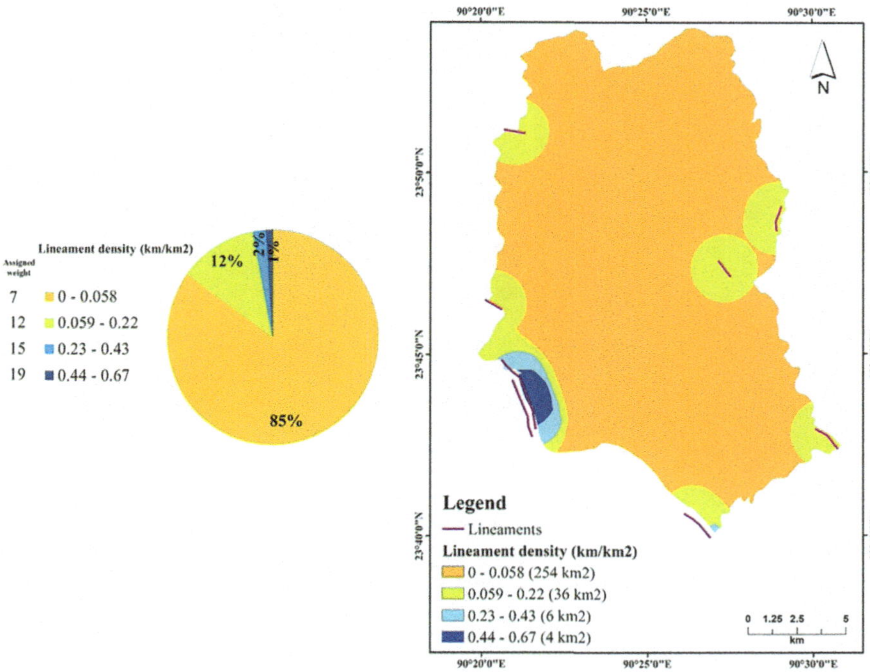

Fig. 5 Lineament density map of Dhaka city, Bangladesh

and very high lineament density (0.44–0.67 km/km^2) occupied the least part (1%) and it is concentrated in the south-western part of the study area.

Land Use/Land Cover

The LULC is an important factor in groundwater recharge as it plays a significant role in the movement and occurrence of groundwater in any region (Singh 2014). In this study, 4 types of LULC were identified from the image analysis and they are (a) built-up area, (b) vegetation, (c) bare land, and (d) water body. From the error analyses, the overall accuracy and Kappa coefficient of the LULC analysis were found 88.89% and 0.83, respectively. The LULC map (Fig. 6) of the study area shows that a major part of the study area is under the built-up area (49%) followed by bare land (33%), water body (11%), and vegetation (7%). Most of the water bodies are located in the south-eastern part of the study area and vegetation area is located in the north-eastern part. During weight assignment, water bodies and vegetated area were assigned a high weightage and followed by the bare land and built-up area (see Table 2).

Slope

The gradient of slope is an important factor for the identification of favorable sites for artificial recharge as it directly influences the infiltration process (Selvam et al. 2015). A higher degree of slope generates less recharge as the steeper slope causes rapid runoff and does not store water easily (Machiwal et al. 2011).

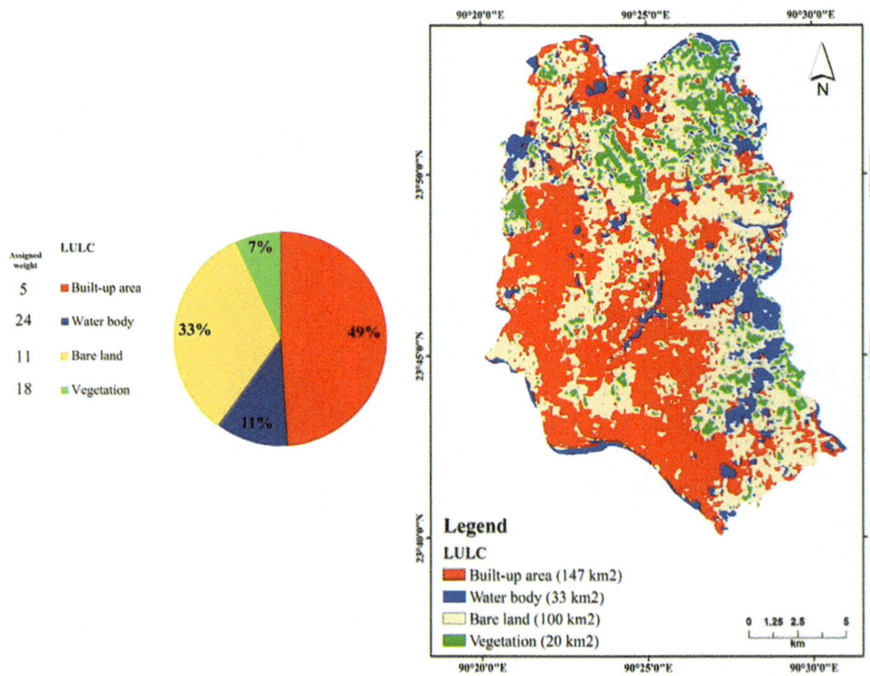

Fig. 6 LULC map of Dhaka city, Bangladesh

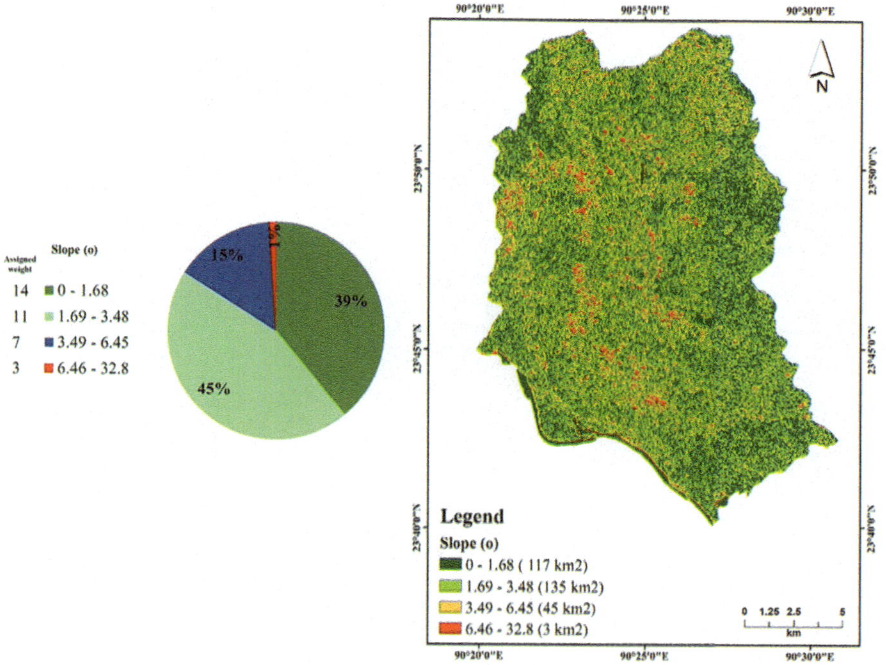

Fig. 7 Slope map of Dhaka city, Bangladesh

From the analysis, the slope of the study area has been classified into 4 groups (as shown in Fig. 7). The slope gradient was found to be relatively flat where approximately about 84% of the total area is under 3.48° and the slope varies from <1.68° to 32.8°.

Lithology

Types of lithology affect groundwater recharge by controlling the infiltration of water flow (El-Baz and Himida 1995). The study area is mainly composed of Madhupur clay residuum, which covers approximately about 41% of the total area and the rest followed by alluvial silt with clay (40%) and marsh clay with peat (19%) (Fig. 8). The Madhupur clay residuum which is not favorable for groundwater recharge, is located at the central part of the study area. The marsh clay with peat is located at the south-eastern part which is moderately favorable for recharge. The alluvial silt with clay which is more favorable for groundwater recharge is located at the western, southern, and to some degree, the north-eastern part of the study area.

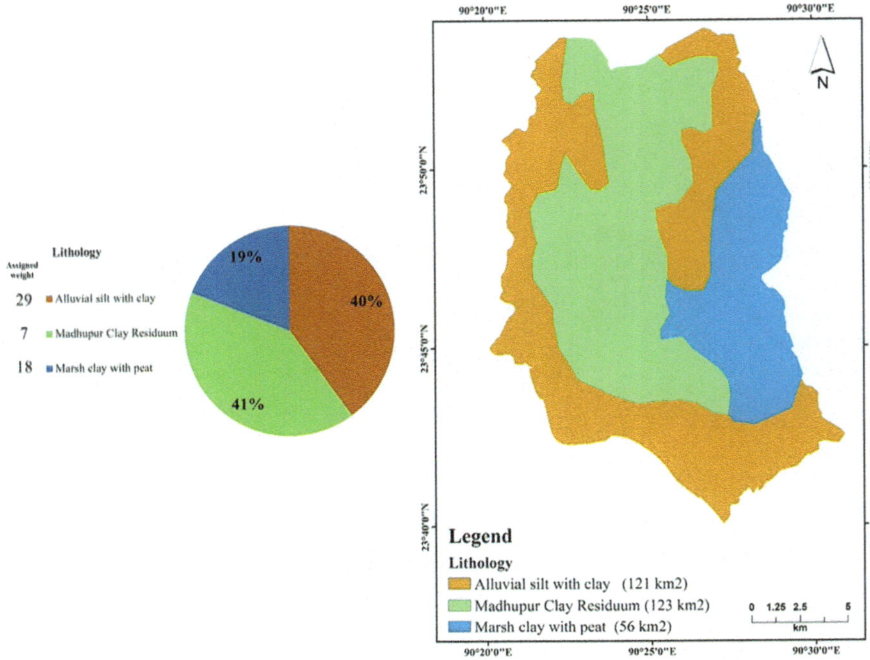

Fig. 8 Lithology map of Dhaka city, Bangladesh

4.3 Deciphering of Groundwater Recharge Potential Zone

Based on the final scores of the integrated map, recharge potential zone (Fig. 9) was classified into 5 categories with groundwater recharge potentiality from poor to excellent. The scores are attributed as: 28–46 (poor), 47–55 (low), 56–64 (moderate), 65–73 (good), and 74–96 (excellent). Analytical results demonstrate that about 11% (32 km^2) area falls under excellent zone followed by 25% (75 km^2), 23% (68 km^2), 22% (67 km^2) and 19% (58 km^2) under good, moderate, low, and poor zone, respectively. As it appears, the south-western part of the city is completely situated in build-up area and the south-eastern part is composed of bare land. However, the south-western, south-eastern, and some area of the north-eastern parts appear to have excellent recharge potentials. It is probably due to the distribution of alluvial silt and clay in their lithological formation of the aquifer, as well as low and moderate drainage density, with the presence of vegetation area.

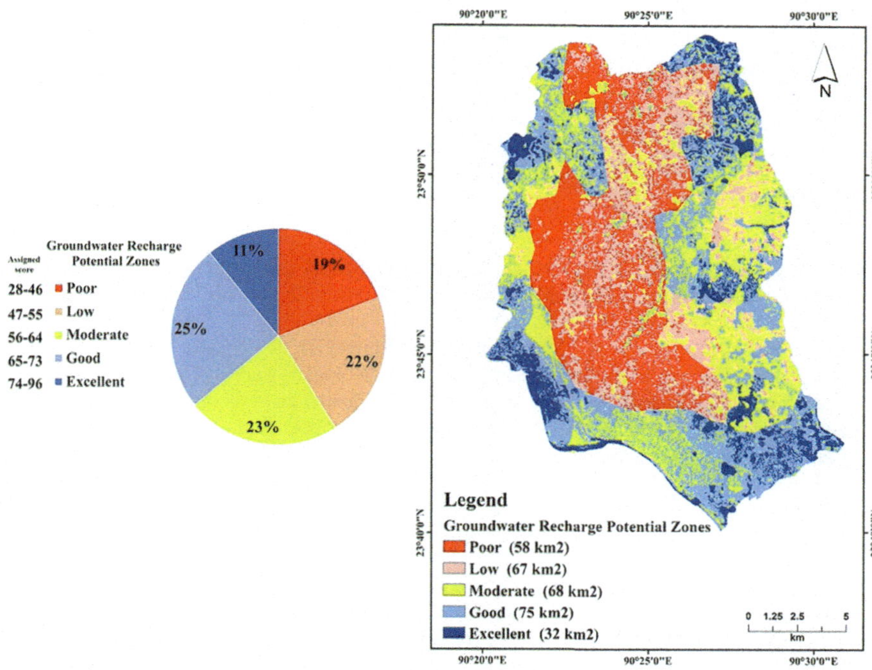

Fig. 9 Artificial groundwater recharge potential zone map of Dhaka city, Bangladesh

5 Conclusion

In this study, potential (artificial) recharge zones in Dhaka, Bangladesh were explored using various RS data and GIS approaches. Five different influencing factors (i.e., drainage density, lineament density, LULC, slope, and lithology) that control the groundwater recharge process were considered, and their thematic maps were prepared. Then, the thematic maps were integrated by weighted overlay analysis method. Based on the analysis, the artificial groundwater recharge potential zone map has been categorized into 5 different zones, namely poor (19%), low (22%), moderate (23%), good (25%), and excellent (11%). However, this analysis is not free from its limitations. For example, the secondary data used in this study was not examined further at field level or cross checked with the real situation. Although the result of the study was not validated with field investigation, the results showed would serve a first-hand information in planning future artificial recharge projects and thus, help in further groundwater exploration analyses. It is expected that the output analyses will be more usable once subjected to execution of a primary survey or field verification.

Acknowledgments The authors would like to thank the Bangladesh University of Engineering and Technology (BUET), Bangladesh, for financially supporting this research under the graduate research assistantship.

References

Akther, H., Ahmed, M. S., & Rasheed, K. B. S. (2009). Spatial and temporal analysis of groundwater level fluctuation in Dhaka City, Bangladesh. *Asian Journal of Earth Sciences, 2*(2), 49–57. https://doi.org/10.3923/ajes.2009.49.57.

Alam, M. K., Hasan, A. K. M., Khan, M. R., & Whitney, J. W. (1990). *Geological map of Bangladesh*. Geological Survey of Bangladesh. US Geological Survey, Dhaka.

Bhuvaneswaran, C., Ganesh, A., & Nevedita, S. (2015). Spatial analysis of groundwater potential zones using remote sensing, GIS and MIF techniques in Uppar Odai sub-watershed, Nandiyar, Cauvery Basin, Tamilnadu. *International Journal of Current Research, 7*(9), 20765–20774.

Chowdhury, A., Jha, M. K., Chowdary, V. M., & Mal, B. C. (2009). Integrated remote sensing and GIS-based approach for assessing groundwater potential in West Medinipur District, West Bengal, India. *International Journal of Remote Sensing, 30*(1), 231–250. https://doi.org/10.1080/01431160802270131.

Cohen, J. A. (1960). A coefficient of agreement for nominal scales. *Educational and Psychological Measurement, 20*(1), 37–46. https://doi.org/10.1177/001316446002000104.

Deepa, S., Venkateswaran, S., Ayyandurai, R., Kannan, R., & Vijay Prabhu, M. (2016). Groundwater recharge potential zones mapping in upper Manimuktha Sub Basin Vellar River Tamil Nadu India using GIS and remote sensing techniques. *Modeling Earth Systems and Environment, 2*(3), 137. https://doi.org/10.1007/s40808-016-0192-9.

El-Baz, F., & Himida, I. (1995). *Groundwater potential of the Sinai Peninsula, Egypt*. United States Agency for International Development. Project Summery, AID, Cairo.

Fashae, O. A., Tijani, M. N., Talabi, A. O., & Adedeji, O. I. (2014). Delineation of groundwater potential zones in the crystalline basement terrain of SW-Nigeria: An integrated GIS and remote sensing approach. *Applied Water Science, 4*(1), 19–38. https://doi.org/10.1007/s13201-013-0127-9.

Fenta, A. A., Kifle, A., Gebreyohannes, T., & Hailu, G. (2015). Spatial analysis of groundwater potential using remote sensing and GIS-based multi-criteria evaluation in Raya Valley, northern Ethiopia. *Hydrogeology Journal, 23*(1), 195–206. https://doi.org/10.1007/s10040-014-1198-x.

Fung, T. (1990). An assessment of TM imagery for land-cover change detection. *IEEE Transactions on Geoscience and Remote Sensing, 28*(4), 681–684. https://doi.org/10.1109/TGRS.1990.572980.

Greenbaum, D. (1985). *Review of remote sensing applications to groundwater exploration in basement and regolith*. British Geological Survey, Report OD 85/8, UK, p. 21.5;

Hoque, M. A., Hoque, M. M., & Ahmed, K. M. (2007). Declining groundwater level and aquifer dewatering in Dhaka metropolitan area, Bangladesh: Causes and quantification. *Hydrogeology Journal, 15*(8), 1523–1534. https://doi.org/10.1007/s10040-007-0226-5.

Kumar, P. K. D., Gopinath, G., & Seralathan, P. (2007). Application of remote sensing and GIS for the demarcation of groundwater potential zones of a River Basin in Kerala, southwest coast of India. *International Journal of Remote Sensing, 28*(24), 5583–5601. https://doi.org/10.1080/01431160601086050.

Machiwal, D., Jha, M. K., & Mal, B. C. (2011). Assessment of groundwater potential in a semi-arid region of India using remote sensing, GIS and MCDM techniques. *Water Resources Management, 25*(5), 1359–1386. https://doi.org/10.1007/s11269-010-9749-y.

Magesh, N. S., Chandrasekar, N., & Soundranayagam, J. P. (2012). Delineation of groundwater potential zones in Theni District, Tamil Nadu, using remote sensing, GIS and MIF techniques. *Geoscience Frontiers, 3*(2), 189–196. https://doi.org/10.1016/j.gsf.2011.10.007.

Selvam, S., Magesh, N. S., Chidambaram, S., Rajamanickam, M., & Sashikkumar, M. C. (2015). A GIS based identification of groundwater recharge potential zones using RS and IF technique: A case study in Ottapidaram Taluk, Tuticorin District, Tamil Nadu. *Environmental Earth Sciences, 73*(7), 3785–3799. https://doi.org/10.1007/s12665-014-3664-0.

Shaban, A., Khawlie, M., & Abdallah, C. (2006). Use of remote sensing and GIS to determine recharge potential zones: The case of occidental Lebanon. *Hydrogeology Journal, 14*(4), 433–443. https://doi.org/10.1007/s10040-005-0437-6.

Singh, A. (2014). Groundwater resources management through the applications of simulation modeling: A review. *Science of the Total Environment, 499*, 414–423. https://doi.org/10.1016/j.scitotenv.2014.05.048.

Souissi, D., Msaddek, M. H., Zouhri, L., Chenini, I., El May, M., & Dlala, M. (2018). Mapping groundwater recharge potential zones in arid region using GIS and Landsat approaches, Southeast Tunisia. *Hydrological Sciences Journal, 63*(2), 251–268. https://doi.org/10.1080/02626667.2017.1414383.

Srinivasa Rao, Y., & Jugran, D. K. (2003). Delineation of groundwater potential zones and zones of groundwater quality suitable for domestic purposes using remote sensing and GIS. *Hydrological Sciences Journal, 48*(5), 821–833. https://doi.org/10.1623/hysj.48.5.821.51452.

Thakur, D., Bartarya, S. K., & Nainwal, H. C. (2018). Mapping groundwater Prospect zones in an Intermontane Basin of the outer Himalaya in India using GIS and remote sensing techniques. *Environmental Earth Sciences, 77*(10), 368. https://doi.org/10.1007/s12665-018-7552-x.

Uddin, A. F. M. A., & Baten, M. A. (2011). *Water supply of Dhaka City: Murky future-the issue of access and inequality*. Unnayan Onneshan-the Innovators.

Wada, Y., Van Beek, L. P. H., Van Kempen, C. M., Reckman, J. W. T. M., Vasak, S., & Bierkens, M. F. P. (2010). Global depletion of groundwater resources. *Geophysical Research Letters, 37* (20), L20402. https://doi.org/10.1029/2010GL044571.

Yeh, H. F., Lee, C. H., Hsu, K. C., & Chang, P. H. (2009). GIS for the assessment of the groundwater recharge potential zone. *Environmental Geology, 58*(1), 185–195. https://doi.org/10.1007/s00254-008-1504-9.

Impact of Climate Change on Water Balance in Lakes of Shchuchinsk-Borovsk Resort Zone of Kazakhstan

Marzhan Saduokasova

1 Introduction

The Shchuchinsk-Borovsk Resort Area (SBRA) is situated in the north of the Republic of Kazakhstan. SBRA represents a unique natural setting with a high recreational potential and is an important strategic water resource of the Republic of Kazakhstan. The lakes Shortan, Burabay, Ulken Shabakti, Keeshi Shabakti, Katarkol, Zhukey, Maybalyk, Tekekol, Karasye and Sulukol are part of SBRA. Long-term hydrological data are available for lakes Shortan, Burabay and Ulken Shabakta (Fig.1). Assessments of climatic and hydrological changes in these lakes is a research gap in this region. To fill this research gap, this study analyzes long-term climatic and hydrological data along with a water balance assessment in changed climatic conditions.

2 Methodology

2.1 Calculation of Total Inflow

The basic water balance equation used in this study is:

$$\Delta W = V + S + P - E, \tag{1}$$

where ΔW – water volume (million m^3), V – total inflow (million m^3), S – snow cover water equivalents on the surface of the lake for the cold period (million m^3),

M. Saduokasova (✉)
Caspian Sea Hydrometeorological Research Department, RSE Kazhydromet, Almaty, Kazakhstan

A. Haque, A. I. A. Chowdhury (eds.), *Water, Flood Management and Water Security Under a Changing Climate*, https://doi.org/10.1007/978-3-030-47786-8_6

Fig. 1 Map of Shchuchinsk-Borovsk resort area. The map shows locations of Burabay lake

P – volume of rainfall on the surface of the lake during without ice period (million m^3), E – evaporation from the lake surface (million m^3).

Total inflow of water to the lake is defined by:

$$V = \Delta W - S - P + E. \quad (2)$$

Inflow of water to the lake in the spring period is defined by analyzing the increase in water level and water volume of the lake. Total inflow to the lakes, as a rule, exceeds its size during the spring period (March, April, May). Only in separate years, spring inflow appears slightly more than the total inflow in a hydrological year. This is apparently in connection with an uneven intake of water in the basin of the lake.

2.2 *Calculation of Evaporation*

To assess evaporation from lake surface, monthly measured data from evaporator (an instrument to measure evaporation with model no. GGI-3000) from June to October is used. A correction factor is applied to the measured data of evaporation so that these data represent evaporation from the lake surface in Northern Kazakhstan.

With this correction factor, the equation that relates evaporation with air temperature is:

$$E = 8.28 * t_{air} + 11.3 \tag{3}$$

E – evaporation, mm
t_{air} – air temperature, °C.
Coefficient of correlation r = 0.99.

According to the meteorological station Shchuchinsk, the evaporation data for the period 1935–2017 represents evaporation without ice period (April to October). Snow cover water equivalents for the cold period (November to March) is defined as the difference of monthly precipitation and evaporation (Baysholanov 2014). For precipitation data, monthly rainfall values from the meteorological station Shchuchinsk is used. Evaporation from snow surface is determined by following the method of Semyonov for the regions of Northern and Central Kazakhstan (Chuntonova et al. 2012). The equation used to calculate evaporation from snow surface is:

$$E = n(0.35d - 0.06) \tag{4}$$

where E – evaporation, mm
n – number of days of the settlement period
d – deficiency of humidity of air, gPa, averaging over n-days.

3 Results

3.1 Precipitation

Precipitation (P, mm) for the last 82 year (1935–2017) in station Shchuchinsk is shown in Fig. 2 and for the last 43 years (1974–2017) in station Burabay is shown in Fig. 3. The figures show that precipitation in these two locations has annual variability with some extreme events. But there is no definite long-term trend of change of precipitation in these locations.

3.2 Air Temperature

Variations of air temperature (T, °C) for the last 82 years in station Shchuchinsk and for the last 43 years in station Burabay are shown in Figs. 4 and 5. The variation of air temperature shown in Figs. 4 and 5 is categorized into 3 time periods spanning over 82 years.

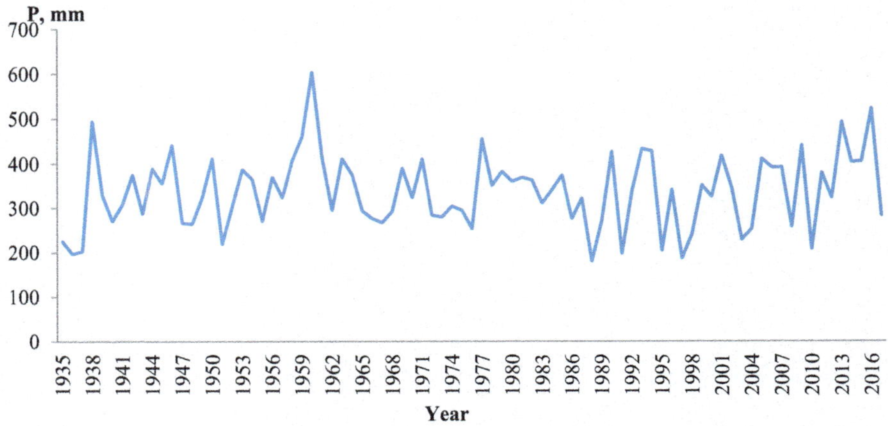

Fig. 2 Variation of precipitation for the last 82 years (1935–2017) in meteorological station Shchuchinsk

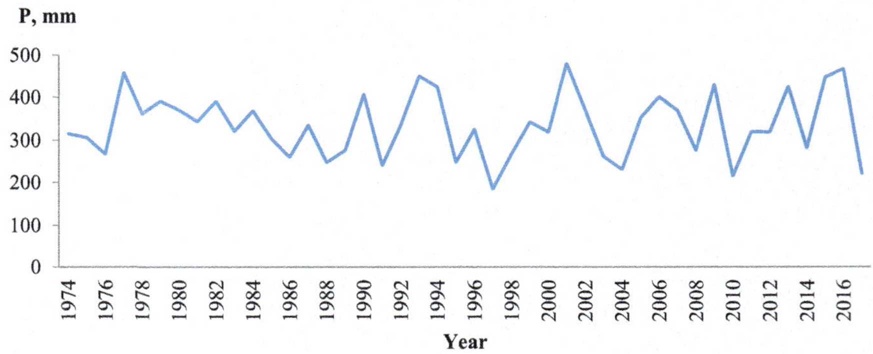

Fig. 3 Variation precipitation for the last 43 years (1974–2017) in meteorological station Burabay

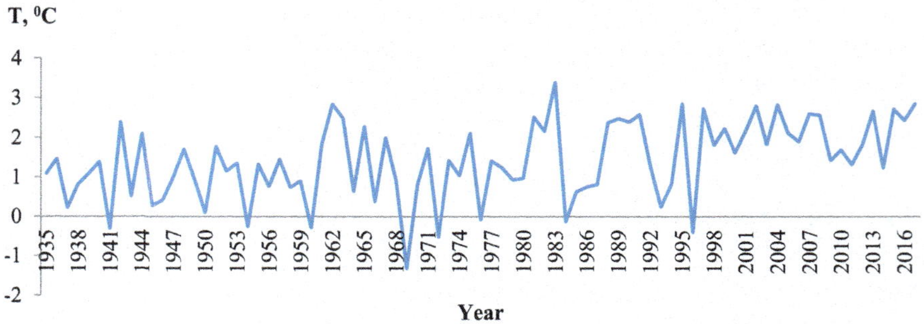

Fig. 4 Variation of air temperature for the last 82 years (1935–2017) in meteorological station Shchuchinsk

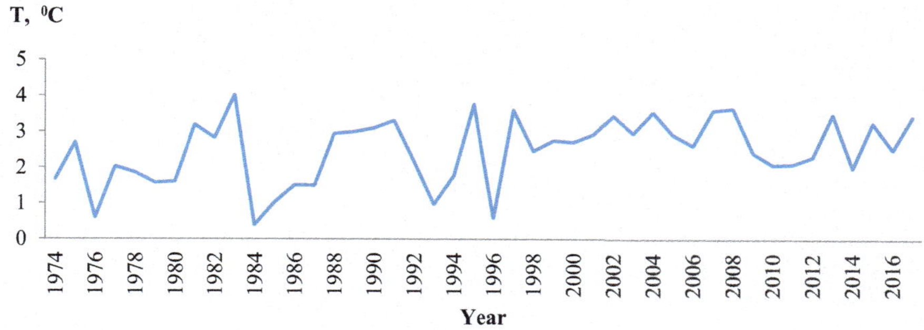

Fig. 5 Variation air temperature for the last 43 years (1974–2017) in meteorological station Burabay

3.2.1 First Period of 38 Years (1935–1973)

During this time period, average annual air temperature is 1.03 °C. This period characterizes climatic conditions of the middle of the twentieth century.

3.2.2 The Second Period of 20 Years (1974–1993)

During this time period, average annual air temperature is 1.47 °C, which shows an increase in air temperature during a transition to the new climatic period.

3.2.3 The Third Period of 23 Years (1994–2017)

During this time period, average annual air temperature is 2.03 °C, which shows an increase in air temperature at the end of the twentieth century towards the beginning of twenty-first century.

The results show a definite trend of increase in average annual air temperature from the middle of twentieth century to the beginning of twenty-first century.

3.2.4 Water Levels

Water level data are available for the lakes Shortan, Burabay and Ulken Shabakti from the beginning of twenty-first century. The measured data for these 3 lakes are shown in Figs. 6, 7, and 8. Overall trend of water level variations for these three stations show a decreasing trend.

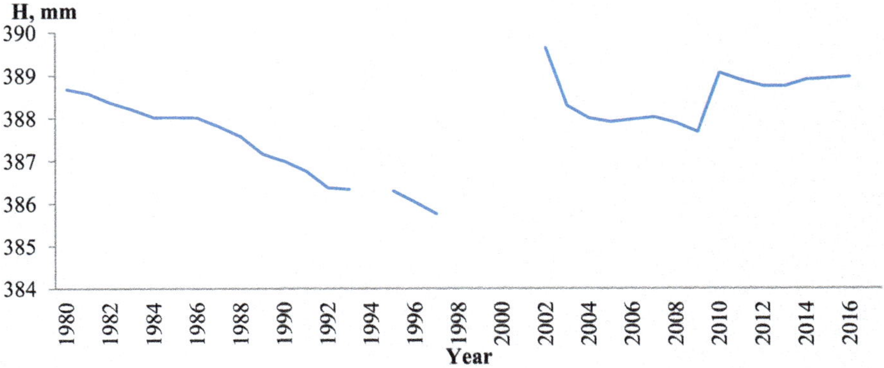

Fig. 6 Variation of average annual water level for the last 36 years (1980–2016) in lake Shortan. Data is missing from 1997 to 2001

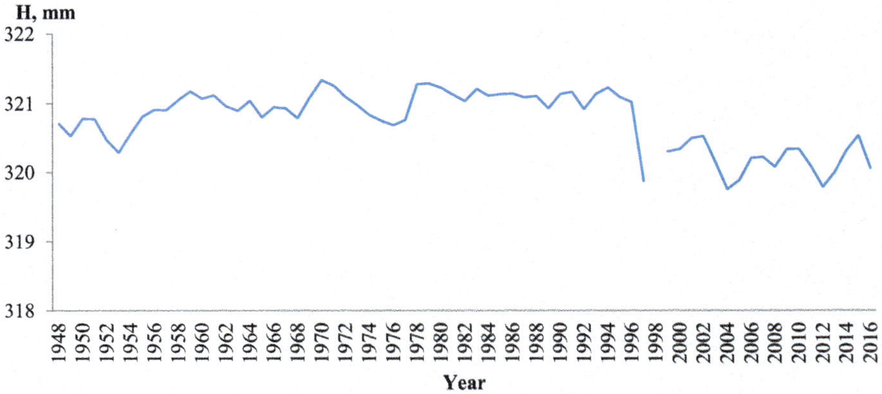

Fig. 7 Variation of average annual water level for the last 68 years (1948–2016) in lake Burabay

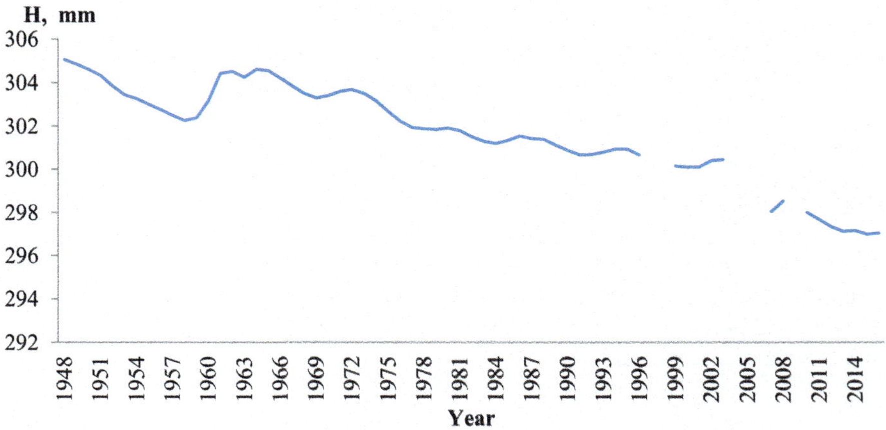

Fig. 8 Variation of average annual water level for the last 68 years (1948–2016) in lake Ulken Shabakti

3.3 Water Volume

Variations and comparisons of observed and calculated water volumes for the three lakes are shown in Figs. 9, 10, and 11. Water volumes in these three lakes are calculated by applying the water balance equation (Eq. 1). Due to unavailability of measured data for all the parameters of Eq. 1, it is not possible to calculate the water volume for the entire period. The observed water volume is infact calculated volume based on measured water level and lake geometry.

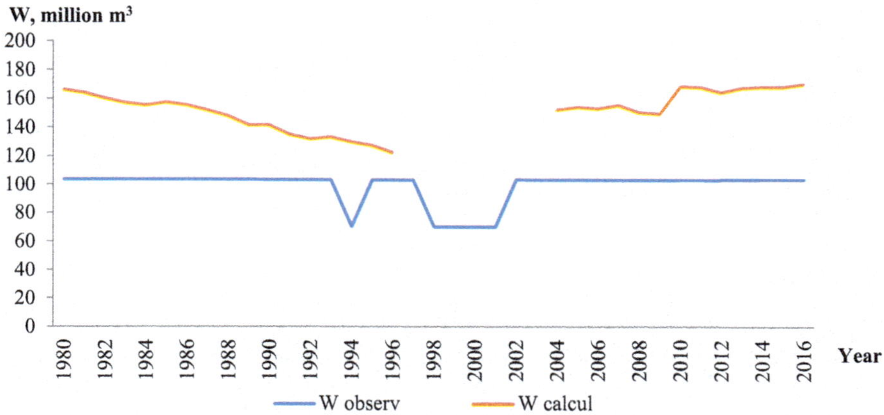

Fig. 9 Variation and comparison of measured and calculated annual water volume for the lake Shortan over the last 36 years (1980–2016)

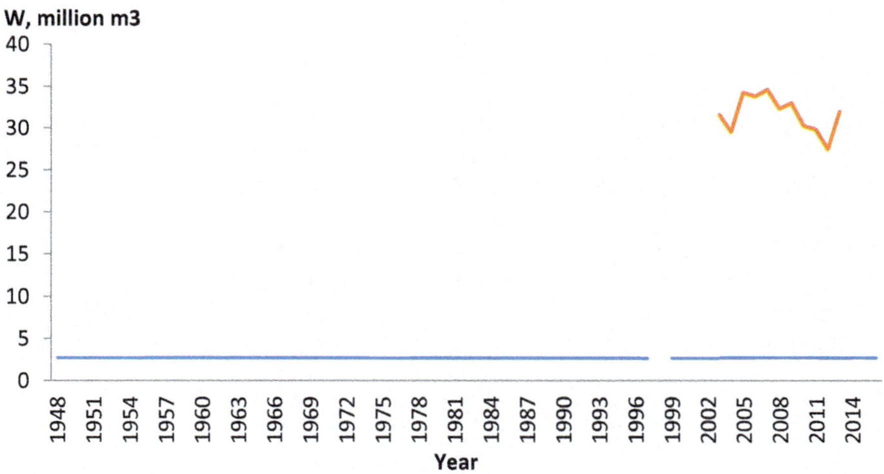

Fig. 10 Variation and comparison of measured and calculated annual water volume for the lake Burabay over the last 68 years (1948–2016)

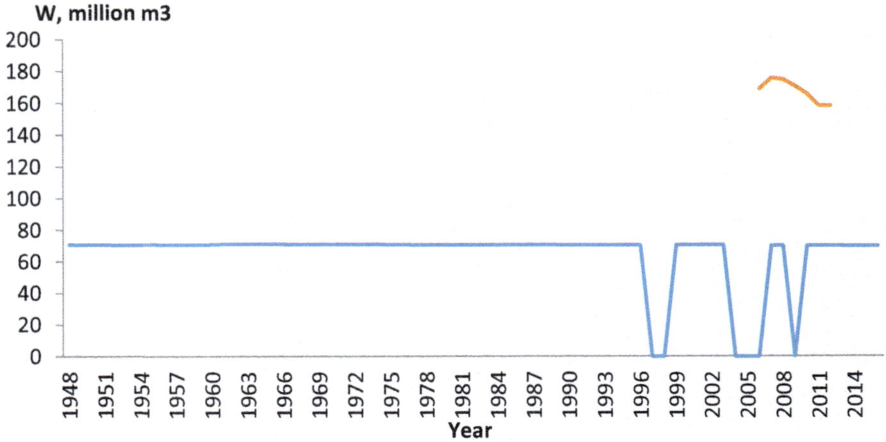

Fig. 11 Variation and comparison of measured and calculated annual water volume for the lake Ulken Shabakta over the last 68 years (1948–2016)

4 Discussion

By applying Eqs. 1, 2, 3, and 4, water balance in three lakes (Shortan, Burabay, Ulken Shabakti) are computed and are presented in Tables 1, 2, and 3. Results are shown in the middle of twentieth century and at the beginning of twenty-first century. These time periods represent two mile-stones to assess impact of climate change determined by the increased temperature (see Sect. 3.2 for details). During these periods there were hardly any change of pattern of precipitation (see Sect. 3.1). So, whatever changes are found in water balance is due to the increasing trend in air temperature (see Sect. 3.2). Except in lake Burabay, inflows of water in other two lakes (determined by water level and water volume) decrease mainly due to increased evaporation (due to increased air temperature which is a climate change impact). Increased evaporation is observed in all three lakes: for lake Shortan increase is from 680 mm to 726 mm, for lake Burabay increase is from 680 mm to 740 mm and for lake Ulken Shabakta increase is from 680 mm to 755 mm. Corresponding change in inflow rates are: for lake Shortan decrease of inflow is 55%, for lake Burabay increase of inflow is 20% and for lake Ulken Shabakta decrease of inflow is 15%. Earlier studies show that total area of 10 lakes in the study area is 99 sq. km (Gidrometeoizdat 1959; Golubtsov et al. 2014). By considering this area, it is calculated that total inflow volume is decreased by 5.64 million m^3.

Table 1 Water balance of lake Shortan in the middle of the twentieth century and at the beginning of the twenty-first century

Water balance parameters	In the middle of the twentieth century		At the beginning of the twenty-first century		Changes	
	Water level mm	Water volume million m^3	Water level mm	Water volume million m^3	Water level mm	Water volume million m^3
Inflow of water to the lake	390	7.25	281	4.00	−109	−3.25
Snow cover water equivalents on the surface of the lake from the beginning of spring snowmelt prior to the beginning of freezing	290	5.40	351	5.00	61	0.40
Evaporation from the lake surface from the beginning of spring snowmelt prior to the beginning of freezing	680	12.65	726	10.1	46	−2.55

Table 2 Water balance of lake Burabay in the middle of the twentieth century and at the beginning of the twenty-first century

Water balance parameters	In the middle of the twentieth century		At the beginning of the twenty-first century		Changes	
	Water level mm	Water volume million m^3	Water level mm	Water volume million m^3	Water level mm	Water volume million m^3
Inflow of water to the lake	390	3.67	450	4.41	60	0.74
Snow cover water equivalents on the surface of the lake from the beginning of spring snowmelt prior to the beginning of freezing	290	2.72	300	2.94	10	0.22
Evaporation from the lake surface from the beginning of spring snowmelt prior to the beginning of freezing	680	6.39	740	7.25	60	0.86

5 Conclusion

Water balance in three lakes of the study area shows that from the middle of twentieth century till the beginning of twenty-first century, inflows of water in two lakes decrease mainly due to increased evaporation. This increased evaporation is caused by increased air temperature which is a climate change impact. To compensate the lost reserve of water in these lakes, it is necessary to feed about 15 million m^3 of water. This additional water is necessary to meet the consumptive need of Shchuchinsk-Borovsk resort area (Krivenko 2003; MEP 2005; Baysholanov 2014).

Table 3 Water balance of lake Ulken Shabakta in the middle of the twentieth century and at the beginning of the twenty-first century

Water balance parameters	In the middle of the twentieth century		At the beginning of the twenty-first century		Changes	
	Water level mm	Water volume million m^3	Water level mm	Water volume million m^3	Water level mm	Water volume million m^3
Inflow of water to the lake	390	8.19	342	6.93	−48	−1.26
Snow cover water equivalents on the surface of the lake from the beginning of spring snowmelt prior to the beginning of freezing	290	6.09	297	5.95	7	−0.14
Evaporation from the lake surface from the beginning of spring snowmelt prior to the beginning of freezing	680	14.28	755	14.5	75	0.22

Acknowledgement These studies were conducted as part of a project that was sponsored by the Ministry of the Republic of Kazakhstan.

References

Baysholanov, S. S. (2014). On climate change in the Shchuchinsko-Borovsk resort area. In *Materials of the international scientific-practical conference "Theoretical and applied problems of geography"* (pp. 192–198). Part II Astana.

Chuntonova, L., Guseva, L., Guseva, E., & Tusupzhanova, A. (2012). *Creation of the hydrographic Atlas of the Shchuchinsko-Borovskoye System of Lakes (ЩБСО) report*. RSE "Kazhydromet", p. 66.

Gidrometeoizdat. (1959). *Surface water resources of the areas of development of virgin and fallow lands*. Issue III. Kokchetav Region of the Kazakh SSR. L., 563 p.

Golubtsov, V. V., Saduokassova, M. T., & Rashenkov, M. N. (2014). About changing in water balance of the lakes of Shuchinsk-Burabay resort zone. *Journal of Hydrometeorology and Ecology, 1*, 45–54.

Krivenko V. G. (2003). Forecast of climate change in Eurasia from the perspective of the concept of its cyclical dynamics. *World conference on climate change: Abstracts* (pp. 514).

MEP. (2005). *Development of a prognostic model of the ecological state of the Territory of the Shchuchinsko-Borovsk resort area for 2005–2007* (227 p.). Astana: MEP. (Typewriting).

Assessment of Cyclone Aila Recovery Progress in Bangladesh: A Comparison Between Rice and Shrimp Farming Villages in Koyra

Md. Shibly Sadik, Hajime Nakagawa, Md. Rezaur Rahman, Rajib Shaw, Kenji Kawaike, and Gulsan Ara Parvin

1 Introduction

The water use and land use conflicts between shrimp aquaculture and rice cultivation are very old debates in Bangladesh (Dove and Khan 1995; Akber et al. 2017). Historically, people living in the southwestern part of Bangladesh delta had been cultivating shrimp by trapping salinity rich tidal water in the rice fields (Akber et al. 2017; Deb 1998). However, with the commercialization of shrimp industries in the mid 1970s the traditional practices had started being replaced by the intensive shrimp aquaculture (Deb 1998). The industry grew in an unregulated way, unsustainably, with policy encouragement from the government, and became a serious threat to the ecology, environment and local livelihood (Deb 1998; Paul and Vogl 2011; Akber et al. 2017). The issue of land regulation to control the expansion of shrimp farms and protect the livelihood of the local community and their rice cultivable lands became a hot topic in research, media talk, and policy

M. S. Sadik (✉)
Center for Environmental and Geographic Information Services (CEGIS), Dhaka, Bangladesh

H. Nakagawa · K. Kawaike
Disaster Prevention Research Institute, Kyoto University, Kyoto, Japan

M. R. Rahman
IWFM, Bangladesh University of Engineering and Technology, Dhaka, Bangladesh

R. Shaw
Keio University, Tokyo, Japan

G. A. Parvin
Independent Researchers, Kyoto, Japan

A. Haque, A. I. A. Chowdhury (eds.), *Water, Flood Management and Water Security Under a Changing Climate*, https://doi.org/10.1007/978-3-030-47786-8_7

making processes (Paul and Vogl 2011; Siddique 2014). In the debate for prioritizing rice or shrimp, environmental issues, ecological issues, economic issues, political and social issues were always considered, and these then reflected in the national policies (Deb 1998; Ali 2006; Hossain et al. 2013; Akber et al. 2017; Parvin et al. 2017). Although shrimp farming is considered to be a reason for deforestation of the mangrove belt along the coast, which eventually caused increase of vulnerability to cyclones and storm surges (Dove and Khan 1995; Deb 1998), the issue was rarely seen from the context of disaster management. However very recently, after observing the sudden increase of switching from shrimp to rice cultivation after cyclone Aila (which struck in 2009) (Barai et al. 2019), the context of disaster impact and management has caught research interest. Social and political challenges at the local level and motivation for people to switch to rice cultivation were always considered as challenges when enforcing land zoning based regulation to prevent unsustainable expansion of shrimp farming (Sadik et al. 2019). Therefore, the reason behind the voluntary switching would be interesting to study.

Most of the research found that reduction of profit, loss of productivity of land, climatic stresses, lack of proper infrastructural supports, local environmental consequence, and loss of employment are the drivers behind the recent trend of switching from shrimp to rice (Paul and Røskaft 2013; Akber et al. 2017; Parvin et al. 2017). Most of the research excluded the context of post-disaster livelihood recovery strategy of an individual or community. However, a recent research on post disaster recovery has found some evidence that suggests the need for post-disaster livelihood recovery might be a driver which motivates local people to switch back to rice cultivation (Sadik et al. 2019). On this line, this research has been designed to compare the livelihood recovery outcomes in Post-Aila shrimp aquaculture and Post-Aila rice cultivation dominated villages. The result thereby would bring the post-disaster recovery context of switching livelihood from shrimp to rice cultivation. Apart from the comparison, this research attempted to measure the recovery progress of different villages subject to the comparison. Measuring the recovery progress is necessary for strengthening the mid-course correction process, evaluation of recovery decision, policy implication, and preparedness for future disasters (Rathfon et al. 2013).

2 Study Area

This paper discusses the case of Cyclone Aila recovery at Koyra, which is one of the most remote coastal upazilas[1] of Khulna district (Fig. 1). The upazila was severely affected by cyclone Aila. Similar to other coastal upazilas of the country, coastal polders (ID 13-14/2 and 14/1) were constructed in Koyra to prevent salinity intrusion and tidal flooding (Fig. 1). On 25th May 2009 when cyclone Aila swept across the southwestern part of the country, the coastal polders overtopped and were breached

[1]Upazila is the third level of administrative unit in Bangladesh Since it functions as a subunit of a district, it can be defined as sub-district.

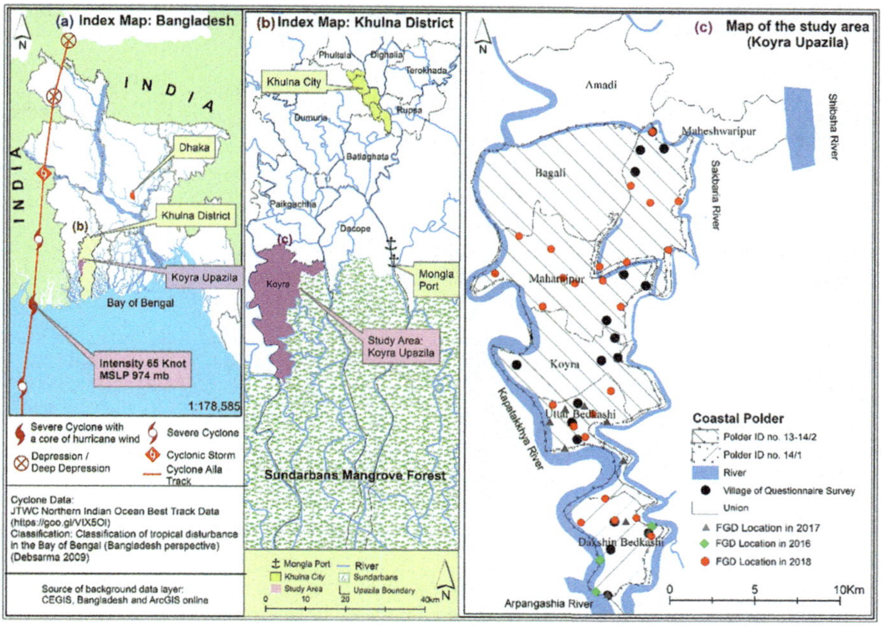

Fig. 1 Map of the study area

at several locations, which resulted in the flooding of almost the entire upazila (Sadik et al. 2018a, b, c). The cyclone affected around 150,000 of the Koyra population and damaged 81 km of embankments (between polder 13–14/2 and 14/1), 680 km of earthen road, 163.5 km of asphalt road, 49 bridge culverts, 42, 440 houses, 9 academic institutions, 192 religious institutions, the crops grown on 11, 500 hectares of land, and 10,364 fish aquaculture farms in the upazila (Koyra Upazila Council 2010).

In addition to the case study, the Uttar Bedkashi area of Koyra was studied for assessing the progress of recovery including the comparison between shrimp aquaculture dominating and rice cultivation dominating villages. Uttar Bedkashi is one of the unions of Koyra where after Aila some of the local villages switched to rice cultivation from shrimp aquaculture, while other parts still remained as shrimp aquaculture dominated areas. Whether a village is dominated by shrimp aquaculture or rice cultivation was decided by the people's perception during the focus group discussions (FGDs). Table 1 contains the list of villages, specifying whether they are dominated by rice cultivation or by shrimp aquaculture.

3 Methodology

This research aimed to compare the recovery progress and outcome between post-Aila shrimp aquaculture dominated and post-Aila rice cultivation dominated villages. In order to do so the research attempted to measure the recovery progress of

Table 1 A comparative analysis between rice cultivation and shrimp farming from the context of BBB

Rice cultivation	Shrimp aquaculture	Sources
It is labor intensive, thereby it generates income opportunity for landless, small farmers and agricultural labors	A large number of labor is only required at the time of pond preparation	FGD, (Deb 1998; Rahman et al. 2008)
It requires small initial capital. Therefore, small and medium farmers can easily resume rice cultivation when the area is protected from tidal floods	It requires large investment, thereby small and medium farmers found it very difficult to renovate/reconstruct their shrimp ponds.	FGD
It allows homestead farming of vegetables and fruits.	Shrimp farming negatively impacts homestead vegetation and ecosystem.	FGD, (Paul and Vogl 2011; Paul and Røskaft 2013)
It allows livestock and poultry rearing	It negatively impacts livestock and poultry by limiting and degrading grazing fields.	FGD, (Barai et al. 2019)
	Increases salinity of drinking water ponds.	FGD
	Reduces sustainability of rural roads by erosion. Roads along any shrimp pond degrades faster due to salinity and wave erosion	FGD, (Sadik et al. 2018a,b,c)
	Reduces sustainability of coastal polders, because shrimp farmers are illegally cutting or installing drainage pipes.	FGD, (Sadik et al. 2018a,b,c)

different sectors i.e. local economy, water supply and sanitation, rural road infrastructures, and housing. Generally measuring the progress requires a wide range of time series data of multi-disciplinary indicators and it needs to be supported by a detailed survey and pre-disaster census data (Tatsuki 2007; Horney et al. 2017). In the absence of any comprehensive and integrated data base, this research adopted a synthetic approach built on peoples' perception (Dutta et al. 2013). The data collection survey included institutional surveys, expert interviews, and focus group discussions.

3.1 *Institutional Survey and Expert Interview*

Institutional surveys followed by expert interviews (in 2016 and 2017) helped to identify institutes involved in Aila recovery in Koyra which eventually helped to understand the overall mechanism of Aila recovery. Since the number of institutions were not recorded in any database or in local government offices, a snowball technique (Goodman 1961) was applied to identify relevant institutes. In this way, a total of 12 institutes were visited which helped to identify 13 Non-Governmental Organizations (NGOs) involved in Aila recovery.

3.2 Focus Group Discussion

With an aim of quantifying the recovery progress 11 informal FGDs were conducted in 11 villages of Uttar Bedkashi. The participants were farmers, agriculture laborers who were direct and indirect beneficiaries of the recovery programs. Villages were selected purposively, considering their accessibility and possibility of being rice cultivation dominating or shrimp aquaculture dominating. This was mainly done by visual observation. While selecting villages a "snow ball sampling" approach was also practiced. The numbers of participants varied depending on location and the time of conducting the FGD. The average number of participants was 12.

The FGDs were administrated by a pre-developed, and pre-tested questionnaire. The questionnaire consisted of a general question about recovery for each of the three sectors (local economy, water supply and sanitation, and housing) – "how was/is the condition of the sector in the following five time periods: before Cyclone Aila, immediately after Aila (on 26 May 2009), one year of Aila (2010), three years of Aila (2012), five years of Aila (2014) and at present (2017/2018)" (Fig. 2). These flagged years corresponded to years when different major recovery programs (e.g. emergency response and relief, rehabilitation of embankment, reconstruction of housing projects, etc.) were ended. Thereby it helped the participants of the FGD to remember from their past experiences. From the experience of field testing the questionnaire, it was understood that local people could easily correlate the past condition to any major recovery events like completion of polder rehabilitation, completion of a road reconstruction, etc.

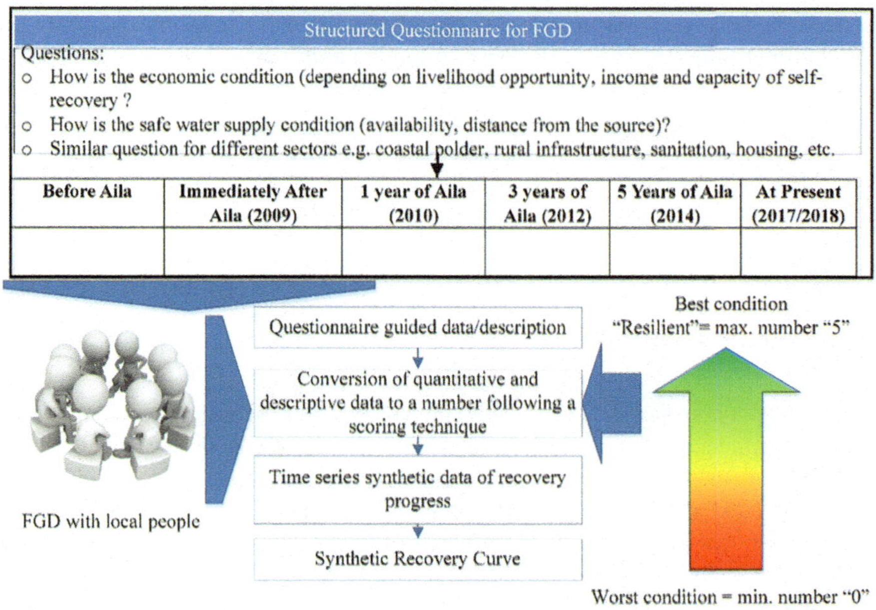

Fig. 2 Illustration of the methodology adopted to measure recovery progress

3.3 People's Perception-Based Scoring Technique for Measuring Recovery

Recovery was measured following a people's perception-based scoring technique (Giupponi et al. 2013; Gain et al. 2015). A numerical scale ranging from the minimum number "0" to the maximum number "5" (Fig. 2) was developed to quantify the sectoral condition with a comparison to pre-Aila condition and an imaginary condition of a resilient future considering Build Back Better (Sadik et al. 2018a, b, c; Sadik 2019). The worst condition "0" means "people are living in emergency" and the best condition "5" means the economy is resilient due to diversified, sustained, safe and certain livelihood opportunities. While conducting the FGD the word scenarios with the numerical scale (Sadik et al. 2018a, b, c) was administered along with the pre-developed questionnaire (Fig. 2) to transform the elicited information and perception of local people into a number quantifying the prevailing condition of a sector in a given year as illustrated in Fig. 2.

3.4 Construction of Synthetic Recovery Curve to Illustrate Recovery Progress

The numbers quantifying the prevailing conditions of different sectors in different years as received from FGDs allowed to construct a time series data of sectoral conditions from the pre-Aila period of 2009 to the year 2017/2018. It was then possible to construct the synthetic recovery curve for each sector from this time series data set. A synthetic recovery curve thereby illustrates recovery progress from 2009 (after cyclone Aila) to 2017/2018. Adoption of this approach of using people's perception to generate synthetic data was motivated by a research conducted in Australia and Japan to quantify future flood impact by people's perceptions (Dutta et al. 2013). Similar approach of quantitative assessment using perception can also be found in vulnerability assessment (Dutta et al. 2011; Dutta et al. 2013) and resilience assessment (Sadik and Rahman 2010; Parvin and Shaw 2011).

4 Cyclone Aila Recovery

Cyclone Aila appeared at a time when the Government with the help of its development partners set its course to bring a massive change in disaster management policies and practices in response to the impact of another super cyclone, Sidr, which struck in 2007 (Mallick and Islam 2014; Kirsten and Uddin 2017). Thereafter, development partners and the government had to change their initiatives of cyclone recovery to include cyclone Aila (Sadik et al. 2017a, b). The post cyclone response from the international community and advocacy of development partners motivated the government to give more priority on cyclone preparedness, enhancing capacity and numbers of cyclone shelter and disaster resilient housing as priority issues in post disaster reconstruction. In this regard, the government along with development partners established multi-sectoral recovery programs to ensure continuation of economic growth and preparedness for any future cyclones (Sadik et al. 2017a, b). From the national level perspective the recovery initiatives appeared to be aligned to BBB principles as it had been claimed (Mallick and Islam 2014). However the story at the local level evidences quite the opposite (Sadik et al. 2018a, b, c).

4.1 *Effort of Government and NGOs at Local Level*

The cyclone Aila recovery is characterized by a new approach of the Government-NGO joint effort (Sadik et al. 2018a, b, c). Soon after the cyclone impact, the Government and international NGOs extended their support of emergency response to the affected community (Sadik et al. 2017a, b). The affected community of Koyra was living in an emergency situation with direct aid from international NGOs and the Government for around 6 months to 3 years (depending on the restoration status of coastal polder to prevent tidal flood). In response to the government's call for recovery support, multi-donor bilateral agencies established early recovery facilities and implemented aid-driven initiatives for housing, infrastructure, coastal polders, water supply, sanitation, and local economic recovery (UNDP, EKN, AusAid, and SDC 2013; EKN and UNDP 2015; Sadik et al. 2018a, b, c). However, proper inclusion of DRR measures from the context of BBB was a serious limitation of such initiatives (Sadik et al. 2018a, b, c). A study conducted on inclusiveness of DRR measures in Aila recovery found that DRR measures related to local economy, water supply, sanitation and hygiene (WASH) and social relations and networks were poorly included while the DRR measures related to physical safety were very poorly included. Most of the implemented recovery initiatives were inclusive of short-term measures while long term measures like hazard-based land use planning, improvement of coastal embankment, improvement of early warning systems, were completely missing. After a systematic characterization of GO and NGO driven recovery initiatives at Koyra, another research concluded that alignment of Aila recovery to the safety aspect of BBB is poor to moderate due to low inclusion of pre-Aila vulnerability reduction and low effectiveness (Sadik 2019). It is to be mentioned here that both the researches agreed that there is a visible change and development of the community but from the context of BBB the recovery goals are yet to achieve.

4.2 *Livelihood Switching: A Strategy for Surviving and Recovery*

Different literature suggests that livelihood diversification, temporal change of occupation and migration were popular choices of recovery strategy after Cyclone

Aila in affected areas (Paul 2013; Abdullah et al. 2016; Saha 2017). Though, whether those choices were recovery strategy or coping strategy can be a topic of argument, it is agreeable that such moves helped the local community to survive in post disaster situations. Similar strategy was also found to be opted by the local people in Koyra. Cyclone Aila and thereafter 3 years of continued tidal flooding due to delayed repair of coastal polder system suspended all kind of agricultural practice in Uttarbedkashi and Dakshin Bedkashi of koyra. People who were involved in either rice cultivation or shrimp aquaculture or related businesses had to change their occupation. A large number of people out migrated for both push and pull factors (IOM 2010). On the other hand, recovery strategy of people who decided to stay in the disaster affected area went through an interesting recovery mechanism. In the years when the entire area was subject to tidal flooding people first changed their occupation to fishing, resources extraction from Sundarbans, wage labor, etc. (Abdullah et al. 2016; Sadik 2019). It is to be noted here that these choices were very hard for them. When Bangladesh Water Development Board (BWDB) repeatedly failed to repair the polder system, NGOs came forward and compartmentalized the flooded area by constructing ring dikes (Sadik et al. 2019). After compartmentalization of the flooded area, few people (mostly the large farmers) started shrimp aquaculture. But small farmers were still unable to resume their shrimp aquacultures due to lack of initial capital. However, after 2–3 years BWDB repaired the polder system and people could resume their shrimp aquaculture or rice cultivation. Interestingly, in the few villages that were pre-dominated by shrimp aquaculture, people switched to rice cultivation. The main reasons behind such switching, as they mentioned during the FGD, were – (i) obvious need of food grain, (ii) requirement of small initial capital to cultivate rice, (iii) labor intensive, thereby opportunity of employment, (iv) experience of negative social and environmental impacts of shrimp (e.g. loss of employment, rise of soil salinity, dying of homestead trees, etc.), (v) opportunity of homestead vegetable cultivation and livestock rearing, etc. Such field finding of switching to rice cultivation aligns to the findings of other researches who conducted satellite image based land use/land cover analysis which prove increase of rice cultivating area (Barai et al. 2019; Mallick 2019). However, such switching to rice cultivation was challenged by the local influential large farmers who have large interest in shrimp aquaculture. Because, majority of the local people of such villages opposed to the development of any shrimp aquaculture to protect their rice cultivable lands and homestead vegetation from salinity. This change was not possible in places where the large farmers could resist such public movement. A land-use based recovery policy which would reduce such conflict and plan the recovery towards a resilient future was missing in this case. The Cyclone Aila created an opportunity of implementing land zoning to regulate unsustainable growth of shrimp farming which was also a source of social vulnerability and vulnerability of coastal polders (Sadik et al. 2017a, b). Unfortunately, such opportunity has not been utilized.

5 The Progress of Aila Recovery at Shrimp and Rice Dominating Villages of Koyra

5.1 Recovery of Local Economic Condition

The Fig. 3 illustrate the prevailing local economic condition (translated to a number) of rice cultivation and shrimp farming dominating villages of Uttarbedkashi in five different millstone years i.e. pre-Aila (2009), immediately after Aila, one year after Ail, 3 years of Aila, 5 years of Aila and in 2017–2018. The figure clearly illustrates how the economic condition changes in post-Aila time which was the outcome of recovery, thereby visualizing the progress of recovery. By the notion "economic condition" this research describes the prevailing local economic status perceived by the local people from the context of their livelihood opportunity, employment opportunity, and self-recovery capacity (economic), after similar disaster.

The synthetic curve shows that the economic condition in both cases of villages was "low (score 2)" to "moderate (score 3)" before Aila. This region was categorized as "very hard-to-reach" due to high poverty and poor condition of water supply and sanitation. Aila severely affected the local economy by damaging all crops, destroying shrimp aquaculture ponds and suspending all livelihood activities. The impact was a little higher in villages where rice cultivation is dominating at present. According to the local people, in both cases, it took around 3 years to come back to an economic status similar to before Aila when the coastal polders were repaired. Thereafter, with the rehabilitation of coastal polders and rural infrastructures (by GO and NGOs), the economic condition started improving. The present economic

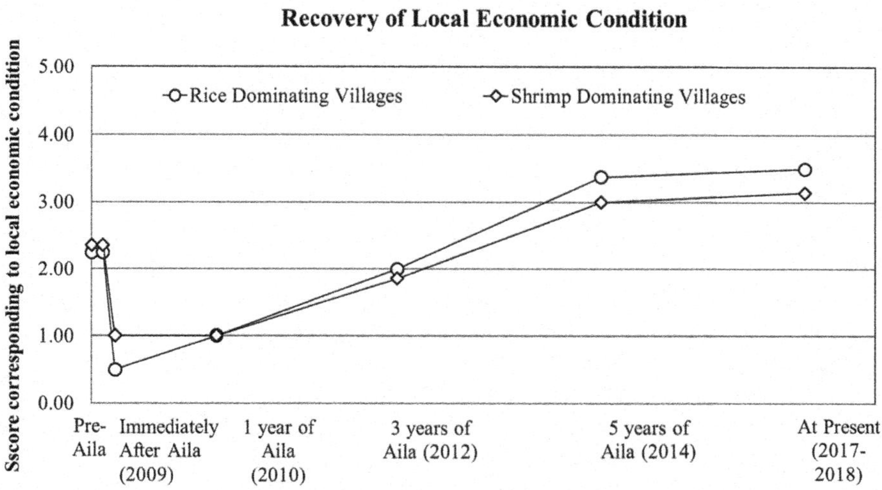

Note: The average score of rice cultivation dominating villages and shrimp farm dominating villages are plotted here. The score is a number quantifying economic condition perceived by local people from the context of their livelihood opportunity, employment opportunity, and self-recovery capacity after a future similar disaster. The

Fig. 3 Recovery of local economic condition

condition is "moderate, (score 3)" in both cases. However, it lacks long-term DRR measures e.g. sustainable agricultural practices, land zoning, sustainability of coastal polders, etc.

The finding shows that in comparison to shrimp farming villages, rice cultivation dominating villages showed better recovery. During FGD local people stated from their individual perspective that shrimp farm owners could economically recover better compared to rice farmers. But from the community perspective, local people apart from large shrimp farm owners (who are very small in number) in shrimp farm dominating villages could not recover their economic condition easily.

5.2 Recovery of Water Supply and Sanitation

Figure 4 illustrates the recovery of water supply and sanitation condition with respect to BBB. High salinity in surface and ground water, lack of appropriate technology, water quality and availability were the pre-existing vulnerabilities. The pre-Aila condition of water supply and sanitation were "low (score 2)" to "moderate (score 3)". Cyclone Aila severely damaged the water supply and sanitation system. Drinking water ponds were inundated, toilets were damaged along with houses, people were living with emergency support of NGOs and GOs for 1–2 years. The condition was gradually improving with interventions of NGOs and GOs. NGOs, and GOs excavated ponds, installed roof top rain water harvesting systems at selected houses. However, the condition started sharply improving after the repair of coastal polder system which prevented every day tidal flood. Within 3 years of Aila both water supply and sanitation conditions improved and became better than pre-Aila conditions. However, the pre-Aila vulnerabilities were mainly the lack of appropriate technology, safety from cyclone and storm surge, poor maintenance, etc.

Similar to economic condition, water supply and sanitation conditions are also better in rice cultivation dominating villages (Fig. 2). According to the local people, in shrimp aquaculture dominating villages most of the re-excavated and newly excavated ponds became saline within 1–2 years due to leaching of salinity from the shrimp aquaculture ponds next to them. Since shrimp aquaculture negatively impact income opportunity of agricultural labor and small farmers, in those villages people have less capacity to maintain sanitation and hygiene. Even most of the sanitary toilets provided by NGOs soon decayed and took bad shape due to a lack of proper maintenance. In general, water supply and sanitation conditions are comparatively better in rice cultivation dominating villages. However, from the context of BBB, water supply and sanitation conditions are still far from being fully resilient, (Fig. 2) mainly due to poor inclusion of DRR (Sadik et al. 2018a, b, c).

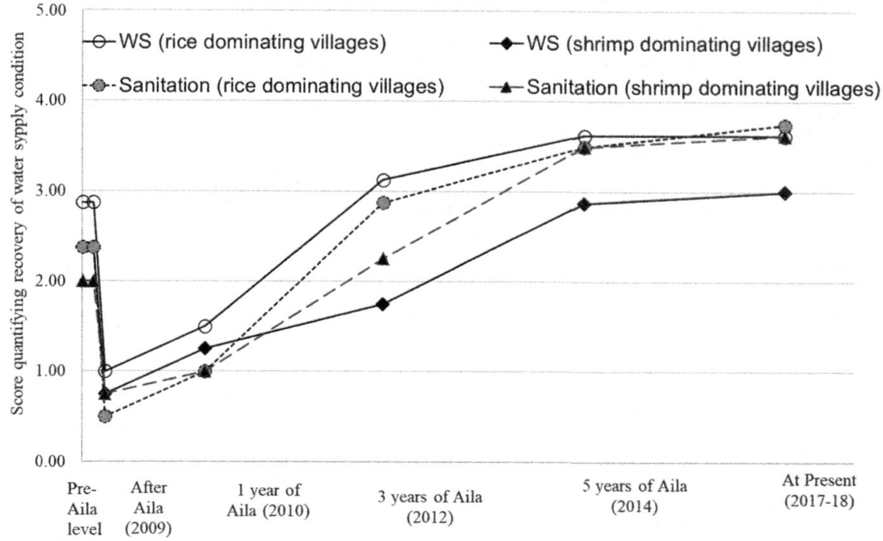

Note: The average score of rice cultivation dominating villages and shrimp farm dominating villages are plotted here. The score is a number quantifying WS & Sanitation condition perceived by local people . The number "0"= worst condition means "living in emergency" and "5"= best condition means "resilient"

Fig. 4 Recovery of water supply and sanitation condition

5.3 *Recovery of Housing*

Figures 5 and 6 illustrate the progress of housing recovery in rice and shrimp farming dominating villages of Uttar Bedkashi. Cyclone Aila and thereafter repeated tidal flooding due to embankment breaching caused severe damages to the housing sector. People started reconstructing their houses after 6 months to 1 year. The area was regularly flooded by tide due to the delay in repair of embankment breaches. After 1 year only 10%–15% houses were rebuilt. After 3 years when the coastal polder was repaired most of the people started rebuilding their houses. When reconstructing or newly constructing their houses people were able to adopt DRR measures. In general, while adopting DRR measures, NGOs and local people considered three types of hazards - wind storm, storm surge and regular tidal flood due to embankment breaching, and depending on their capacity adopted DRR measures against such hazards. Therefore, to illustrate recovery from the context of BBB, three criteria have been considered, which are:

(i) number of houses constructed considering DRR measures against cyclonic wind similar to Aila (Fig. 5),
(ii) number of houses constructed considering DRR measures against tidal flood (if embankment breaching happens) (Fig. 6)
(iii) number of houses constructed considering DRR measures against storm surge similar to Aila (Fig. 6).

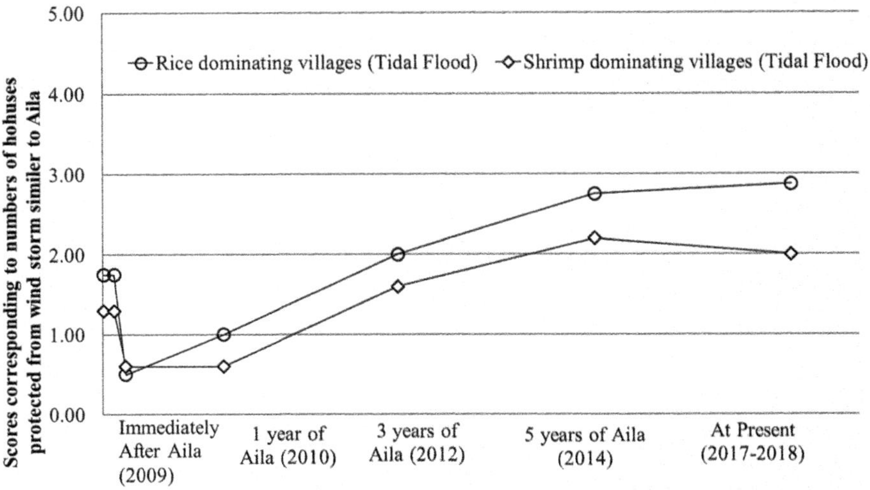

Note: The average score of rice cultivation dominating villages and shrimp farm dominating villages are plotted here. The score is a number quantifying housing condition perceived by local in the context of BBB. The number "0"= 0% houses, "1"= 20% houses, and "5"= 100% houses with DRR measurese

Fig. 5 Recovery of houses considering DRR against Windstorm of Cyclone Similar to Aila

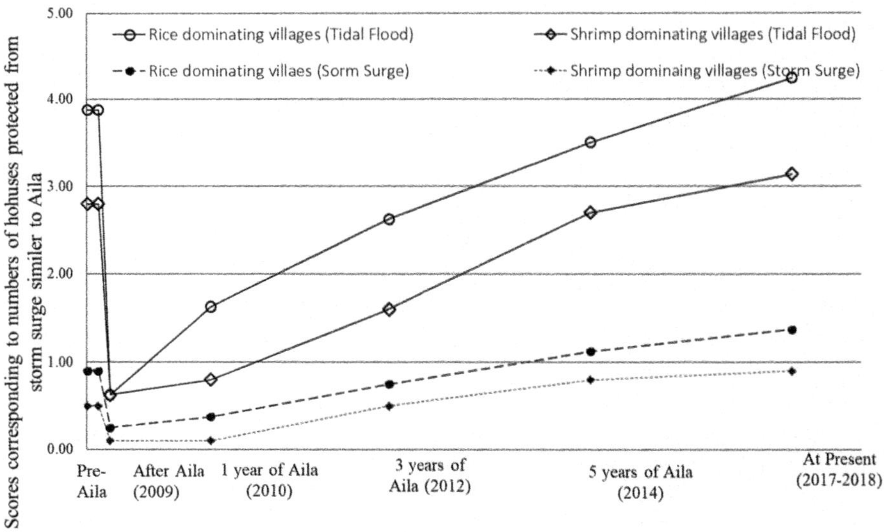

Note: The average score of rice cultivation dominating villages and shrimp farm dominating villages are plotted here. The score is a number quantifying housing condition perceived by local in the context of BBB. The number "0"= 0% houses, "1"= 20% houses, and "5"= 100% houses with relevent DRR measurese

Fig. 6 Recovery of houses considering DRR against Storm Surge and Tidal Flood

The result shows that in all three aspects, rice cultivation dominating villages exhibit better recovery.

There is a general trend of improvement in comparison with pre-Aila conditions. However, from the context of BBB, the progress is yet to achieve the safety against storm surge of BBB (Fig. 6). While reconstructing or newly constructing houses people adopted special protection against wind storms and regular tidal floods e.g. especial roof fitting, concrete pillar, T-footing, corrugated iron (CI) sheet replacing thatched roofs, CI sheet or thatched fence replacing earthen wall, and plinth level above the regular tidal flood (Sadik et al. 2018a, b, c). However, a small number of houses have been constructed adopting DRR measures against storm surges (Fig. 6).

6 Discussion: Why Rice Cultivation Dominating Villages Perform Better Recovery

According to the local people shrimp farming only benefits the shrimp farm owners whose pond size is large and large farmers who lend their land to shrimp farmer. Apart from these groups, shrimp farming does not bring benefits to small and medium farmers and agricultural laborers. The estimation of Department of Agricultural Extension (DAE) of Koyra says, in Uttar Bedkashi large farmers are only 4.6% and around 23.5% farmers are land less and in general around 60.5% families live by agriculture labor. Therefore, it is obvious that whichever (between rice and shrimp farming) benefits the majority (agricultural labor, land less, and small and medium farmers), will contribute more in community recovery. Table 1 clearly demonstrates reasons why instead of shrimp farming rice cultivation may be more beneficial for the majority of the people.

7 Concluding Remarks

Hazard based land use planning is considered to be the center of post-disaster recovery programs. This research brings out an example that even in the absence of land use based decision making in government and NGO initiatives, local people adopted their livelihood recovery strategy by changing their land use. This research indicates that such change of land use, i.e. switching from shrimp to rice helped them to recover in better ways than other villages where such change did not take place. The reason behind better recovery performance of rice cultivation dominating villages can be understood from Table 1. Local people of the surveyed villages in Uttar Bedkashi apprehended those benefits and switched to rice cultivation. Despite the reasons being valid this trend of switching was not followed everywhere. Another researcher found a tendency of switching from rice to shrimp after cyclone Aila (Mallick 2019) in another village of the southwest coast. There are several factors including public motivation, unity, support from local influential people, effectiveness of coastal polder to prevent tidal flood, productivity, land suitability, capital, etc. that play an important role when making these decisions (Kulsum et al. 2017). As such it is very difficult to predict such changes. There was an initiative to understand individual decision towards a choice between rice or crop, especially in changing environment, where the researchers tried to develop a mental model to understand the decision of a farmer (Kulsum et al. 2017). They have found that an individual farmer might analyze 25 triggering concepts in their mind before making a choice, which suggests that the decision-making process at the community level might be more complex to understand. Again, the social process involved in the implementation of a community decision is also important in this case. In this study area, local people shared that their local political leaders had supported their decision of switching from shrimp to rice which helped them to implement their decision. Another case in Maheshwaripur of Koyra was studied, which showed that the local people of Maheshwaripur united for preventing shrimp farming after Aila but were unable to implement their decision due to lack of support from local influential people. Such complex social processes call for effective regulation. At the international level there is now a growing interest for developing necessary regulation to build the resilience of coastal communities (World Bank Group and GFDDR 2010). The National Coastal Zone Policy, the Disaster Management Plan, and the Bangladesh Delta Plan 2100 recommend land-use based zoning to regulate shrimp farming for ensuring sustainability of coastal zones, coastal infrastructure and resilience of coastal community. However, the bottleneck of implementing such a plan is believed to be the acceptance of such land zoning by the local people. The example of Uttar Bedkashi shows us that in post-disaster situations implementation of such land zoning is an opportunity and easier to implement when there is public motivation. However, unfortunately, such land-use based planning was not applied during the post-Aila recovery. Even though the opportunity is missed, community driven Aila recovery strategy of land use change informs the policy makers about the dynamic of the land-use conflict of the post-disaster period which is yet to be considered in the current process of finalizing the draft *Agricultural Land Protection and Land Use Bill* of the government and the land zoning in the coastal areas.

References

Abdullah, A. N. M., Zander, K. K., Myers, B., Stacey, N., & Garnett, S. T. (2016, September). A short-term decrease in household income inequality in the Sundarbans, Bangladesh, following cyclone Aila. *Natural Hazards, 83*(2), 1103–1123.

Akber, M. A., Islam, M. A., Ahmed, M., Rahman, M. M., & Rahman, M. R. (2017, May). Changes of shrimp farming in southwest coastal Bangladesh. *Aquaculture International, 25*(5), 1883–1899.

Ali, A. M. S. (2006). Rice to shrimp: Land use/land cover changes and soil degradation in southwestern Bangladesh. *Land Use Policy, 23*(4), 421–435.

Barai, K. R., Harashina, K., Satta, N., & Annaka, T. (2019, January). Comparative analysis of land-use pattern and socioeconomic status between shrimp- and rice- production areas in southwestern coastal Bangladesh: A land-use/cover change analysis over 30 years. *Journal of Coastal Conservation*, 1–12.

Deb, A. K. (1998, October). Fake blue revolution: Environmental and socio-economic impacts of shrimp culture in the coastal areas of Bangladesh. *Ocean and Coastal Management, 41*(1), 63–88.

Dove, M. R., & Khan, M. H. (1995, May). Competing constructions of calamity: The April 1991 Bangladesh cyclone. *Population and Environment, 16*(5), 445–471.

Dutta, D., Wright, W., & Rayment, P. (2011, April). Synthetic impact response functions for flood vulnerability analysis and adaptation measures in coastal zones under changing climatic conditions: A case study in Gippsland coastal region, Australia. *Natural Hazards, 59*(2), 967–986.

Dutta, D., Wright, W., Nakayama, K., & Sugawara, Y. (2013, February). Design of synthetic impact response functions for flood vulnerability assessment under climate change conditions: Case studies in two selected coastal zones in Australia and Japan. *Natural Hazards Review, 14* (1), 52–65.

EKN and UNDP. (2015). *Final Narrative Report of Building Community Resilience Through Integrated Water Management (IWM) Project in bangladesh*. Dhaka, Bangladesh: The Embassy of the Kingdom of the Netherlands and United Nations Development Programme.

Gain, A. K., Mojtahed, V., Biscaro, C., Balbi, S., & Giupponi, C. (2015, December). An integrated approach of flood risk assessment in the eastern part of Dhaka City. *Natural Hazards, 79*(3), 1499–1530.

Giupponi, C., Giove, S., & Giannini, V. (2013, June). A dynamic assessment tool for exploring and communicating vulnerability to floods and climate change. *Environmental Modelling and Software, 44*, 136–147.

Goodman, L. A. (1961, March). Snowball sampling. *Annals of Mathematical Statistics, 32*(1), 148–170.

Horney, J., Dwyer, C., Aminto, M., Berke, P., & Smith, G. (2017, January). Developing indicators to measure post-disaster community recovery in the United States. *Disasters, 41*(1), 124–149.

Hossain, M. S., Uddin, M. J., & Fakhruddin, A. N. M. (2013, September). Impacts of shrimp farming on the coastal environment of Bangladesh and approach for management. *Reviews in Environmental Science and Bio/Technology, 12*(3), 313–332.

IOM. (2010). *Joint Position Paper on: Cyclone Aila: Priorities for Action*. Dhaka: International Organization for Migration.

Kirsten, L., & Uddin, A. M. K. (2017). *World resources report case study: Bangladesh's comprehensive approach to disaster management*. Washington, DC.

Koyra Upazila Council. (2010). *At a glance of the Cyclone Aila Damage in Koyra (এক নজরে কয়রায় ঘূর্ণঝিড় 'আইলায়' ক্ষয়ক্ষতির ববিরন)*. Koyra: Koyra Upazila Parishad and Office of the Upazila Nirbahi Officer.

Kulsum, U., et al. (2017). Modelling Farmers' Livelihood Decision Making to Understand Their Adaptation Response to Changeing COnditions in Southwest Coastal Bangladesh In the Ganges Delta. In: *6th INternational COnference on Water and Flood Management 2017* (pp. 397–406).

Kumar Paul, S. (2013). Post-cyclone livelihood status and strategies in coastal Bangladesh. *Rajshahi University Journal of Life & Earth and Agricultural Sciences, 41*, 1–20.

Mallick, B. (2019, June). The Nexus between socio-ecological system, livelihood resilience, and migration decisions: Empirical evidence from Bangladesh. *Sustainability, 11*(12), 3332.

Mallick, F., & Islam, A. (2014). Post-Aila community recovery innovations and planning. In R. Shaw (Ed.), *Disaster recovery: Used or Missused development opportunity, disaster risk reduction* (pp. 241–264). Tokyo: Springer.

Parvin, G. A., & Shaw, R. (2011). Climate disaster resilience of Dhaka City corporation: An empirical assessment at zone level. *Risk Hazards Crisis Public Policy, 2*(2), 121–150.

Parvin, G. A., Ali, M. H., Fujita, K., Abedin, M. A., Habiba, U., & Shaw, R. (2017). Land use change in southwestern coastal Bangladesh: Consequence to food and water supply. In *Land use Management in Disaster Risk Reduction* (pp. 381–401). Tokyo: Springer.

Paul, A. K., & Røskaft, E. (2013, December). Environmental degradation and loss of traditional agriculture as two causes of conflicts in shrimp farming in the southwestern coastal Bangladesh: Present status and probable solutions. *Ocean and Coastal Management, 85*, 19–28.

Paul, B. G., & Vogl, C. R. (2011). Impacts of shrimp farming in Bangladesh: Challenges and alternatives. *Ocean and Coastal Management, 54*(3), 201–211.

Rahman, M. M., Flitner, M., Krause, G., & Maniruzzaman, M. (2008). Socioeconomic assessment of shrimp farming in relation to local livelihoods in the south-west coastal Bangladesh. *Bangladesh Journal of Fisheries Research, 12*(1), 109–120.

Rathfon, D., Davidson, R., Bevington, J., Vicini, A., & Hill, A. (2013, April). Quantitative assessment of post-disaster housing recovery: A case study of Punta Gorda, Florida, after hurricane Charley. *Disasters, 37*(2), 333–355.

Sadik, M. S. (2019). *Characterization, diagnostic analysis and assessment of Progress of community recovery after cyclone Aila in Bangladesh.* Kyoto University.

Sadik, M. S. & Rahman, M. R. (2010). Community engagement in analyzing their livelihood resilience to climate change induced salinity intrusion in Sundarbans mangrove forest. In *International Symposium on Coastal Zones and Climate Change: Assessing the Impacts and Developing Adaptation Strategies.*

Sadik, M. S., et al. (2017a). Systematic study of cyclone Aila recovery efforts in Koyra, Bangladesh highlighting the possible contribution to vulnerability reduction. *Journal of Japan Society for Natural Disaster Science, 36*(Special), 107–119.

Sadik, M. S., Nakagawa, H., Shaw, R., Rahman, M. R., Kawaike, K., & Fujita, K. (2017b). A study on the humanitarian aid driven cyclone Aila (2009) recovery in Koyra Upazila of Bangladesh. *The Disaster Prevention Research Institute Annual, 60*(B), 746–756.

Sadik, M. S., et al. (2018a). Humanitarian aid driven recovery of housing after cyclone Aila in Koyra, Bangladesh: Characterization and assessment of outcome. *Journal of Japan Society for Natural Disaster Science, 37*(Special), 73–91.

Sadik, M. S., et al. (2018b). Measuring Progress of recovery from People's perception: A study on cyclone Aila recovery in Koyra, Bangladesh. *Annual disaster prevention research institute, 61* (B), 637–646.

Sadik, M. S., Nakagawa, H., Rahman, R., Shaw, R., Kawaike, K., & Fujita, K. (2018c, March). A study on cyclone Aila recovery in Koyra, Bangladesh: Evaluating the inclusiveness of recovery with respect to Predisaster vulnerability reduction. *The International Journal of Disaster Risk Science, 9*(1), 28–43.

Sadik, M. S. et al. (2019). Recovery of Coastal Polder after Cyclone Aila in Koyra, Bangladesh. In *People at Risk*. Disaster Research Training and Management Center, Dhaka University. p. accepted.

Saha, S. K. (2017, July). Cyclone Aila, livelihood stress, and migration: Empirical evidence from coastal Bangladesh. *Disasters, 41*(3), 505–526.

Siddique, A. New policy to restrict shrimp farming. *Dhaka Tribune*, Dhaka, Bangladesh, 03-Feb-2014.

Tatsuki, S. (2007, December). Long-term life recovery processes among survivors of the 1995 Kobe earthquake: 1999, 2001, 2003, and 2005 life recovery social survey results. *Journal of Disaster Research, 2*(6), 484–501.

UNDP, EKN, AusAid, and SDC. (2013). *Early recovery facility (ERF) project annual Progress report 2013*. Dhaka: United Nations Development Programme.

World Bank Group and GFDDR. (2010). *Building regulation for resilience: Managing risks for safer cities*. Washington, DC.

Salinity Intrusion in Southwest Coastal Bangladesh: An Insight from Land Use Change

Md. Hasibul Hasan, Mohammad Jobayer Hossain, Md. Arif Chowdhury, and Maruf Billah

1 Introduction

All over the world, natural, as well as, anthropogenic salinity intrusion is considered to be one of the critical issues especially for the coastal regions (Shameem et al. 2015). Coastal areas of Bangladesh are diversified and competing regarding land use change; shrimp cultivation is one of the noticeable land use practices due to increasing features of salinity intrusion (Islam 2006; Shameem et al. 2015). Besides increasing aspects of shrimp cultivation, land use induced changes also appear in the aspects of agriculture practice, alteration of vegetation cover and water bodies, including rising of social conflicts, etc. (Majid and Gupta 1997; Mahmuduzzaman et al. 2014).

Globally, around 13% of the urban dwellers in the entire population live near coastal areas and more than 75% of those live in Asia (Rosenzweig et al. 2011). Furthermore, sea level may rise up to 0.2–0.3 m by 2100, for which salinity intrusion will also penetrate in the coastal areas at an alarming rate (Rosenzweig et al. 2011; Trieu and Phong 2015). Based on the impact of climate change on salinity intrusion, Trieu and Phong (2015) summarized that, salt water will enter about 70–80 km which will cover about 63% of the total space of the Mekong Delta for which farming area and availability of freshwater will be reduced by about 11% (Trieu and Phong 2015). In addition, saltwater intrusion has an impact on groundwater contribution, livelihood, and agriculture practice of people living in coastal areas of Bangladesh (Naidu et al. 2013). Shrimp cultivation is established at large scale in many coastal lowland areas of tropical and subtropical regions due to the availability of suitable climate; where it has different forms of impacts on the environment and

M. H. Hasan (✉) · M. J. Hossain · M. A. Chowdhury · M. Billah
Institute of Water and Flood Management (IWFM), Bangladesh University of Engineering and Technology (BUET), Dhaka, Bangladesh

A. Haque, A. I. A. Chowdhury (eds.), *Water, Flood Management and Water Security Under a Changing Climate*, https://doi.org/10.1007/978-3-030-47786-8_8

livelihood patterns of people (FAO 1999; Páez-Osuna 2001). Furthermore, due to saline water intrusion and shrimp cultivation, an emerging burden appeared on the environment too, with major impacts on the agricultural sectors (Alonso-Pérez et al. 2003).

Furthermore, though the coastal zone of Bangladesh covers 32% of total land area and 30% of cultivable land, it supports a huge population of around 743 people per km^2 with diversified livelihood activities (WARPO 2004). In 2001, coastal people of this country were about 35.1 million which will be growing to about 57.9 million in 2050 (Islam 2004). Directly or indirectly about 60% of the population in this country are dependent on agriculture for their livelihood contributing 20% of its GDP (BBS 2011). But in the coastal areas, the net-cropped land shows a decreasing trend because of some climatic factors e.g. salinity intrusion, increasing rate of the cyclone, flood, etc. (Islam 2004). The effects of climate change on agriculture are not only a global concern but also for Bangladesh, due to the dependency on agriculture, it is becoming a great threat to national food security (Sikder and Xiaoying 2014). Directly or indirectly, 19 districts of Bangladesh are being affected by climate change-related hazards like tidal waters, storm surges, cyclone (Aila, Sidr, etc.), which causes the frequent inflow of saline water. In addition, anthropogenic reasons like extensive shrimp cultivation are causing the increase of salinity intrusion into the coastal areas (Haque 2006). According to the Soil Resources Development Institute, per year increasing rate of salinity intrusion is around 0.74% (SRDI 2010). From the beginning of 1985, saltwater dominated shrimp aquaculture was introduced to the southwest region from the economic perspective in fallow that gradually helped to change the local landscape, and negatively affected surface and groundwater resources resulting from the increase in soil salinity and decrease in paddy production (Ali 2006; Azad et al. 2009; Datta et al. 2010). Nowadays, it has become a common practice to rotate land use from agriculture to shrimp farming, where it contributes to up to 77% of loss of crop fields come from shrimp culture (Sarwar 2005).

Research on land use change is a prerequisite to formulate national documents including policy, plans and programs to take proper measures on the basis of present scenarios, perception of people and future demands of communities and environment. Coastal areas of Bangladesh are facing different forms of natural and man-made calamities where needs of people differ from community to community (Parvin et al. 2008; Rwanga and Ndambuki 2017). In coastal areas of Bangladesh people are mostly dependent on resources and in such areas salinity intrusion induced land use change is imposing different sorts of impacts on their livelihoods and daily life (Parvin et al. 2008). Following the background, this research attempts to understand the relationship between salinity intrusion and land use changes in the perspective of different forms of land use pattern. There are several studies available on the southwest region of Bangladesh that address climate change and its impacts on agriculture, land cover change and soil degradation, aquaculture and its ecological consequence (Islam 2004; Sarwar 2005; Ali 2006; Haque 2006; Parvin et al. 2008; Azad et al. 2009; Datta et al. 2010; SRDI 2010; Rosenzweig et al. 2011; Kabir and Iva 2014; Rwanga and Ndambuki 2017). However, research work on the impact of salinity intrusion in the perspective of land use changes in the southwest part of

Bangladesh especially in Bagerhat is not sufficient. The study results will help different kinds of stakeholders including government and non-government sectors to understand the impacts of salinity intrusion on land use pattern that may contribute to better agricultural development planning in the country.

2 Methodology

2.1 *Study Area*

As a part of the Bengal basin, Bangladesh is one of the largest geo-synclines in the world and has a large coastline of 710 km (Islam 2004). Due to the geophysical settings, coastal Bangladesh are vulnerable to climate change and salinity intrusion is one of the burning issues for agriculture in such locales. In Bangladesh, the coastal area is categorized into three regions (southwest, south-central and southeast) with 19 coastal districts. Some districts are considered as exposed coasts, whereas, some are known as interior coasts according to their threshold limit (salinity level, movement of tide and cyclone risk) (ICZMP 2003). The scope of this study was limited to the southwest coastal district named Bagerhat; the Southwestern part of Bangladesh is located in between 21°49′ to 22°59′ northern latitudes and 89°32′ to 89°98′ eastern longitudes. It is the coastal district of Khulna division, consisting a total area of 3959.11 square kilometers and 1,549,031 population. This district is bounded by Barguna, Gopalganj, Pirojpur district on the east, Khulna on the west, Narail on the north and Bay of Bengal on the south side. Bangra, Baleshwar, Daratana, Haringhata, Panguchi, Madhumati, Pasur, Mongla and Goshairkhali are the main rivers of this district. Mongla, one of the main ports of Bangladesh, is located here. Mollahat, Fakirhat, Chitalmari, Bagerhat Sadar, Rampal, Kachua, Morrelganj, Mongla, Sharankhola are the nine upazillas in Bagerhat, amongst which Rampal and Fakirhat are famous for their huge production of shrimp (BBS 2011) (Fig. 1).

2.2 *Data Collection*

The study has considered both primary and secondary sources of data to fulfill the objective of the research. Satellite images were collected from the United States Geological Survey (USGS) to identify the land use change from 1980 to 2018 for Bagerhat district, using remote sensing and GIS technique. For accuracy assessment of image analysis, overall accuracy, producer's and user's accuracy; the commission & omission error and Kappa statistics have been considered, as discussed in Sect. 2.4. Data on river salinity and cultivated agricultural area in Bagerhat district were collected from the Bangladesh Water Development Board (BWDB) and Bangladesh Bureau of Statistics (BBS), respectively.

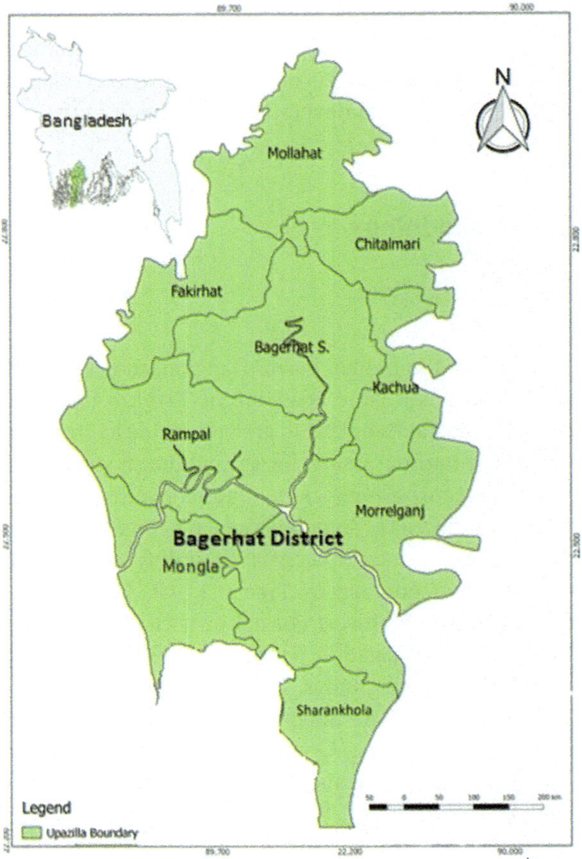

Fig. 1 Location of Bagerhat, Khulna, Bangladesh (Study area)

In addition, in case of the qualitative part, timeline analysis was done to investigate the land use change from the field and for this 27 Key Informant interviews (KIIs) were conducted by using a structured checklist within the whole district, 3 in each Upazila. KIIs were done with the farmers who were knowledgeable and have more than 35 years of involvement in agriculture or fisheries. The informants (e.g. shrimp farmer, agriculturist, local elite, Union parishad chairman etc.) were randomly chosen based on their availability, mainly to understand the historical development of shrimp farming along with other agricultural practices and resource management. Local perceptions on salinity intrusion in these regions have also been identified thoroughly. After collecting data from KIIs, all information was checked and analyzed on the basis of the objectives of this research. This field investigation also aids to validate the information obtained from the image classification of the year 2018.

2.3 Image Analysis

Through the use of various satellite images (e.g. Landsat, Spot, Ikonos, Orbview, Quickbird), freely available Landsat TM images with 30 m resolution were acquired from USGS. Only dry season images (January–February) with less than 5% cloud coverage were considered because of the occurrence of cloud-free sky and less rainfall in this period, this aided in the actual detection of the classified areas. Five images with 10 years interval (1980–2018) were geo-rectified with WGS 84 and zone 45 and consecutively projected with Universal Transverse Mercator's (UTM) projection system, which were classified by using supervised classification methods considering false color composition with QGIS. Where, in total 307 sample pixels for the image of 2018, that are representative of specific classes (As images were classified into four land classes: agriculture, vegetation, river water, and aquaculture), were selected for the supervised image classification. In addition, 278, 301, 297 and 288 sample pixels were used respectively, for the years 2010, 2000, 1990 and 1980. The overall error of the supervised classification is 17.86%.

2.4 Accuracy Assessment

Accuracy assessment is an important step in the image classification process, as the objective of this assessment is to assess how effectively the pixel were sampled from the satellite image for the land cover classification (Rwanga and Ndambuki 2017). The image captured in February 2018 has been chosen for the accuracy assessment as it is the most recent image of our study. A total of 56 points (land cover locations) were chosen for the accuracy assessment.

Google earth was used to identify the river water and vegetation, and field GPS points were demarcated to locate the agriculture and aquaculture, which were used as reference sources. Table 1 presents the error matrix of the land cover classified image of 2018, which illustrates the relationship between the resembling classified data and ground truth data. The columns of Table 1 show produced values which represent the accuracy from the map maker's point of view and also shows the more real features on the ground, correctly shown on the classified map. On the other hand, the rows show observed value that represent the accuracy from map user's point of view,

Table 1 Error matrix of Land Use classification

Classification	River water	Aquaculture	Agriculture	Vegetation	Total	Correct sampled
River water	7	1	0	0	8	7
Aquaculture	2	16	2	0	20	16
Agriculture	0	3	14	1	18	14
Vegetation	0	0	1	9	10	9
Total	9	20	17	10	56	46

Table 2 Accuracy assessment of different accuracy indicators

Class	Sensitivity	Specificity	Commission error	Omission error	UA	PA
River water	0.78	0.98	0.02	0.22	0.88	0.78
Aquaculture	0.80	0.89	0.11	0.20	0.8	0.8
Agriculture	0.82	0.90	0.10	0.18	0.78	0.82
Vegetation	0.90	0.98	0.02	0.10	0.9	0.9

which shows how often the land class will accurately be presented on the map with the ground. This study considered several classification accuracy indicators derived from the error matrix to narrate accuracy through overall accuracy, producer's and user's accuracy, commission and omission error, and Kappa statistics followed by Rwanga and Ndambuki (2017). If only the overall accuracy of classification or the kappa coefficient was indicated, it would have been an incomplete description. It is better to indicate the commission and omission errors as there is a possibility of high overall accuracy, whereas, individual class will contain an unusual amount of errors. On the other hand, only producer accuracy or user accuracy values will not be enough indication of classification accuracy since the full picture will not be found here (Fleiss et al. 1981; Mchugh 2012; Rwanga and Ndambuki 2017). Using the above formulas, various accuracy evaluating parameters were computed and tabulated shown in Table 2.

3 Results

3.1 Accuracy Assessment of Image Classification

Results derived from the accuracy assessment showed that the overall accuracy obtained from the random sampling process was 82.14%. Both user's and producer's accuracy ranged from 78% to 90% (Table 2). In addition, sensitivity depicts the accuracy of prediction of the particular category, where user accuracy expresses reliability of the classification to the user that reflects a more apropos measure of the classification's real effectiveness in the study area. Vegetation was fabricated to be more feasible with 90% of user accuracy. The commission error, called overestimation, reflects a feature that is incorrectly included in the category being evaluated. For instance, the commission error is highest in case of aquaculture (0.11), which means that some points which should not fall under this category were classified as aquaculture.

Similarly, the omission error, known as underestimation, reflects the number of a feature that is left out of the category being evaluated. The omission error in the case of water body was more (0.22), which should fall in this category but is not being categorized in this class. While the overall accuracy obtained was 82.14%, Kappa coefficient was measured at 75.20%, which is rated as excellent according to Fleiss (Fleiss et al. 1981). The calculated overall classification accuracy showed that each parameter gives more detailed accuracy in performance assessment of land use classification for the particular class or category of its field of interest.

3.2 *Salinity Intrusion*

The measuring station of Bangladesh Water Development Board (BWDB) at Daratona station of Alaipur Khal in Bagerhat district showed an increasing trend of river water salinity from 2000 to 2014 (Fig. 2a). Besides that, an occurrence of cyclone-induced storm surge (that is considered as an extreme event) has suddenly increased the salinity level tremendously, it will take several years to return to its prior condition. When extreme events (e.g. Sidr, Aila, Komen etc.) frequently occurred before the salinity level had reached its previous concentration, it led to the increased salinity level itself being set as the new base level of salinity, as a result of which the salinity level has been gradually increasing on a daily basis (Fig. 2b).

In case of qualitative findings, the local perceptions have highlighted that dysfunctional water control structures (e.g. sluices, embankment), its poor operation and maintenance, ailing water governance system, unstructured and inactive local water management institutions, questionable involvement of local government in water management and illegal power practices by local power holders have been intensifying salinity intrusion over the years. In addition, past activities led by the local government and powerful people regarding the operation of water control structure (e.g. sluice gate operation and embankment cutting) were also accelerating the saline water intrusion in the agriculture land inside the polder.

3.3 *Land Use Change: Image Classification*

In the studied 38 years, significant land use change in Bagerhat district has been detected, shown in Fig. 3 and quantified in Table 3. Within that period, the river of

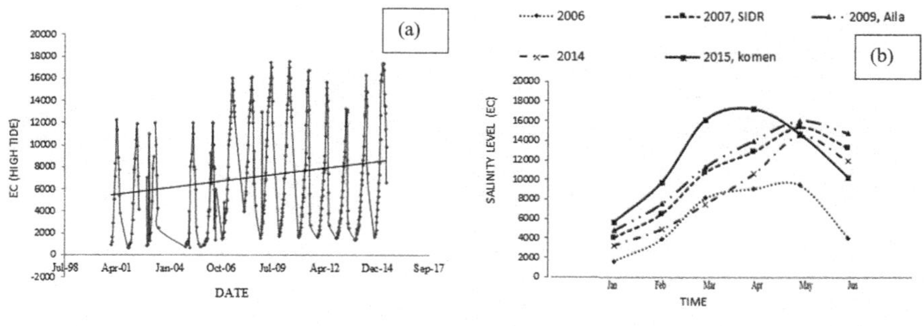

Fig. 2 (**a**) Change of river salinity from 2000 to 2015 at Daratona station of Alaipur Khal and (**b**) Salinity level after extreme events

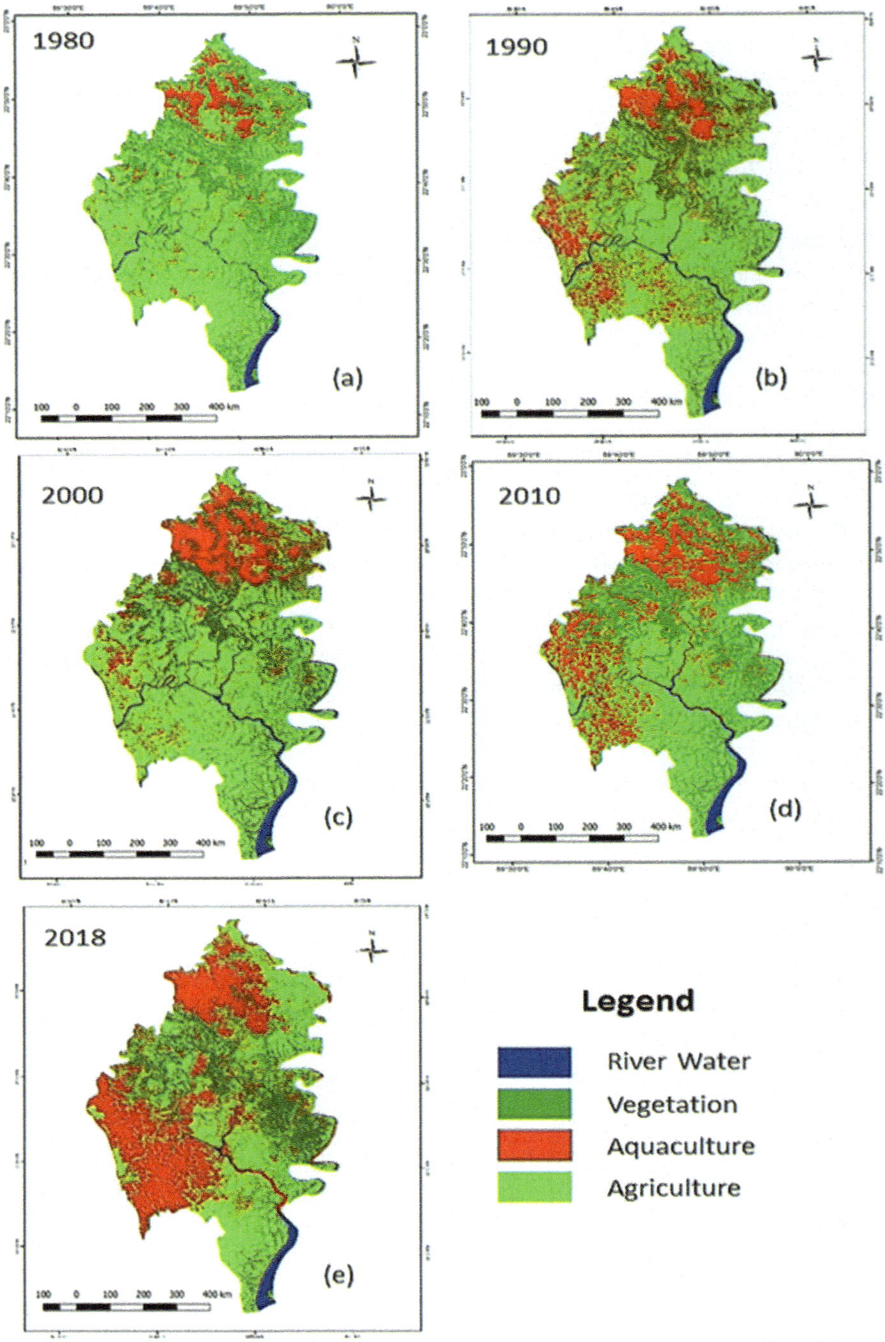

Fig. 3 Classified land use map of Bagerhat district in the year (**a**) 1980, (**b**) 1990, (**c**) 2000, (**d**) 2010, (**e**) 2018

Table 3 Status of land use in Bagerhat districts in different years

	River Water		Aquaculture		Agriculture		Vegetation	
Year	Km^2	%	Km^2	%	Km^2	%	Km^2	%
1980	61.31	2.98	147.94	7.20	1299.7	63.33	543.1	26.46
1990	73.74	3.59	292.84	14.27	1280.63	62.40	404.85	19.72
2000	69.23	3.37	315.21	15.36	1217.51	59.33	450.13	21.93
2010	66.50	3.24	499.14	24.32	1038.45	50.60	447.99	21.83
2018	43.85	2.13	633.90	30.89	996.01	48.53	378.33	18.43

the district was almost the same but aquaculture, vegetation and agriculture have shown fluctuation year by year. Aquaculture was increased by 23.7%, from 7.2% in 1980 to 30.9% in 2018, in respect to the total land of the district. This resulted to a decrease of 14.8% of agricultural land and 8% of vegetation. The increase of aquaculture mostly (9%) triggered in the first decade of this century.

From the image analysis of the year 1980 (Fig. 3a), aquaculture was found in Fakirhat and Mollarhat Upazila (sub-district) and some parts of Bagerhat Sadar and Chitalmari Upazila. In other areas, very few scattered aquacultures were detected. Besides this, vegetation and agriculture were almost equally identified in the northern part of the district while only agriculture was dominating the southern part in that year. In the classified image of 1990 (Fig. 3b), aquaculture was also initiated in the western part of the district, namely Rampal and Mongla Upazila. Along with this, there was an increase in its concentration at Fakirhat, Mollarhat, Chitalmari, and Bagerhat Sadar Upazila, compared to the 1980s. On the other hand, agriculture was found in the middle and south-eastern part of the district. Vegetation was highlighted in the upper middle part.

The whole northern area of the district covering Fakirhat, Mollarhat, Chitalmari, and Bagerhat Sadar Upazila had undergone huge aquaculture, found in the classified image of the year 2000 (Fig. 3c). Vegetation coverage was still concentrated on the upper middle part of the district but the whole southern part was dominated by agriculture. Some parts of Rampal and Mongla Upazila was found under aquaculture, which indicates the continuing process started in 1990 in these areas.

In 2010 (Fig. 3d), the increase of aquaculture was continuing and had spread almost all over Fakirhat, Mollarhat, Chitalmari and Bagerhat Sadar Upazila in the northern part and Rampal and Mongla Upazila in the western part of the district. On the other hand, the southeastern part was still dominated by agriculture, while vegetation was detected in the middle areas of the district.

Finally, in the year 2018 (Fig. 3e), tremendous coverage of aquaculture was seen in the western part, particularly in Rampal and Mongla Upazila. Besides that, the areas which were not identified as aquaculture in 2010, such as Fakirhat, Mollarhat and upper part of Bagerhat Sadar Upazila, had turned into aquaculture dominated areas in 2018. On the contrary, most of the areas of Chitolmari Upazila were detected as agricultural land, indicating a shift from aquaculture in 2010 to agriculture in 2018. The middle part of the district, including Kochua, Bagerhat Sadar, some parts of Fakirhat and Rampal Upazila, was vegetation dominated while the southeastern part was dominated by agriculture as usual.

3.4 Land Use Timeline Analysis

Before the 1980s, in the aftermath of poldarization, Amon was the dominant crop in the wet season followed by some Robi crops (i.e. mainly Mustard, Pea) in the dry season all over Bagerhat. Additionally, capture fishery was a very common practice at that time in canals and wetlands (locally known as Beel), which were open to all and well connected to the river through tidal flow. But in the early 1980s, culture fishery, mainly shrimp farming (both saline and fresh water shrimp), was first introduced in a few areas of the northern parts of the district (Fakirhat, Mollahat, Chitalmari and Bagerhat Sadar Upazila). This was due to the availability of saline water inside the polders through sluices and embankment cutting. Since the northern parts of the district were more than 80 km from the sea, where the tidal flow was distributed through the Baleshwari river system, the river salinity in these upper parts of the district were minimal which enabled the growth of both saline and fresh water shrimp in the same aquaculture pond (locally known as Gher). This system was a bit complex. Firstly, saline-water shrimp (locally known as Baghda) was cultured during January–February, which was then grown and sold till May–June, when the river water remained saline. Secondly, culturing from March–April, fresh-water shrimp (locally known as Golda) was mainly grown after May, when river salinity gradually declined with regular rainfall until the next dry period.

For consequences, such as higher economic return than the earlier practice, capture fisheries and Robi crops' land were gradually converted to Gher. Since then, this aquaculture practice has slowly been spreading all over the northern parts till the very first decade of this century. Within this period, a shift from solely shrimp farming to mixed farming, which combined brackish and freshwater shrimp in relatively low saline water along with Boro and Robi crops (vegetables) on the peripheral embankment of the Gher in the dry season, was started in some parts of Chitalmari and Bagerhat Sadar Upazila. Later, this cropping pattern had gained more popularity, still persisting till today, due to the lower economic return from sole aquaculture. This is happening due to loss of production from repeated virus attacks, fall of shrimp price in the market and increasing cost of production. Besides this, the concentration of river salinity has been gradually lowered due to river siltation down-stream, making the condition unfavorable for sole shrimp farming, as in Chitalmari upazila where currently Boro rice and vegetables are dominant crops. Moreover, the low economic return from aquaculture increased the vulnerability of farmers regarding economic and food safety, which led them to shift their focus to agriculture in order to be self-sufficient in food production, considering the higher uncertainty of shrimp production.

In the western part (Rampal and Mongla Upazila) culture fisheries initiated as saline shrimp aquaculture started in the late 1980s. As these areas were very close to the sea, river salinity was much higher than the northern part, due to which mixed shrimp cultivation is not feasible. Consequently, saline shrimp farming had tremendously spread all over the western part from the early years of this century. In addition, at present almost all areas of Rampal and Mongla Upazila are under the coverage of saline shrimp farming along with some saline water fishes.

Within the eastern parts (Kachua, Morrelganj and Sarankhola Upazila), Aman (during the wet season) along with Boro and/or Robi crops (during the dry season) have been practiced since the 1980s. In the late 2010s, very some scattered shrimp farming was initiated in these areas, including the southern parts of Bagerhat Sadar Upazila and upper western parts of Morrelganj Upazila, which has since then shown gradual increase on a daily basis.

3.5 *Agricultural Land in Bagerhat District*

According to the agricultural census of the years 1983–1984 (BBS 1984, 1996, 2008) conducted by Bangladesh Bureau of Statistics (Fig. 4), agricultural land was gradually increased in Chitalmari, Sarankhola, Kachua, and Bgerhat Sadar Upazila. On the other hand, in Fakirhat, Rampal, Mollahat, and Morrelganj Upazila, a significant decline of agricultural land throughout the years have been observed.

4 Discussion

Land use in Bangladesh is basically stated by geology, climatic events and land features (Sarwar 2005). In Bangladesh, land use patterns in coastal areas are varied and ambitious; land in such areas are used for different purposes e.g. agriculture, shrimp cultivation, fish and salt production, ship-breaking yard, industrial activities, and human habitats, etc. (Asib 2011; Rwanga and Ndambuki 2017).

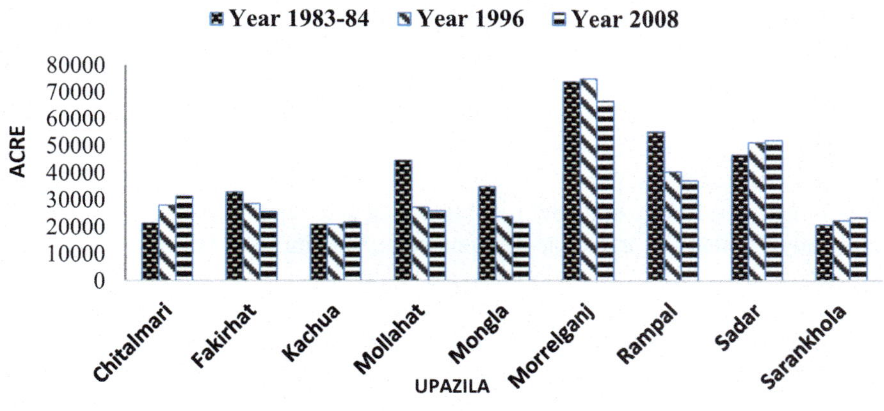

Fig. 4 Agricultural production of different Upazilas in Bagerhat district

Landsat 8 OLI_TIRS, 2018 satellite images, which were used for the accuracy assessment, shows a satisfactory result. Our research depicts that sensitivity in the accuracy statistical parameters is 0.79, which is almost close to the findings of (Rwanga and Ndambuki 2017), which were 0.87. Furthermore, all of the findings in the accuracy statistical parameters, e.g. specificity, commission error, omission error, user accuracy and producer accuracy, etc., are almost similar to the findings of (Rwanga and Ndambuki 2017). This justified the accuracy of our research on salinity intrusion induced land use changes regarding aquaculture, agriculture, vegetation and water bodies.

Salinity intrusion is one of the important issues in Bangladesh, where coastal areas are highly vulnerable and climate change is accelerating the salinity intrusion scenario and impacting land use and livelihood pattern (Hossen et al. 2019). Mahmuduzzaman et al. (2014) described that the issues related to salinity intrusion is exacerbated by natural, socioeconomic and political phenomena where all these systems are interlinked with one another. From this study, it depicts that salinity intrusion is happening at a significant rate, where an increasing thread is being noticed in coastal areas from 2000 to 2014. In addition, local people's perception on salinity intrusion also indicate natural phenomenon (e.g. cyclones, sea level rise) and ineffective governance approaches as reasons for salinity intrusion in coastal areas of Bangladesh. Furthermore, Shamsuddoha and Chowdhury (2007) stated that 5 ppt saline water will penetrate 40 km inland at the freshwater pocket from the shoreline if there is a rise of sea level up to 88 cm, which ultimately indicates sea level rise as an issue when it comes to salinity intrusion. Moreover, Flaherty et al. (2000) also summarized that shrimp cultivation can lead to an extreme consequence of saline water intrusion in agricultural land and this statement supports the qualitative findings of this research work.

In the last few decades, conversion of agriculture land to shrimp fields have led to a mainland cover change in the southern part of Bangladesh. Despite the climate-induced factors, e.g. frequent disasters, sea level rise, salinity intrusion etc., several socio-economic factors, such as agriculture production rate, profit from other practice, labor expenses etc., are also lying behind the changes of land (Turner and Ali 1996). In the instance of land use change pattern in Bagerhat district, which is situated in the southwest part of Bangladesh, this research found that agricultural land is being decreased (about 15%). This is where aquaculture has flourished in practice over the past 38 years and the decrease in agricultural land has mostly occurred due to salinity intrusion. Similarly, research conducted by Ali (2006) in a village in the southwest part of Bangladesh found that about 79% of the agriculture land turned into shrimp cultivation fields, which is a large percentage compared to this study. Present research done in Damarpota, Satkhira shows the reduction of the quantity of agriculture land as a noticeable figure.

Furthermore, shrimp cultivation is now considered as one of the major sources of GDP. The DoF (2003) has identified this sector as a contributor of 5.2% of the overall GDP. The timeline analysis in this research derived that the cultivation of shrimp farming started in 1980s, whereas, Akber et al. (2017) previously stated that the practice had started in coastal areas before the 1960s. People adapted shrimp farming against the changing scenarios of salinity intrusion and in most areas they are currently engaged in mixed cultivation approaches (Martínez-Porchas et al. 2010; Johnson et al. 2016). This is also valid for some Upazilas (sub-district) of Bagerhat. Among the different upazilas in Bagerhat district, farmers in Chitalmari returned to mixed culture after practicing shrimp cultivation for a few years, whereas, aquaculture practice is being increased in Rampal and Bagerhat Sadar Upaliza. Moreover, earlier research identified that within Rampal Upazila of the same district more than 60% of the rice field were turned into rice-shrimp farming during 1975 and 1999, which coincides with the findings of this research. This is happening because agricultural practice is not profitable at desired level due to saline water intrusion and farmers have found that shrimp cultivation is easy to execute with saline water while requiring less manpower and accruing more profit. In addition, mixed culture or shrimp cultivation increased dramatically among agriculture farmers in the last few decades as majority of the farmers or stakeholders, e.g. owner of the shrimp "Gher (a locally termed used to describe shrimp cultivation farm)", accommodated new practice of production management (Mitro et al. 2014). On the other hand, in the southern parts of the Bagerhat district, such as in Sharankhola, agriculture practice is noticed since saline water is not available to the same extent as other areas of Bagerhat district.

Furthermore, due to the change in land use regarding agriculture, shrimp and mixed culture in the southwestern parts of Bangladesh, crop cultivation pattern and quantity are also being changed, as discussed in this study. Transformation of high yielding paddy field into shrimp farms also reduced the production of rice in Bangladesh especially in coastal areas (Islam 2006). Furthermore, based on the land suitability maps developed by Islam (2006), increasing salinity intrusion is a positive catalyst for Bagda shrimp, whereas, salinity intrusion has harmful impacts on the production of T. aman (Transplanted paddy). Along with agriculture and aquaculture aspects, land use changes have also been identified in the perspective of vegetation and river water in Bagerhat but not in a similarly large scale. Correspondingly, several studies described that due to the practice of shrimp farming there has been a reduction in the vegetation cover, increase in the salinity intrusion and declination in the drinking water availability and grazing ground for livestock (Alauddin and Tisdell 1998; Maniruzzaman 1998).

Along with the impacts on land use change, due to salinity intrusion, and changes in agriculture, shrimp cultivation, vegetation pattern etc., there is a significant impact on the environment and social cohesion. Mixed production practice has an impact on environment due to diversity of crops in production practice and the move from one ecosystem to another needs considerable revolution of the characteristics of ecosystem (Ali 2006). In addition, Haque (2017); Karim and Stellwagen (1998); Islam and Koudstaal (2003) and Firoze (2003) have discussed that shrimp cultivation induced land use changes in coastal areas generated large scale conflicts in the community, alteration of environmental components and destruction of natural resources.

5 Conclusion

The land use of Bagerhat district was directly influenced by saline water intrusion which was triggered by cyclone induced storm surge and continual increase of river salinity, imposing differential risks to land and livelihood. Agriculture land was being decreased in Bagerhat, whereas, aquaculture was flourishing due to the combined effect of poor water management, natural calamities, and upstream intervention. Furthermore, some of the Upazilas in Bagerhat returned to mixed culture practice after practicing aquaculture for a few years. Whereas, people from some areas, e.g. Sharankhola, are fully practicing agriculture due to the availability of fresh water. To ensure the sustainable practice of different forms of land use, it is important to understand the requirements of people, as needs of people differ from community to community. In addition, impacts on livelihood and the environment should be considered to implement sustainable livelihood patterns in the coastal areas of Bangladesh. Therefore, policymakers should draw attention to salinity issues and formulate a sustainable land use policy to minimize the potential impact of salinity intrusion in the coastal areas of Bangladesh. Following this research, further study can be done in polder scale as water management varies from polder to polder in coastal areas, which will help understand the different scenarios in detail.

Acknowledgments Authors are thankful to Key Informant persons for their help during data collection in Bagerhat district.

References

Akber, M. A., Islam, M. A., Ahmed, M., Rahman, M. M., & Rahman, M. R. (2017). Changes of shrimp farming in southwest coastal Bangladesh. *Aquaculture International, 25*(5), 1883–1899.

Alauddin, M., & Tisdell, C. (1998). *The environment and economic development in South Asia: An overview concentrating on Bangladesh*. Springer.

Ali, A. M. S. (2006). Rice to shrimp: Land use/land cover changes and soil degradation in Southwestern Bangladesh. *Land Use Policy, 23*(4), 421–435.

Alonso-Pérez, F., Ruiz-Luna, A., Turner, J., Berlanga-Robles, C. A., & Mitchelson-Jacob, G. (2003). Land cover changes and impact of shrimp aquaculture on the landscape in the Ceuta coastal lagoon system, Sinaloa, Mexico. *Ocean and Coastal Management, 46*(6–7), 583–600.

Ahmed, A. (2011). Some of the major environmental problems relating to land use changes in the coastal areas of Bangladesh: A review. *Journal of Geography and Regional Planning, 4*(1), 1–8.

Azad, A. K., Jensen, K. R., & Lin, C. K. (2009). Coastal aquaculture development in Bangladesh: Unsustainable and sustainable experiences. *Environmental Management, 44*(4), 800–809.

BBS. (1984). *Bangladesh Bureau of Statistics, census of agriculture*. Dhaka: Bangladesh Bureau of Statistics.

BBS. (1996). *Bangladesh Bureau of Statistics, census of agriculture*. Dhaka: Bangladesh Bureau of Statistics.

BBS. (2008). *Bangladesh Bureau of Statistics, census of agriculture*. Dhaka: Bangladesh Bureau of Statistics.

BBS. (2011). *Population census*. Dhaka: Bangladesh Bureau of Statistics.

Datta, D. K., Roy, K., & Hassan, N. (2010). Shrimp Culture: Trend, Consequences and Sustainability in the South-Western Coastal Region of Bangladesh. In *Management and sustainable development of coastal zone environments* (pp. 227–244). Springer, Dordrecht.
DoF (Department of Fisheries). (2003). *Fishery statistical yearbook of Bangladesh: 2002–2003* (p. 41). Dhaka: Department of Fisheries.
FAO. (1999). *Aquaculture production statistics 1988±1997* (Fisheries Circular No. 815 Rev. 11). Rome: FAO.
Firoze, A. (2003). The southwest coastal region: Problems and potentials. *The Daily Star, 25.*
Flaherty, M., Szuster, B., & Miller, P. (2000). Low salinity inland shrimp farming in Thailand. *Ambio*, 174–179.
Fleiss, J. L., Levin, B., & Paik, M. C. (1981). The measurement of interrater agreement. *Statistical Methods for Rates and Proportions, 2*(212–236), 22–23.
Haque, A. E. (2017). *Sanitary and phyto-sanitary barriers to trade and its impact on the environment* (No. id: 12353).
Haque, S. A. (2006). Salinity problems and crop production in coastal regions of Bangladesh. *Pakistan Journal of Botany, 38*(5), 1359–1365.
Hossen, M. A., Chowdhury, M., Hans, A., Tagoe, C. A., Allan, A., Nelson, W., & Das, S. (2019). Governance challenges in addressing climatic concerns in coastal Asia and Africa. *Sustainability, 11*(7), 2148.
ICZMP (Integrated Coastal Zone Management Plan). (2003). *Delineation of the coastal zone.* WP005, Dhaka.
Islam, M. R. (2004). *Where land meets the sea: A profile of coastal zone of Bangladesh.*
Islam, M. R. (2006). 18 managing diverse land uses in coastal Bangladesh: Institutional approaches. *Environment and Livelihoods in Tropical Coastal Zones, 237.*
Islam, M. R., & Koudstaal, R. (2003). *Coastal zone management: an analysis of different policy documents* (Working Paper 09, p. 52). Program Development Office for ICZM.
Johnson, F. A., Hutton, C. W., Hornby, D., Lázár, A. N., & Mukhopadhyay, A. (2016). Is shrimp farming a successful adaptation to salinity intrusion? A geospatial associative analysis of poverty in the populous Ganges–Brahmaputra–Meghna Delta of Bangladesh. *Sustainability Science, 11*(3), 423–439.
Kabir, M. H., & Iva, I. J. (2014). Ecological consequences of shrimp farming in southwestern Satkhira District of Bangladesh. *Australian Journal of Earth Sciences, 1*(1), 7.
Karim, M., & Stellwagen, J. (1998). *Shrimp aquaculture.* Fourth Fisheries Project, Department of Fisheries: Final report, Dhaka, 6: 101.
Mahmuduzzaman, M., Ahmed, Z. U., Nuruzzaman, A. K. M., & Ahmed, F. R. S. (2014). Causes of salinity intrusion in Coastal Belt of Bangladesh. *International Journal of Plant Research, 4*(4A), 8–13.
Majid, M. A. & Gupta, M. V. (1997). Research and Information Needs for Fisheries and Development and Management. In *Proceedings of National Workshop on Fisheries Resources Development and Management in Bangladesh*, 29 October–1 November 1995. MOFL/BOBP/FAO/ODA (pp. 160–177).
Maniruzzaman, M. (1998). *Intrusion of commercial shrimp farming in three rice-growing villages of southern bangladesh: Its effects on poverty, environment and selected aspects of culture.* Unpublished PhD thesis, University of the Philippines, Diliman, Quezon City.
Martínez-Porchas, M., Martínez-Córdova, L. R., Porchas-Cornejo, M. A., & López-Elías, J. A. (2010). Shrimp Polyculture: A potentially profitable, sustainable, but uncommon aquacultural practice. *Reviews in Aquaculture, 2*(2), 73–85.
Mchugh, M. L. (2012). Interrater reliability: The kappa statistic. *Biochemia Medica: Biochemia Medica, 22*(3), 276–282.
Mitro, S. H. E. A. T. U., Khatun, R., & Baten, M. A. (2014). Socio-economic and environmental impacts of shrimp culture in some selected areas of Bagerhat District. *Journal of Environmental Science and Natural Resources, 7*(1), 265–269.

Naidu, L. S., VVSe, G. R., Rao G, T., Mahesh, J., Padalu, G., Sarma, V. S., Prasad, P. R., Rao, S. M., & BM, R. R. (2013). An integrated approach to investigate saline water intrusion and to identify the salinity sources in the central Godavari Delta, Andhra Pradesh, India. *Arabian Journal of Geosciences, 6*(10), 3709–3724.

Páez-Osuna, F. (2001). The environmental impact of shrimp aquaculture: A global perspective. *Environmental Pollution, 112*(2), 229–231.

Parvin, G. A., Takahashi, F., & Shaw, R. (2008). Coastal hazards and community-coping methods in Bangladesh. *Journal of Coastal Conservation, 12*(4), 181–193.

Rosenzweig, C., Solecki, W. D., Hammer, S. A., & Mehrotra, S. (Eds.). (2011). *Climate change and cities: first assessment report of the urban climate change research network.* Cambridge/New York: Cambridge University Press.

Rwanga, S. S., & Ndambuki, J. M. (2017). Accuracy assessment of land use/land cover classification using remote sensing and GIS. *International Journal of Geosciences, 8*(04), 611.

Sarwar, M. G. M. (2005). *Impacts of sea level rise on the coastal zone of Bangladesh.* See http://static.weadapt.org/placemarks/files/225/golam_sarwar.pdf

Shameem, M. I. M., Momtaz, S., & Kiem, A. S. (2015). Local perceptions of and adaptation to climate variability and change: The case of shrimp farming communities in the coastal region of Bangladesh. *Climatic Change, 133*(2), 253–266.

Shamsuddoha, M., & Chowdhury, R. K. (2007). *Climate change impact and disaster vulnerabilities in the coastal areas of Bangladesh.* Dhaka: COAST Trust.

Sikder, R., & Xiaoying, J. (2014). Climate change impact and agriculture of Bangladesh. *Journal of Environment and Earth Science, 4*(1), 35–40.

SRDI. (2010). *A report of saline soil of Bangladesh*, ARMAF Project, Ministry of Agriculture, Government of the People's Republic of Bangladesh.

Trieu, T. T. N., & Phong, N. T. (2015). The impact of climate change on salinity intrusion and Pangasius (Pangasianodon Hypophthalmus) farming in the Mekog Delta, Vietnam. *Aquaculture International, 23*(2), 523–534.

Turner, B. L., & Ali, A. S. (1996). Induced intensification: Agricultural change in Bangladesh with implications for Malthus and Boserup. *Proceedings of the National Academy of Sciences, 93* (25), 14984–14991.

WARPO (Water Resources Planning Organization). (2004). *Living in the coast: People and livelihoods.* Integrated Coastal Zone Management Project (ICZMP) (pp 3–17). Simon Centre (5th Floor), House 4A, Road 22, Gulshan Dhaka 1212, Bangladesh.

Groundwater Vulnerability Mapping to Salinity Intrusion Using GALDIT Method: A Case Study of the South-Western Coastal Region of Bangladesh

Md. Rezaul Hasan, Md. Esraz-Ul Zannat, A. F. M. Afzal Hossain, S. M. Shah-Newaz, and M. Monowar Hossain

1 Introduction

Bangladesh lies in the northeastern part of South Asia, has 710 km coastline, which consists of 32% of the land area (MoWR 2005). Islam (2001) classified the coast of Bangladesh into three distinct regions namely southeast, central and southwest. The south-western coastal region is characterized by the Ganges tidal flood plains with low relief, crisscrossed by rivers, tidal marshes, and swamps. Although groundwater (GW) is abundant in the region, saline water intrudes into the aquifer system due to reduction of upstream freshwater flow, sea level rise and over-abstraction of groundwater, which makes the condition worse. The intrusion of pollutants to groundwater alters the water quality and reduces its value to the consumers (Melloul and Collin 1994).

MPO (1987) studied that the main source of potable water is groundwater in the coastal zones because of high surface water salinity. SRDI (2009) reported that high

M. R. Hasan (✉)
Institute of Water Modelling (IWM), Dhaka, Bangladesh
e-mail: mrh@iwmbd.org

M. E.-U. Zannat
Department of Urban and Regional Planning, Khulna University of Engineering & Technology, Khulna, Bangladesh

A. F. M. A. Hossain
P&D, IWM, Dhaka, Bangladesh

S. M. Shah-Newaz
IRM Division, IWM, Dhaka, Bangladesh

M. M. Hossain
IWM, Dhaka, Bangladesh

A. Haque, A. I. A. Chowdhury (eds.), *Water, Flood Management and Water Security Under a Changing Climate*, https://doi.org/10.1007/978-3-030-47786-8_9

salt content is limiting crop intensification in the coastal zones of Bangladesh. Salt enters inland through rivers and channels during the latter part of the dry season when the upstream freshwater flow becomes low. River water causes an increase in salinity of groundwater and creates an unfavorable environment of crop production. The dry season flow of the Gorai River, which influences the dry season salinity of the southwest region, especially to the Nabaganga-Rupsa-Passur systems (Halcrow 1993) significantly dropped after the Farakka Barrage became operational. In addition, the possible impact of climate change induced sea level rise on lateral seawater intrusion is being recognized increasingly (World Bank 2010). On the other hand, Khulna is also vulnerable to over exploration in upstream areas that may lead to intrusion of saline water into groundwater resources (Adhikary et al. 2012). In Bangladesh, the recommended chloride (Cl) concentration in drinking water is 150–600 ppm (ECR 1997). However, higher values (up to 1000 ppm) are considered acceptable in problem areas including the coastal belt, such as the Khulna area (DPHE 2006). Spatial distribution of groundwater salinity with varying degrees of salinity concentration has been noticed. In the backdrop of this scenario, determination of groundwater vulnerability owing to salinity has become crucial.

IWM (2013) has conducted a comprehensive study covering 19 coastal districts of Bangladesh. In this study, aquifer properties with varying depths along with water table measurements have been conducted in more than 600 locations. Besides, water quality parameters were analyzed both for surface water and groundwater samples. During the dry season, in above areas the salinity of surface water increases 10–20 times of the salinity level during the wet season. As for groundwater, the study measured salinity from nested wells and line wells. In Khulna, salinity was found at different depths varying from 0.6 gl^{-1} to 2.1 gl^{-1} during the wet season. During that time in Jashore, it was measured to be between 0.3 gl^{-1} and 1.6 gl^{-1}. Moreover, in Satkhira, it was measured to be between 1.6 gl^{-1} and 5.5 gl^{-1}. During the dry period, salinity increased in most parts of these locations.

There are several methods for assessing groundwater vulnerability to contamination, such as using DRASTIC (Aller et al. 1987), GOD (Foster 1998) and GALDIT (Chachadi and Lobo-Ferreira 2001). Both DRASTIC and GOD methods are designed on the grounds that contaminants are introduced from the ground surface. Besides, GALDIT approach is such that the bottom of the aquifer(s) lies below mean sea level so that the seawater can move groundwater laterally. This is more appropriate to address salinity risk in the coastal aquifers. This method was originally developed by Chachadi and Lobo-Ferreira (2001) to assess the vulnerability of the Indian coastal aquifer systems due to sea water intrusion. The application of the method was studied for the assessment of aquifer vulnerability to seawater intrusion in Portugal (Chachadi and Lobo-Ferreira 2005) and in Greece (Kallioras et al. 2011).

In the GALDIT method, the most important factors controlling seawater intrusion were found to be the following:

- **G**roundwater occurrence (unconfined, confined and leaky aquifer);
- **A**quifer hydraulic conductivity;
- Depth to groundwater **L**evel above the sea;

- **D**istance from the shore (distance inland perpendicular from the shoreline);
- **I**mpact of the existing status of seawater intrusion in the area, and
- The **T**hickness of the aquifer, which is being mapped.

The acronym GALDIT is calculated from the highlighted letters of the parameters for ease of reference. These factors, in combination, are determined to include the basic requirements needed to assess the general seawater intrusion potential of each hydro-geologic setting. GALDIT factors represent measurable parameters for which data are generally available from a variety of sources without detailed examination. According to the GALDIT method, each of the six parameters has a pre-determined fixed relative weight that reflects its relative importance to vulnerability (Table 1). Thus, one can use the hydrologic setting as a mappable unit to define the area of interest, choose corresponding ratings and calculate seawater intrusion using the GALDIT index. Once the GALDIT index has been computed, it is possible to identify three types of vulnerable areas which are more likely to be susceptible to seawater intrusion, relative to one another as given in the Table 2 (Chachadi and Lobo-Ferreira 2005). The maps for each parameter can be prepared using GIS and then superimposed to get a vulnerability score map.

In the southwest region, most of the research works to date have been carried out mainly on irrigation, drainage, flood control and diversion of the Ganges flow during

Table 1 GALDIT parameters, weights and ratings for each parameter

Parameter	Weight	Parameter variables	Rating
G (groundwater occurrence)	1	Confined aquifer	10
		Unconfined aquifer	7.5
		Leaky confined aquifer	5
		Bounded aquifer	2.5
A (aquifer hydraulic conductivity, m/day)	3	High (>40)	10
		Medium (10–40)	7.5
		Low (5–10)	5
		Very low (<5)	2.5
L (groundwater level above sea level, m)	4	High (<1.0)	10
		Medium (1.0–1.5)	7.5
		Low (1.5–2.0)	5
		Very low (>2)	2.5
D (distance from shoreline, m)	4	Very small (<500)	10
		Small (500–750)	7.5
		Medium (750–1000)	5
		Far (>1000)	2.5
I (impact status of sea water intrusion)	1	High (>2)	10
		Medium (1.5–2)	7.5
		Low (1–1.5)	5
		Very low (<1)	2.5
T (aquifer thickness, m)	2	Large (>10)	10
		Medium (7.5–10)	7.5
		Small (5–7.5)	5
		Very small (<5)	2.5

Table 2 GALDIT vulnerability classes

Sl no	GALDIT index (range)	Vulnerability classes
1	10 ($\geq$7.5)	High vulnerability
2	7.5 (5–7.5)	Moderate vulnerability
3	5 (<5)	Low vulnerability

the dry season. Very few studies were undertaken to study groundwater salinity in that area. In this context, the Institute of Water Modelling (IWM) has taken up a research study to map groundwater vulnerability through hydro-chemical investigation and participatory rapid appraisal (PRA). While hydro-chemical investigation indicates actual salinity of the groundwater, sociological indicators are also useful in understanding the impacts of salinity.

2 Methodology

2.1 *Study Area*

The study area is bounded on the North by three GW observation wells of Bangladesh Water Development Board (BWDB) and on the other sides by rivers namely Kobadak, Shibsha, and Rupsa. The study area covers eight Upazilas: Dumuria, Batiaghata, Phultala and Paikgacha Upazilas of Khulna district, Abhaynagar, Keshabpur and Manirampur Upazilas of Jashore district and Tala Upazila of Satkhira district, with a total area of about 1534 km^2 (Fig. 1).

2.2 *Study Design*

The study involved hydro-chemical investigations and qualitative surveys, which were conducted in April 2013.

2.3 *Hydro-chemical Investigation*

Groundwater samples were collected from hand tube-wells with screening depths varying from 35–120 m (Table 3). The number of tube wells were selected from each union (Sub-Upazila) by consulting the list of tube-wells available with the Department of Public Health Engineering (DPHE). A total number of tube-wells tested were 186, of which 161 were used for the present study due to consistency of data.

Groundwater salinity parameters like temperature, pH, EC, Total Dissolved Salts (TDS) and chloride concentration were determined using a field test-kit (Hach Salinometer model EC5DL which could not show salinity reading below 300 gl^{-1}). Water samples of eight wells were analyzed for EC and Cl at laboratories

Fig. 1 Map showing study area

of Bangladesh University of Engineering and Technology (BUET), Bangladesh Council of Scientific and Industrial Research (BCSIR) and DPHE. The results obtained from DPHE and BUET were found to be consistent. Therefore, DPHE laboratory was selected for carrying out the laboratory analysis. Since the field test-kit results do not show a high level of accuracy, the field values were modified according to relations derived from the lab test results. In addition, duplicate samples from 22 tube-wells were tested at the DPHE Central Laboratory for getting the values of carbonates such as CO_3 and HCO_3 required for GALDIT parameter.

Table 3 Distribution of the tubewell samples of salinity and carbonates by Upazila

District	Upazila	Salinity parameters	Carbonates
Khulna	Dumuria	49	6
	Batiaghata	18	3
	Phultala	5	1
	Paikgacha	53	5
Jashore	Abhaynagar	5	1
	Keshabpur	16	2
	Manirampur	16	–
Satkhira	Tala	24	4
Total		186	22

2.4 *Qualitative Survey*

For collecting qualitative data on salinity aspect, PRA approach was followed including focus group discussion (FGD) and key informant interviews (KII). Data was collected by FGDs and KIIs using a checklist. The FGDs were conducted at six unions with local level public representatives and people of different professional groups including farmers, businessmen, service holders, shopkeepers, teachers, housewives, etc. Eight to twelve people participated in each of the FGD sessions. The KIIs were held with staff of DPHE based in the study area.

To verify the results of the qualitative survey in relation to groundwater salinity in the reported less saline areas, field kit tests were carried out at most of the PRA locations. Samples were taken from at least three tube-wells at each location. Before taking the water samples, sufficient water was pumped out to ensure that the sample was represented that of the surrounding groundwater. The information from the qualitative social survey thus verified by field test-kit was used and modified as mentioned earlier. The field values of all the salinity data collected were later used to get GALDIT parameters.

2.5 *Measurement of GALDIT Parameters*

Groundwater Occurrence (G) The extent of salinity intrusion is dependent on the basic nature of groundwater occurrence in layers such as confined, unconfined or leaky confined. In the present study, the weightage and rating of parameter G was taken as 10 and 2 respectively considering unconfined aquifers.

Aquifer Hydraulic Conductivity (A) Hydraulic conductivity indicates the ability of the aquifer to transmit water, which determines the rate of flow of contaminant material within the groundwater system. The hydraulic conductivity map was obtained from the IWM study (IWM. 2013).

Depth to Groundwater Level Above the Sea (L) This is the most important factor in the evaluation of seawater intrusion in an area because it determines the hydraulic pressure availability to push back the seawater front. The study considered the values pertaining to minimum groundwater levels above sea (pre-monsoon during April), as this would provide the highest possible vulnerability risk. Groundwater level data of particular date during the pre-monsoon was taken from the study findings (IWM. 2013).

Distance from the Shore (D) The magnitude of the impact of seawater intrusion generally decreases as one moves inland at right angles to the shore and the creek. Since the nearest surface water salinity is the river which is connected with the Bay of Bengal, data for parameter D was computed using the satellite map of the area.

Impact of Existing Status of Saline Water Intrusion in the Area (I) The existing imbalance in the saline water-freshwater interface is considered with this parameter which is very significant in mapping aquifer vulnerability to salinity intrusion. The ratio of Cl/[HCO_3+ CO_3] as a criterion to identify the extent of seawater intrusion into the coastal aquifers (Revelle 1941). Chloride is the dominant ion in the seawater and it is only available in small quantities in groundwater while bicarbonate (HCO_3), which is available in large quantities in groundwater, occurs only in very small quantities in the seawater. This ratio was used while assigning the importance rating for the GALDIT parameter I.

Thickness of the Aquifer (T) Aquifer thickness or saturated thickness of an unconfined aquifer plays an important role in determining the extent and magnitude of seawater intrusion in the coastal areas. The aquifer thickness in the area was obtained from lithological logs (IWM. 2013).

3 Results and Discussion

3.1 *Calibration of the Field Test-kit*

The relationship between the EC measured using the field test-kit and in the BUET laboratory was strong and the polynomial relationship ($y = -0.0058* x^2 + 1.023*x - 0.0345$), $R^2 = 0.999$ (Fig. 2a) was near perfect. Salinity is the

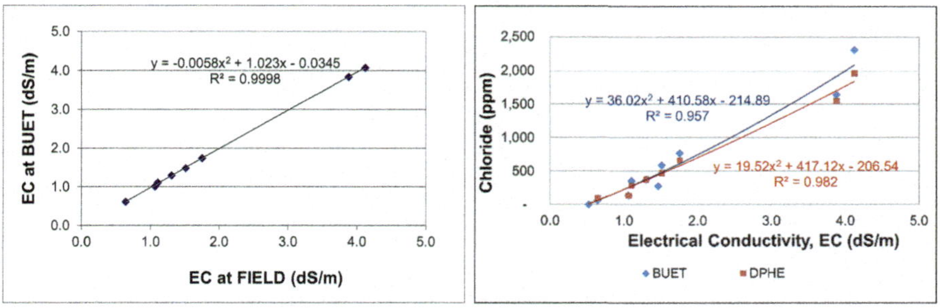

a) field kit ECvs BUET- EC **b) field kit EC vs Chloride concentration**

Fig. 2 Calibration of the field test-kit against laboratory values from BUET and DPHE measure of the concentration of dissolved salts in water, such as chloride anions (Cl), and can be determined from the EC relationship curve. Relationships between field EC and chloride concentration determined at BUET and DPHE are in Fig. 2b. In case of EC > 4 dS/m the BUET equation has been considered due to its consistency with field data; while for the lower EC values, the DPHE equation was used for Cl data analysis.

3.2 *Groundwater Quality*

Around 50% of the hand tube-wells had chloride concentrations exceeding 1000 ppm in the study Upazilas of Khulna (Table 4). However, The Cl concentration of the tube-wells in Jashore and Satkhira Upazilas were below 600 ppm and thus suitable for drinking at all locations. In case of Paikgacha Upazila, the adjacent areas of Kobadak River had low Cl concentration and other areas have very high values. This indicated that in some parts of this Upazila the available groundwater is suitable for drinking while in others it is not.

3.3 *Findings of the Qualitative Survey*

The study area covers both saline and less saline aquifers. While, change in groundwater salinity is not prominently perceived, there is still a growing apprehension of

Table 4 Range (median) of groundwater parameters

District	Upazila	EC (dSm^{-1})	Cl (gl^{-1})	TDS (gl^{-1})	CO_3 (mgl^{-1})	HCO_3 (mgl^{-1})
Jashore	Abhaynagar	1.2–1.9 (1.4)	0.3–0.67 (0.4)	0.6–1.0 (0.7)	2.46	525
	Keshabpur	0.6–1.9 (0.8)	0.3–0.47 (0.15)	0.3–0.9 (0.4)	1.26–1.27 (1.265)	340–540 (440)
	Manirampur	0.6–2.7 (0.9)	0.3–2.0 (0.17)	0.3–1.4 (0.4)	–	–
Khulna	Dumuria	0.6–6.1 (2.5)	0.3–2.1 (0.95)	0.3–3.0 (1.3)	0.56–1.66 (0.97)	265–560 (450)
	Phultala	0.9–5.6 (1.5)	0.3–1.9 (0.45)	0.5–3.5 (1.0)	1.51	640
	Batiaghata	1.1–8.5 (2.7)	0.3–3.4 (0.98)	0.5–4.6 (1.4)	1.45–1.62 (1.5)	775–1090 (800)
	Paikgacha	0.6–12.9 (3.6)	0.3–12.9 (1.5)	0.3–17.0 (1.9)	0.64–1.04 (0.79)	365–880 (605)
Satkhira	Tala	0.5–1.6 (0.9)	0.3–0.5 (0.3)	0.3–0.8 (0.5)	0.78–1.27 (0.98)	285–680 (355)

Note: 1 gl^{-1} = 1000 mgl^{-1} = 1000 ppm

increase. It is perceived that salinity is increasing in groundwater for a few decades. In case of Paikgacha Upazila, two distinct features are observed with low and relatively high values of salinity indicating that in some parts of this Upazila shallow aquifer is suitable for drinking purposes while in the other parts the shallow aquifers are not suited for providing drinking water. This may be due to the fact that historically in the Kobadak system sweet water used to flow from upstream thereby making shallow aquifer of adjacent areas less saline. But for the rest of the area of Paikgacha Upazila the water is not suitable for drinking.

3.4 Computation of Index Map

In the preparation of the vulnerability map following the GALDIT model, separate maps of the study area were prepared considering each of the 6 individual parameters (Fig. 3). A final map showing the vulnerability index was prepared by superimposing the individual index maps (Fig. 4). It reveals that the groundwater in the southern part is highly vulnerable to contamination. The moderate vulnerability risk class is situated in the north, the east and central parts of the area. Based on the analyses performed using this method, about 17.0% of the study area was identified as high vulnerability risk, 62.4% as moderately vulnerability risk and 20.6% as low vulnerability risk classes.

4 Conclusions

This paper attempted to develop a vulnerability map showing groundwater vulnerable areas with varying degrees of salinity. The vulnerability has been assessed using GALDIT index owing to salinity intrusion. The assessment was based on the outcomes of the limited data collected in the month of April 2013 as a pilot programme project from eight Upazilas under three Districts of the South-Western coastal regions of Bangladesh. Both quantitative and qualitative approaches were adopted in the present study. This study revealed that in regard to drinking water, the tube-wells of the Upazilas of Khulna had more groundwater salinity than that of the tube-wells of Jashore and Satkhira. The GALDIT model showed that around 17.0% of the study area is high vulnerability risk, 62.4% is moderately vulnerability risk and 20.6% is low vulnerability risk classes. As the moderate vulnerability risk class is situated in the north, the east and the central part of the area, it can be said that these parts should be taken care from the contamination of groundwater. The southern part falls under the high vulnerability class.

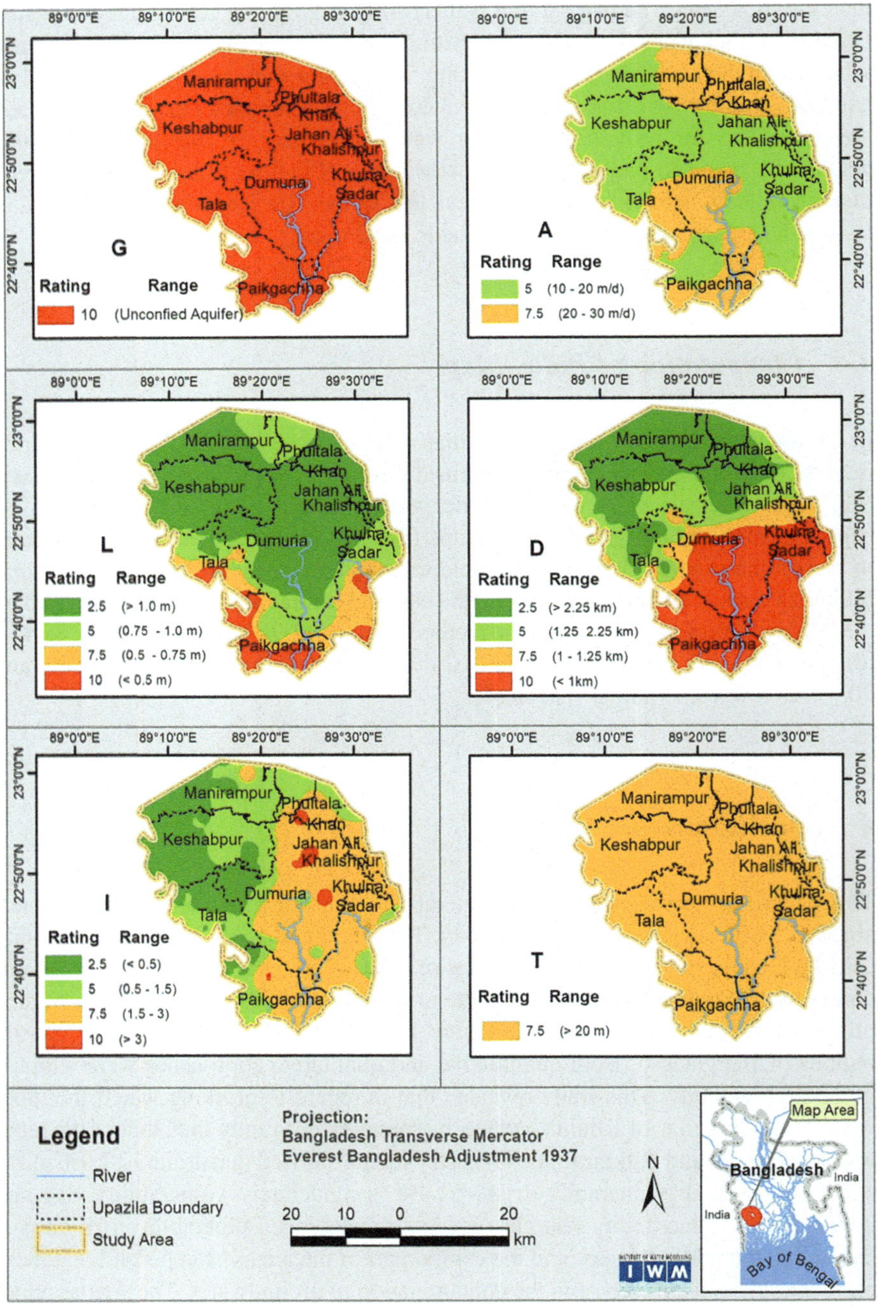

Fig. 3 Index of GALDIT parameters

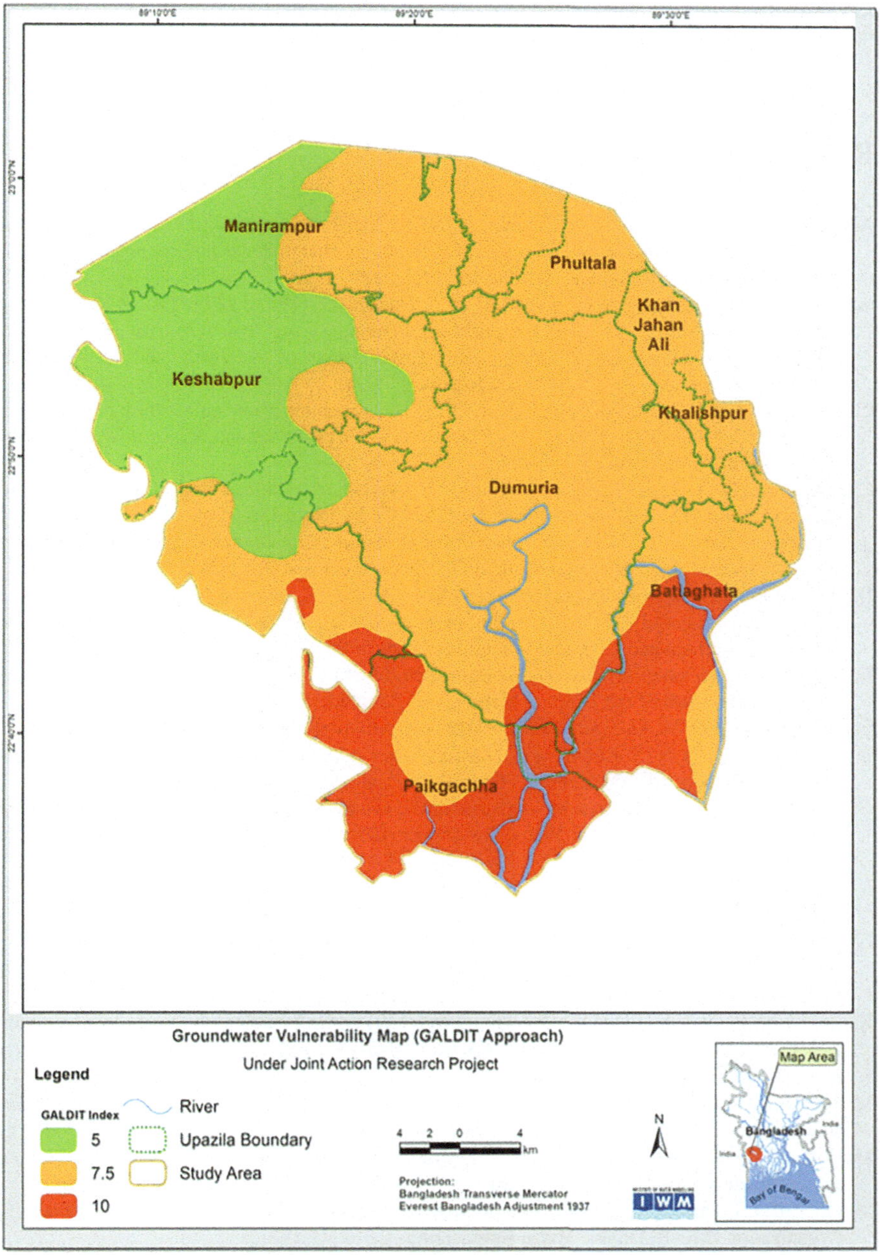

Fig. 4 Groundwater vulnerability map (GALDIT method)

Acknowledgement The authors would like to thank all the field staff of IWM at the Khulna Project office who provided the data required for carrying out the study.

References

Adhikary, S. K., Elahi, M. M., & Hossain, A. M. I. (2012). Assessment of shallow groundwater quality from six wards of Khulna City Corporation, Bangladesh. *International Journal of Applied Sciences and Engineering Research, 1*(3), 488–498.

Aller, L., Bennet, T., Lehr, J. H., Petty, R. J., & Hachet, G. (1987). *DRASTIC: A standardised system for evaluating groundwater pollution potential using hydrogeologic settings (EPA 600/2-87)* (Environmental Research Laboratory, Office of Research and Development, U.S. Environmental Protection Agency Report. p. 622).

Chachadi, A. G., & Lobo-Ferreira, J. P. (2001). Sea water intrusion vulnerability mapping of aquifers using GALDIT method. In: *Proceedings Workshop on modelling in hydrogeology* (pp.143–156) Anna University, Chennai.

Chachadi, A. G. & Lobo-Ferreira, J. P. (2005). Assessing aquifer vulnerability to sea-water intrusion using GALDIT method: Part 2 – GALDIT Indicator Descriptions. IAHS and LNEC, *Proceedings of the 4th The Fourth Inter Celtic Colloquium on Hydrology and Management of Water Resources*, held at Universidade do Minho, Guimarães, Portugal, July 11–13.

DPHE. (2006). *Development of Deep Aquifer Database and Preliminary Deep Aquifer Map*, Final Report. Department of Public Health Engineering (DPHE), Ministry of Local Government and Rural Development and Cooperatives, Government of the People's Republic of Bangladesh.

ECR. (1997). *Environmental conservation rules, Bangladesh Gazette*. Ministry of Environment and Forest, Government of the People's Republic of Bangladesh.

Foster, S. S. D. (1998). Groundwater recharge and pollution vulnerability of British aquifers: a critical review. In N. S. Robins (Ed.), *Groundwater pollution, aquifer recharge and vulnerability* (Vol. 130, pp. 7–22). London,. Special Publications: Geological Society.

Halcrow. (1993). *Final report on soil and agriculture, southwest area water resources management project*, FAP-4. Ministry of Water Resources, Government of the People's Republic of Bangladesh.

Islam, M. S. (2001). *Sea-level changes in Bangladesh: The last ten thousand years*. Asiatic Society of Bangladesh.

IWM. (2013). *Establishment of monitoring network and mathematical model study to asses salinity intrusion in groundwater in the coastal area of bangladesh due to climate change* (Package III, Final Report, Vol. I). Dhaka: Institute of Water Modelling (IWM).

Kallioras, A., Pliakas, F., Skias, S., & Gkiougkis, I. (2011). Groundwater vulnerability assessment at SW Rhodope aquifer system in NE Greece. *Advances in the Research of Aquatic Environment, 2*, 351–358.

Melloul, A., & Collin, M. (1994). Water quality factor identification by the principal components statistical method. *Water Science and Technology, 34*, 41–50.

MoWR. (2005). *Coastal Zone Policy (CZPo)*. Ministry of Water Resources (MoWR), Government of the People's Republic of Bangladesh.

MPO. (1987). Technical Report No. 5 on Groundwater Resources of Bangladesh, Master Planning Organization (MPO), Dhaka, Harza Engineering.

Revelle, R. (1941). Criteria for recognition of sea water in ground-waters. *Transactions of the American Geophysical Union, 22*, 593–597.

SRDI. (2009). *Saline water quality assessment, land resources Database& its use in agricultural development activities, February 2009*. Dhaka: Soil Research Development Institute (SRDI).

World Bank. (2010). *Implication of climate change for fresh groundwater resources in coastal aquifers in Bangladesh*. Technical report, 2010, World Bank Office, Dhaka.

On the Importance of Typhoon Size in Storm Surge Forecasting

Md. Rezuanul Islam and Hiroshi Takagi

1 Introduction

The global occurrence of destructive typhoons (TY) over the last several years has urged the development of a more reliable system for real-time prediction of typhoon-induced storm surges. Major typhoons, which caused more than 1000 fatalities, such as the 2005 Hurricane Katrina, the 2007 cyclone Sidr, the 2008 cyclone Nargis, and the 2013 typhoon Haiyan, all occurred in the past 15 years and many of these causalities were the direct result of the coastal surge (Esteban et al. 2015). However, this is the fifth decade in which the surface wind speed-based TY scale (i.e., Saffir-Simpson hurricane scale (SSHS)) has been used by meteorological scientists and emergency planners as a convenient tool for forecasting storm surge as well as wind intensity. Although there are many advantages to using wind speed-based scales through simple categorization of TY, the reliance on this scale as an indicator of potential storm surge has also led to significant misconceptions within the public and scientific communities. The primary disadvantage of the wind speed-based scale for TY surge warning is its inability to account for some of the important factors such as TY size, forward speed, and the track that directly influence storm surge generation (Needham and Keim 2011).

Correlating maximum surge with TY meteorological conditions in the 1950s through the 1970s, previous studies suggested that the size of hurricane is weakly correlated with peak surge and SSHS may be a reasonable surge indicator (Conner et al. 1957; Hoover 1957; Harris 1963; Jelesnianski 1972). Consequently, most storm surge studies, for both forecasting and coastal protection design, have relied heavily on either central pressure deficit or the related maximum wind speed (i.e. SSHS) as the determining factors for surge response (Berke et al. 1984). Prior

M. R. Islam (✉) · H. Takagi
School of Environment and Society, Tokyo Institute of Technology, Tokyo, Japan

A. Haque, A. I. A. Chowdhury (eds.), *Water, Flood Management and Water Security Under a Changing Climate*, https://doi.org/10.1007/978-3-030-47786-8_10

to recognizing the limitations of the SSHS, Dolan and Davis proposed a new hurricane intensity scale in the early 1990s (Dolan and Davis 1992). Nevertheless, no conclusion could be drawn in regard to the influence of hurricane size on surge for a given storm intensity. After the 2005 Hurricane Katrina caused catastrophic storm surge at the Gulf Coast of the United States, Irish et al. claimed that typhoon size plays a key role in storm surge generation, particularly over mildly sloping bathymetry (Irish et al. 2008). Later, using long-term observation data for Gulf Coast, Needham and Keim revealed that hurricane size and storm surge are statistically correlated (Needham and Keim 2014). In this manner, few recent studies have advanced storm surge climatology particularly in the Gulf Coast of the United States. However, a thorough literature review on this topic reveals that no studies have been conducted on semi-enclosed bays particularly. Despite the importance of surface wind forcing, several recent studies have attempted to quantify the influence of hurricane track, forward speed, and landfall location to determine storm surge behavior (Zhong et al. 2010; Sebastian et al. 2014) in semi-enclosed bays. However, none of these studies have considered the relationship between hurricane size and storm surge.

This study aimed to investigate the general influence of typhoon size in generating surge, considering Tokyo Bay as a study area. Tokyo Bay was chosen for two main reasons. As confirmed from Fig. 1, the bay has intricate coastlines that can either defend the megacity composed of Tokyo, Kanagawa and Chiba Prefecture or render them particularly vulnerable because it may trap and amplify storm surges

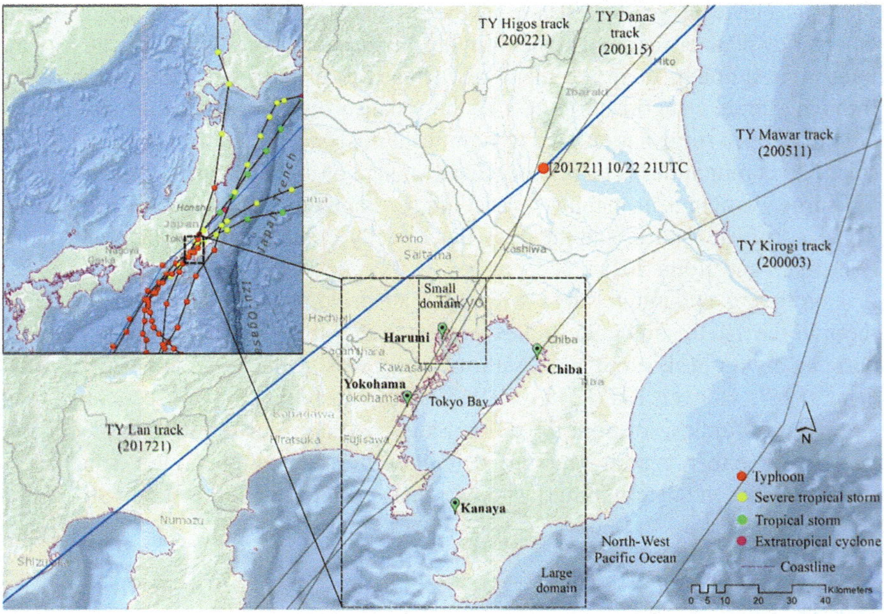

Fig. 1 Track of Typhoon Lan in 2017 as it approaches Tokyo Bay (Japan Meteorological Agency 2019a, b); Green marks indicate measurement stations used in this study

(Hirano et al. 2014). The other reason is that the frequency of typhoon landfall is not remarkably high compared to the western south part of Japan. Hence, the storm surge risk for Tokyo Bay has not been sufficiently understood. Considering the growing concern for unprecedented storm surge in the present-day, Tokyo Bay is a coastal system well suited for conducting the sensitivity of storm surge to typhoon size.

2 Methodology

The authors performed numerical simulations by employing the parametric typhoon model developed by Takagi and Wu (2016); Takagi et al. (2016, 2017, 2018), coupled with the fluid dynamics model Delft3D-FLOW (DELTARES 2011). The accuracy of the model has been validated for recent typhoons such as Haiyan in 2013, Goni in 2015, and Hato in 2017, all of which occurred in the North-West Pacific basin (Takagi and Wu 2016; Takagi et al. 2016, 2017). The typhoon model calculates both the pressure and wind fields using parameters obtained from the Japan Meteorological Agency (JMA) best track dataset, which includes the central position of the typhoon, pressures, and 50-knot (26 m/s) wind radius (R_{50}) at every recording period. Delft3D-FLOW was then used to simulate the movement of a storm surge from the deep sea to shallow water.

In this study, two different domains (Fig. 1) were used. First, the simulation was performed for a wide area (grid size: 250 m) encompassing parts of Tokyo, Kanagawa, and Chiba prefecture in order to assess potentially affected areas. Then, a more detailed simulation was performed for inner bay locations (grid size 25 m) where the mean bathymetry is shallower than 5 m. Using this grid configuration, numerical simulations were performed for four different bottom slopes, varying from 1:2050 to 1:117. The bathymetry data (50 m resolution) over the target area and astronomical tide data were collected from the Japan Coast Guard and the TPXO7.1 Global Tide Model (Egbert and Erofeeva 2002), respectively. The numerical settings used in this study are shown in Table 1.

Table 1 Numerical model settings

Typhoon path	JMA typhoon best track https://www.jma.go.jp/jma/jma-eng/jma-center/rsmc-hp-pub-eg/trackarchives.html
Typhoon model	Pressure: Empirical estimation by Myers formula; wind: Gradient winds considering super-gradient wind effect
Fluid dynamics model	Delft3D flow ver. 6.02
Domain	Spherical coordinate system, large domain: 250 m; small domain: 25 m
Bathymetry	50 m grid spacing

Source: Japan Coast Guard

In order to assess the validity of the storm surge model, simulated results for the 2017 Typhoon Lan were compared with observations. A total of 4 tide observation stations was selected to determine the storm surge response of the bay. These stations are located at the upper bay (i.e., Harumi, Chiba), the middle bay (i.e., Yokohama), and the lower bay (i.e., Kanaya) (Fig. 1).

A series of numerical sensitivity experiments was conducted considering the Typhoon Lan as a reference typhoon. On September 22, 2017, at 1800 UTC, Typhoon Lan made landfall (Fig. 1) about 125 km southwest of Tokyo Bay with a minimum central pressure of 950 hPa and maximum sustained wind speed of 40 m/s. One of Lan's most unusual characteristics is that the overall size ($R_{50} = 389$ km at landfall) of the typhoon system was the largest in recorded history (Islam et al. 2017) and generated the highest storm tide in Yokohama and other coastal areas (e.g. Mera) of Tokyo Bay in recent decades. The average R_{50} for typhoons made landfall in Tokyo Bay is 187 km (Japan Meteorological Agency 2019a, b). Therefore, considering Lan was a very large typhoon, additional four cases were analyzed to investigate the sensitivity of the storm surge by altering the original size.

Maximum wind radius (R_{max}) is a spatial parameter that directly represents typhoon size. However, R_{max} is not recorded in the JMA best track. Hence, R_{50} data were converted into a mean value of R_{max} using the simplified equation, $R_{max} = 0.23 \times R_{50}$ (Takagi and Wu 2016). Five different radii of R_{max} (during landfall) ranging from 13 to 89 km were selected by referring to previous major typhoons in Tokyo Bay (Fig. 1): R_{max}: Danas (2001) = 13 km; Mawar (2005) = 26 km; Higos (2002) = 43 km; Kirogi (2000) = 55 km; Lan (2017) = 89 km). A plausible maximum rise of sea water level respective to each R_{max} was calculated, while other parameters such as typhoon track, forward speed and intensity remained same as the Typhoon Lan had. In this study, the impacts of waves on resultant water levels were neglected to simplify the analyses. The justification for this is twofold. First, this study is not aimed at improving the precise estimation of storm surge, but rather to segregate the effect of storm size on surge levels. Second, the contribution of the wave set up particularly at inner bays is negligible. Thus, the effect would not change the overall results of this study.

3 Results and Discussion

3.1 Model Validation

Figure 2a–c represent the comparison of the simulated and measured storm tides at three tide stations (Harumi, Chiba, and Yokohama) (Japan Meteorological Agency 2019a, b), all of which are located within the computational domain (Fig. 1). The model validation could not be carried out for Kanaya because observed storm tide data during Typhoon Lan is not available for this site. The coefficient of determination R^2 and the root mean square on the average of the three stations are 0.90 and 0.21 m, respectively. Overall, the time series of the simulated water levels agree well

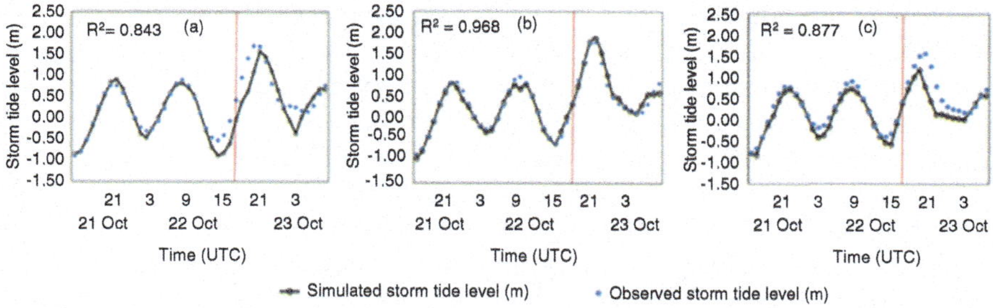

Fig. 2 Comparison of observed and simulated storm tides at Harumi (**a**), Chiba (**b**) and Yokohama (**c**). The red colored line indicates the time of landfall

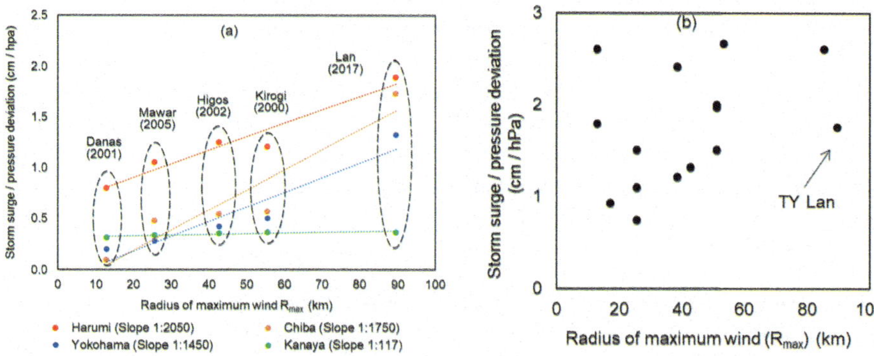

Fig. 3 Simulated peak storm surge normalized by central pressure deficit vs typhoon size (R_{max}). (**a**) Simulated peak storm surge normalized by central pressure deficit vs typhoon size (R_{max}) at landfall. (**b**) Observed peak storm surge normalized by central pressure deficit vs typhoon size (R_{max}) at typhoon landfall

with the recorded tides. However, the predictions had slightly underestimated peak water levels at the Harumi and Yokohama stations.

3.2 *Sensitivity Analysis*

Figure 3a illustrates the obtained relationship between storm size (R_{max}) and surge response in Tokyo Bay. Here, the influence of typhoon intensity was separated by dividing peak surge by the central pressure deficit Δp from mean sea level pressure (i.e. 1013 hPa). The rough estimation of Δp is proportional to the square of the maximum wind speed, corresponding to the Saffir-Simpson scale (Simpson 1974). This figure reveals that storm surge height tends to increase towards the inner Tokyo Bay as the typhoon size (R_{max}) increases. In addition, with the increment of storm size, the linear trend becomes steeper as the shelf slope becomes smaller, indicating the significance of typhoon size over flat seabed areas. Figure 3b demonstrates that the historically observed data support the numerical findings presented in Fig. 3a. Data for Chiba, Yokohama, and Kanaya are not presented in Fig. 3b because the database of past storm surge for mean sea level at these sites is not available. Furthermore, the role of typhoon size near open coast areas (i.e., Kanaya) could not be explained properly because the wave-induced set up was neglected in the numerical model. The size of a storm affects both the fetch and duration of the generating waves. Therefore, wave heights tend to increase with increasing storm size, which would lead to higher wave set up and coastal surges along open coasts (Irish et al. 2008). Nevertheless, the findings presented in this study are consistent with the results of a storm surge study by Irish et al. (2008), in which they also demonstrated that increases in storm size increase storm surge and that this relationship becomes more significant as shelf slope becomes milder.

The results of this study suggest that if typhoon parameters including intensity, track and forward speed remain constant, sea water level in a semi-enclosed bay varies with typhoon size. As an example, the simulated wind speed and water surface elevation from mean sea level (before and after 6 h of landfall) associated with each typhoon size for Harumi station (innermost part of the bay) are presented in Fig. 4. This statistic explains the previous figure as well, particularly for a very large typhoon (i.e., $R_{max} = 89$ km) compared to an average sized ($R_{max} \sim 43$ km in Tokyo Bay),

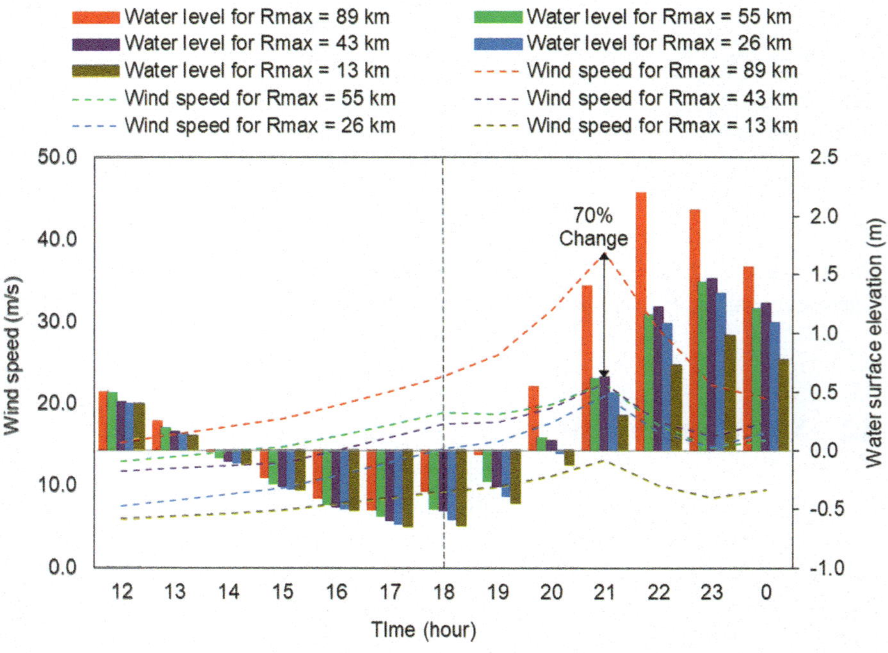

Fig. 4 Time series (before and after 6 h of typhoon landfall) simulated wind speed and water surface elevation varying with typhoon size at Harumi observation station. The dash black colored line indicates the time of the landfall

corresponding to an increase in wind speed of up to 70%. Therefore, even though it makes landfall, the resultant large swath of strong wind affects a larger area of the sea, bringing more water into motion and elevates approximately 0.7 m of water level over that by a typhoon with $R_{max} = 43$ km causes. Furthermore, Fig. 4 reveals that a larger storm (i.e. $R_{max} = 89$ km) along with strong wind (≥ 20 m/s) affects a certain location for a longer period—it takes longer time for the whole storm to pass over a certain point because of the size and the time (hour) stretches over more than triple that of an average sized typhoon ($R_{max} \sim 43$ km). Thus, the time series distribution of strong winds and water levels would also change with the change of typhoon size. This numerical finding is partially explained that why Typhoon Lan raised water level approximately 2 m in upper Tokyo Bay, even though it made landfall far away from the Bay. It is also evident from Fig. 4 that water level more than 1 m stands considerably larger in duration when the typhoon is very large (i.e. $R_{max} = 89$ km). The results show that the impact of the storm surge, particularly in the upper bay, appears to be limited when a comparatively smaller typhoon (i.e. $R_{max} \leq 26$ km) passes over Tokyo Bay (Fig. 4). However, it should be noted that a smaller but powerful typhoon could have a very strong impact on a particular location as the pressure gradient tends to be steep, resulting in stronger winds near the typhoon eye.

Illustrating the importance of typhoon size, the numerical simulations in this study revealed that physical phenomena governing surge generation in a semi-enclosed bay depends not only on wind intensity but also on typhoon size, which could exacerbate storm surge, particularly in the upper bay. The variations in many recorded storm surges could be explained by the differences in storm size. For instance, TY Lan (2017) having a R_{max} 3.5 times larger than TY Mawar (2005) ($R_{50} = 111$ km; $R_{max} = 26$ km) (Fig. 5) produced a storm surge 4 times (1.2 m) that

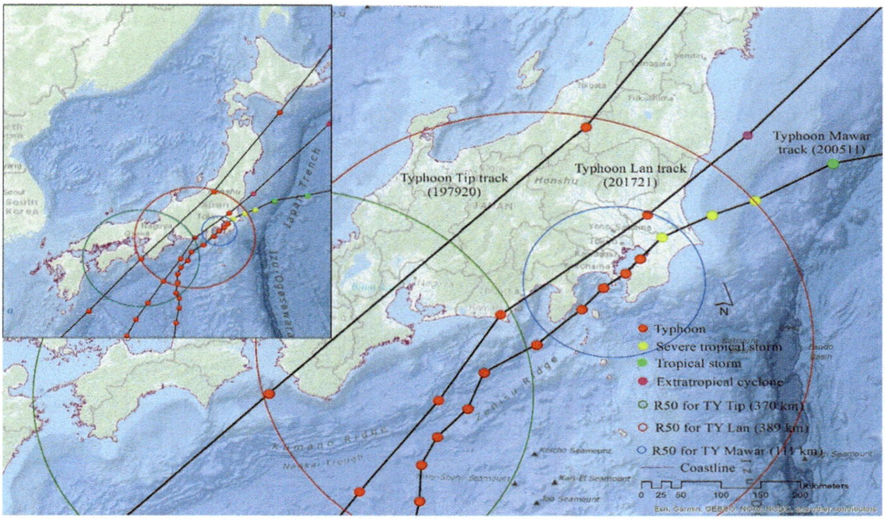

Fig. 5 Track and 50-kt (26 m/s) wind radius (R_{50}) of Typhoon Tip in 1979, Typhoon Mawar in 2005 and Typhoon Lan in 2017 (Japan Meteorological Agency 2019a, b) as it approaches Japan

generated by Typhoon Mawar (0.3 m) at upper Tokyo Bay (Japan Meteorological Agency 2019a, b). Although the maximum sustained wind speed (35 m/s) and track for both typhoons are comparable, TY Mawar made landfall directly over Tokyo Bay (Miura peninsula) (Japan Meteorological Agency 2019a, b). A similar type of incident was observed when 1.25 m of storm surge engulfed the upper Tokyo Bay during TY Tip (1979) (Fig. 5) that made landfall approximately 430 km southwest of the bay (Japan Meteorological Agency 2019a, b). It also had a very large wind field with a R_{max} of 85 km (R_{50} = 370 km;) (Japan Meteorological Agency 2019a, b). Such statistics imply that the passage of a typhoon far away from a semi-enclosed bay does not necessarily imply the occurrence of correspondingly small storm surges. Nevertheless, upper bay areas are more vulnerable to intense storm surges when typhoons make landfall with large swaths of strong wind. This is contrary to the conventional perception that the storm surge disaster risk of any given coastal place lessens as the distance from the track increases.

4 Conclusion

This study performed numerical sensitivity analysis on the relationship between typhoon size and maximum storm surge heights by taking Typhoon Lan as a reference case. Despite the lack of systematic study of the potential impact of typhoon size in semi-enclosed bays, the analyses demonstrate that peak storm surge height tends to intensify as the size of typhoons (R_{max}) increases. Storm size plays a key role in generating surge at upper bay, particularly for very large typhoons making landfall in upper bay areas, which is also evidenced from historical records. Thus, the findings suggest the incorporation of the role of storm size into the wind speed-based TY scale (i.e., SSHS), which has historically provided the basis of estimating storm surge height.

The findings presented in this paper would be most advantageous in two main areas. First, meteorologists and oceanographers could re-evaluate the importance of typhoon size for predicting peak surge heights precisely. Secondly, depending on the storm size, disaster management practitioners could determine the proper geographical extent in order to warn citizens of potential coastal surges. Considering that typhoon intensity is not the only factor affecting the generation of storm surges, the findings of this study will significantly contribute to improved typhoon categorization and planning measures to mitigate disasters related to storm surges.

Acknowledgement This research was funded by a grant awarded to Tokyo Institute of Technology (Japan Society for the Promotion of Science, 16KK0121 and 19 K04964).

References

Berke, P., Larsen, T., & Ruch, C. (1984). A computer system for hurricane hazard assessment. *Computers, Environment and Urban Systems, 9*, 259–269.

Conner, W. C., Kraft, R. H., & Harris, D. L. (1957). Empirical methods for forecasting the maximum storm tide due to hurricanes and other tropical storms. *Monthly Weather Review, 85*, 113–116.

DELTARES. (2011). Delft3D-FLOW – Simulation of multi-dimensional hydrodynamic flows and transport phenomena, including sediments, user manual Delft3D-FLOW, The Netherlands, pp. 690.

Dolan, R., & Davis, R. E. (1992). An intensity scale for Atlantic Coast Northeast storms. *Journal of Coastal Research, 8*, 840–853.

Egbert, G. D., & Erofeeva, S. Y. (2002). Efficient inverse modeling of barotropic ocean tides. *Journal of Atmospheric and Oceanic Technology, 19*(2), 183–204.

Esteban, M., Takagi, H. & Shibayama, T. (2015). Lessons from the last 10 years of coastal disasters. In *Handbook of* Coastal Disaster Mitigation for Engineers and Planners (pp. xxv–xxx). Elsevier.

Harris, D. L. (1963). *Characteristics of the Hurricane Storm Surge* (U.S. Weather Bureau Tech. Paper 48, pp. 139).

Hirano, K., Bunya, S., Murakami, T., Lizuka, S., Nakatani, T., Shimokawa, S., & Kawasaki, K. (2014). Prediction of typhoon storm surge flood in Tokyo Bay using unstructured model ADCIRC under global warming scenario. In *Proceedings of the ASME 2014 4th Joint US-European Fluids Engineering Division Summer Meeting*.

Hoover RA (1957) Empirical relationships of the central pressure in hurricanes to the maximum surge and storm tide, vlo 85. Monthly Weather Review, pp. 167–174.

Irish, J. L., Resio, D. T., & Ratcliff, J. J. (2008). The influence of storm size on hurricane surge. *Journal of Physical Oceanography, 38*, 2003–2013.

Islam, M. R., Takagi, H., Anh, L. T., Takahashi, A. & Bowei, K. (2017). Typhoon Lan Reconnaissance Field Survey in Coasts of Kanto Region, vol *74*(2). *Japan, Journal of Japan Society of Civil Engineers*, Ser. B3 (Ocean Engineering).

Japan Meteorological Agency. (2019a). *JMA tide observation data site*. http://www.data.jma.go.jp/kaiyou/db/tide/genbo/index.php

Japan Meteorological Agency. (2019b). *RSMC best track data site*. https://www.jma.go.jp/jma/jma-eng/jma-center/rsmc-hp-pub-eg/trackarchives.html

Jelesnianski, C. P. (1972). SPLASH (Special Program to List Amplitudes of Surges from Hurricanes): I. Landfall Storms. NOAA Tech. Memo. NWS TDL-46, pp. 52.

Needham, H. F. & Keim, B. D. (2011*) Storm surge: Physical processes and an impact scale, in recent hurricane research—Climate, dynamics, and societal impacts* (E. Lupo, Ed.) Croatia: Intech Open Access.

Needham, H. F. & Keim, B. D. (2014). *An empirical analysis on the relationship between Tropical Cyclone Size and Storm Surge Heights along the U.S.* Gulf Coast, Vol. 18. Earth Interactions, pp. 1–14.

Sebastian, A. G., Proft, J. M., Dietrich, J. C., Du, W., Bedient, P. B., & Dawson, C. N. (2014). Characterizing hurricane storm surge behavior in Galveston Bay using the SWAN + ADCIRC model. *Coastal Engineering, 88*, 171–181.

Simpson, R. H. (1974). *The hurricane disaster potential scale*, Vol. 27. Weatherwise, pp. 169, 186.

Takagi, H., & Wu, W. (2016). Maximum wind radius estimated by the 50 Kt Radius: Improvement of storm surge forecasting over the Western North Pacific. *Natural Hazards and Earth System Sciences, 16*, 705–717.

Takagi, H., DE Leon, S. L. I., Esteban, M., Mikami, M. T., Matsumaru, R., Shibayama, T., & Nakamura, R. (2016). Storm surge and evacuation in urban areas during the peak of a storm. *Coastal Engineering, 108*, 1–9.

Takagi, H., Esteban, M., Shibayama, T., Mikami, T., Matsumaru, R., Leon, M. D., Thao, N. D., Oyama, T., & Nakamura, R. (2017). Track analysis, simulation, and field survey of the 2013 Typhoon Haiyan storm surge. *Journal of Flood Risk Management, 10*(1), 42–52.

Takagi, H., Xiong, Y., & And Furukawa, F. (2018). *Track analysis and storm surge investigation of 2017 Typhoon Hato: Were the warning signals issued in Macau and Hong Kong timed appropriately?* (Vol. 12, pp. 297–307). Georisk: Assessment and Management of Risk for Engineered Systems and Geohazards.

Zhong, L., Li, M., & Zhang, D. L. (2010). How do uncertainties in hurricane model forecasts affect storm surge predictions in a semi-enclosed bay? *Estuarine, Coastal and Shelf Science, 90*(2), 61–72.

Investigation of Flash Flood Producing Rainstorm in Northeast Bangladesh Using WRF Model

Mohan K. Das, A. K. M. Saiful Islam, Samarendra Karmakar, Jamal Uddin Khan, Khaled Mohammed, G. M. Tarekul Islam, and Sujit Kumar Bala

1 Introduction

The wetland regions of the north-eastern part of Bangladesh, typically referred to as the Haor region, experiences frequent flash floods during the pre-monsoon season. Unlike monsoon flooding, pre-monsoon flash floods tend to last for a relatively shorter period of time, but can occur at a rapid rate, usually in the timescale of a day (Basher et al. 2017; Roy et al. 2019). These flash floods are often the result of heavy rainfall accompanied by severe thunderstorms/Mesoscale Convective Systems (MCSs) within a short span of time in the hilly areas of the Haor region (IMD 1944; Karmakar et al. 2017; Nowreen et al. 2014). Typically, the formation of these systems coincide with the rice harvesting period of this region resulting in massive loss of crops (CEGIS 2012). The incidence of flash flooding becomes even more critical when it tends to occur prior to harvesting of the primary and only crop (Boro rice) of that region. Accurate flash flood forecasting with sufficient lead time (Webster et al. 2010) can reduce damage to these crops. Therefore, it is imperative

M. K. Das (✉) · A. K. M. S. Islam · G. M. T. Islam · S. K. Bala
Institute of Water and Flood Management (IWFM), Bangladesh University of Engineering and Technology (BUET), Dhaka, Bangladesh
e-mail: akmsaifulislam@iwfm.buet.ac.bd; tarek@iwfm.buet.ac.bd; bala@iwfm.buet.ac.bd

S. Karmakar
National Oceanographic and Maritime Institute (NOAMI), Dhaka, Bangladesh

J. U. Khan
LEGOS, Université de Toulouse, CNES, CNRS, IRD, UPS, Toulouse, France
e-mail: jamal.khan@legos.obs-mip.fr

K. Mohammed
Department of Civil Engineering, University of Sherbrooke, Sherbrooke, Canada
e-mail: khaled.mohammed@usherbrooke.ca

A. Haque, A. I. A. Chowdhury (eds.), *Water, Flood Management and Water Security Under a Changing Climate*, https://doi.org/10.1007/978-3-030-47786-8_11

to understand these systems to develop an efficient early warning system (EWS) for forecasting flash floods in this area. It has become essential to improve the accuracy of rainfall forecasting during the pre-monsoon season using any mesoscale models (Abhilash et al. 2007; Das and Debsarma 2012; Houze et al. 2017) such as the widely-used community model Weather Research and Forecasting (WRF). WRF is a mesoscale model, which is capable of solving the compressible, non-hydrostatic Euler equations in 3-dimensions (Skamarock et al. 2008, 2019) and designed for both operational applications and atmospheric research. WRF is selectively used in research and operational institutions in Bangladesh for academic research and weather forecasting including high-impact weather simulations. Understanding of the large scale processes for occurrence of torrential rain (Das et al. 2017; Dimri et al. 2017; Hasan and Islam 2018; Rasmussen and Houze 2012, Rajeevan et al. 2010) over the northeast area of Bangladesh is an important aspect of this study. It is expected that NWP models would be useful for obtaining guidelines in the forecasting of flash flood.

2 Methodology

2.1 Heavy Rainfall Events

Two consecutive rainfall events resulting in flash floods were selected for investigation based on surface synoptic observations. On 17 April 2010, the accumulated rainfall over a 24-h period at Cherapunji and Sylhet were found to be 97 mm and 160 mm respectively. However, the next day, the 24-h accumulated rainfall found at Cherapunji and Sylhet were 103 mm and 61 mm respectively. This heavy (44—88 mm day^{-1}) to very heavy (>89 mm day^{-1}) rainfall was responsible for the occurrence of flash floods over the region. This type of flash flooding occurs every few years during the pre-monsoon and early monsoon period over northeastern Bangladesh.

2.2 Weather Research and Forecasting (WRF) Model

Numerical Weather Prediction (NWP) is the state-of-art technology for seamless weather forecasting at all spatial and temporal scales. The NWP model has the ability to predict micro as well as mesoscale weather phenomena.

The Advanced Research Weather Research and Forecasting model (ARW), version 3.7.1 (Skamarock et al. 2008) has been used in this study. WRF is a three-dimensional, fully compressible, non-hydrostatic model. In the study, the domain is considered with a nested domain with 27, 9 and 3 km horizontal spatial resolution as shown in Fig. 1a along with the northeast region of Bangladesh and the Meghna basin with stations locations Fig. 1b. Figure 1c shows daily rainfall over Sylhet,

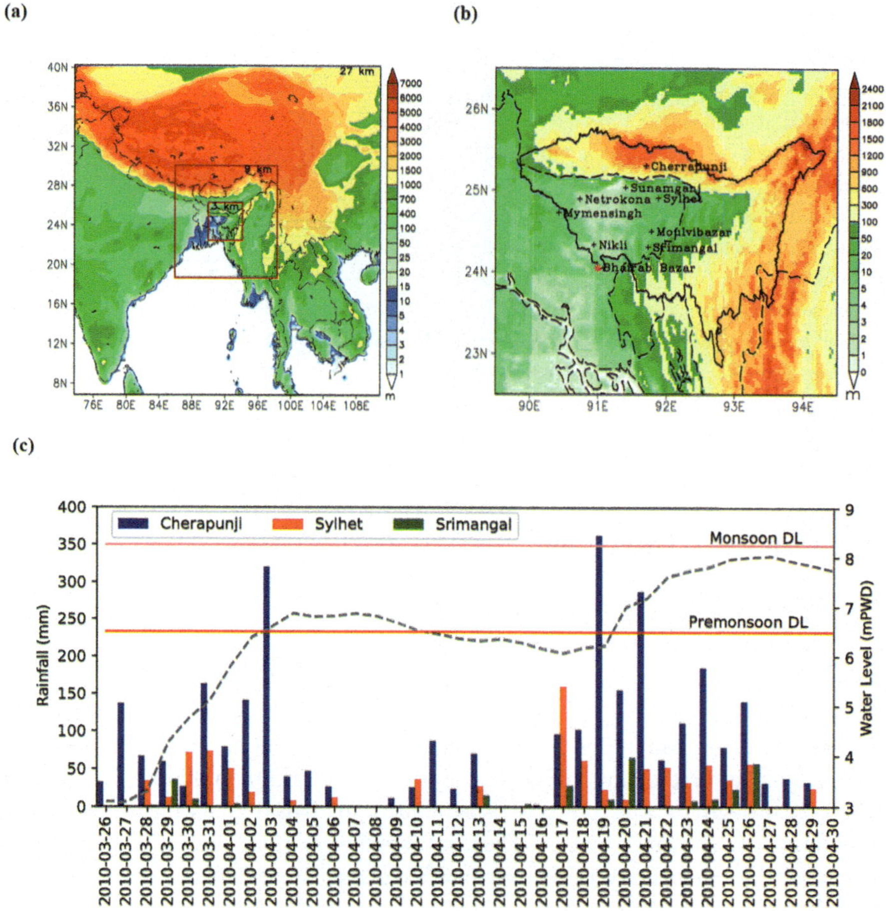

Fig. 1 Geographic map of the study area along with the terrain heights (**a**) triple nested domains, (**b**) northeastern Bangladesh with the Meghna Basin. The solid line shows the basin area and the dashed line shows the borders and shorelines. The location of stations (+) and the Meghna basin outlets Bhairab Bazar (∗) are shown. (**c**) The daily rainfall time series along with water level of Sunamganj station (SW 269) and the Pre-monsoon danger level (DL) and monsoon DL

Srimangal and Cherrapunji along with water level and pre-monsoon and monsoon danger level (DL). The 1.0° × 1.0° gridded NCEP FNL (Final) Operational Global Analysis and Global Forecast System (GFS) data are used as initial and Lateral Boundary Conditions (LBC) of the model. The main features of the model employed for this study are summarized in Table 1.

Table 1 Features of numerical model configurations

Model	WRF Version 3.7.1
Map projection	Mercator
Horizontal resolution	Nest: 27, 9 and 3 km
Vertical levels	40
Land cover	USGS
Time integration	Semi implicit
Vertical differencing	Arakawa's energy conserving scheme
Convection	Kain-Fritsch (new Eta) scheme (Kain 2004)
Planetary boundary layer	Yonsei University Scheme (YSU)
Cloud microphysics	WRF Single-Moment 6-Class (WSM6) (Hong and Lim 2006)
Surface layer	Monin-Obukhov
Radiation	RRTM (LW), SW (Dudhia 1989)
Land surface processes	Unified NOAH land surface model
Horizontal grid scheme	Arakawa C-grid

3 Result and Discussion

3.1 *Large Scale Synoptic Process*

The large scale synoptic process and forcing with sea level pressure and wind flow at 10 m is shown in Fig. 2a. The geopotential height in meter and wind flow patterns at 200 hPa are presented in Fig. 2b. The analyses are based on the National Centers for Environmental Prediction (NCEP) reanalysis (NCEP 2000) on 0000 UTC 17 April 2010. A trough of low pressure is extended towards the northeastern part of Bangladesh as shown in Fig. 2a. There is a southwest moist flow at the 10 m level over the country and converging over sub-Himalayan West Bengal with an extended trough towards the northeast (Fig. 2a).

The wind speed is about 20 ms^{-1} over the north BoB. In Fig. 2b, a ridge of the westerly jet is present over the north Arabian Sea, Pakistan, and north India at 200 hPa level. There is a strong trough of both geopotential height and wind circulation over northern Bangladesh and further north, extending over the Meghalaya region. The geopotential height over the northeast region is 12,250 m, which is lower relative to the surrounding areas and is responsible for the formation of thunderstorms. The sub-tropical westerly jet (SWJ) is embedded in the upper level at 200 hPa; this SWJ also favors the formation of rainstorms.

3.2 *Radar-Derived Rainfall Rate and Model Simulated Reflectivity*

The Weather radar is being used worldwide for observation of various extreme weather events like MCSs, rainstorms, squalls, thunderstorms, tornados, hail storms,

(a)

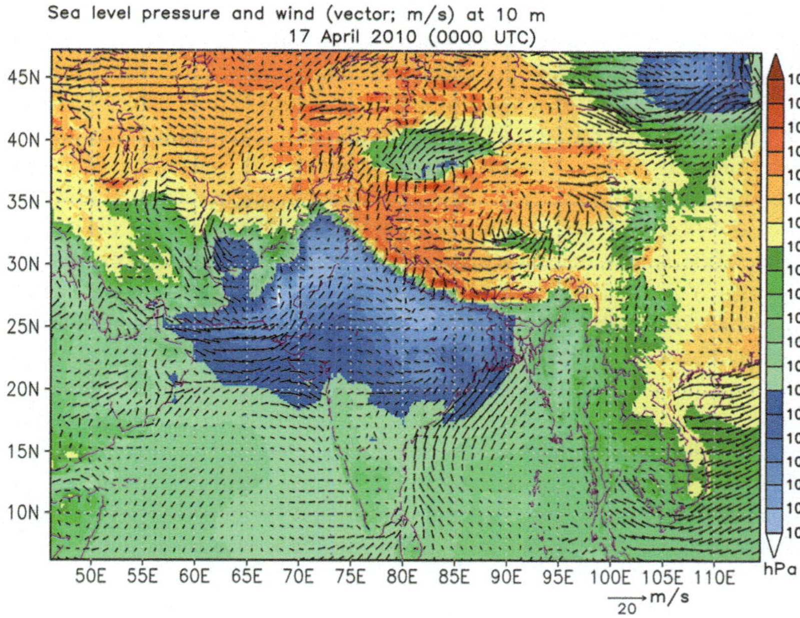

(b)

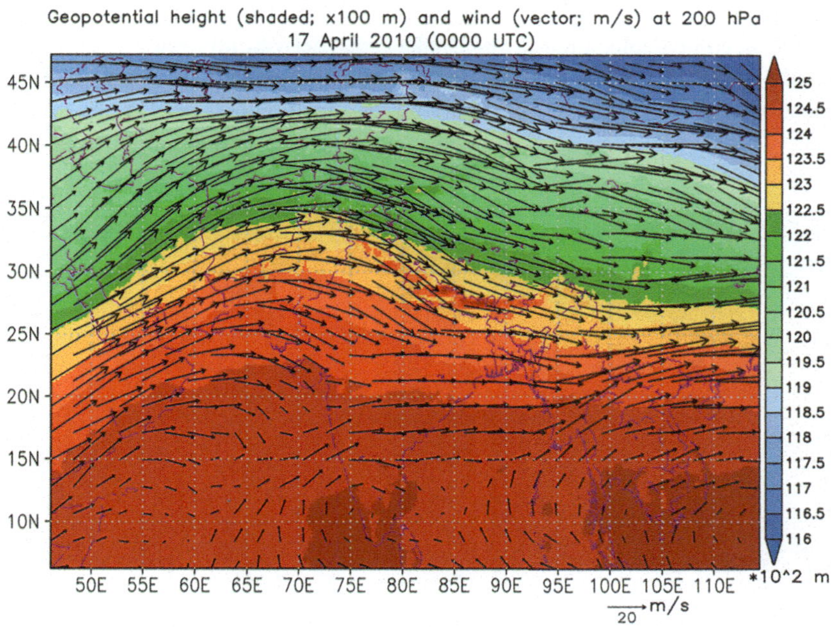

Fig. 2 Large scale process based on NCEP reanalysis on 0000 UTC 17 April 2010 (**a**) Sea Level Pressure (shaded; hPa) and wind vector (m s^{-1}) at 10 m. (**b**) Geopotential height (shaded; $\times 100$ m) and wind vector (m s^{-1}) at 200 hPa

the direction of movement of thunderstorms and cyclones. To estimate the rainfall on the real-time basis, the radar has a great potential to enhance the proficiencies of researchers and scientists (Chatterjee et al. 2008; Das et al. 2015; Pradhan and Sinha 2005). Radar can also measure the speed at which rain or hail is moving away from or towards the radar. From a radar volume scan (series of 360° sweeps, each tilting a little higher than the last); radar meteorologists can get a complete look at organizations and movements of storms close to the radar. The Bangladesh Meteorological Department (BMD) operates a radar system consisting of three Doppler Weather Radars (DWRs) and two conventional radars at different places of the country. The nearest DWR is Moulvibazar (MLV) DWR (Fig. 1b) which is located at 24° 29′ 8″ N, 91° 46′ 30″ E. DWR of Moulvibazar is not operated continuously and there is no other DWR nearby Sylhet and Meghalaya. Therefore, the conventional radar Dhaka retrieved rainfall rate are collected through the SAARC STORM project (Das et al. 2014).

Rain-gauge data cannot recognize the fact that strong rainfall cores are surrounded by light rainfall regions (Islam et al. 2005). To understand the dimension, forms, extent, propagation as well as the life cycle of the rainfall, the radar analysis provides useful insights. In this study, DWR reflectivity is correlated with rainfall. Dhaka radar derived rainfall indicates rain rate over Sylhet and nearby region >60 mm h^{-1} at 0300 UTC on 17 April 2010 (Fig. 3a). The model simulated reflectivity of hydrometeor in the atmosphere at 0300 UTC on 17 April 2010 are presented in Fig. 3b. It is found that the reflectivity becomes maximum and distinct over the place of occurrence as shown in Fig. 3a. The model simulated reflectivity value is more than 50 dBZ, indicating severe thunderstorm and intensive rainfall which is comparable (Fig. 3a) to Dhaka radar derived rain rate. While the model results show the minor shifting of the areas of precipitation both in time and locations, the intensities of precipitation rates are simulated very well.

(a) **(b)**

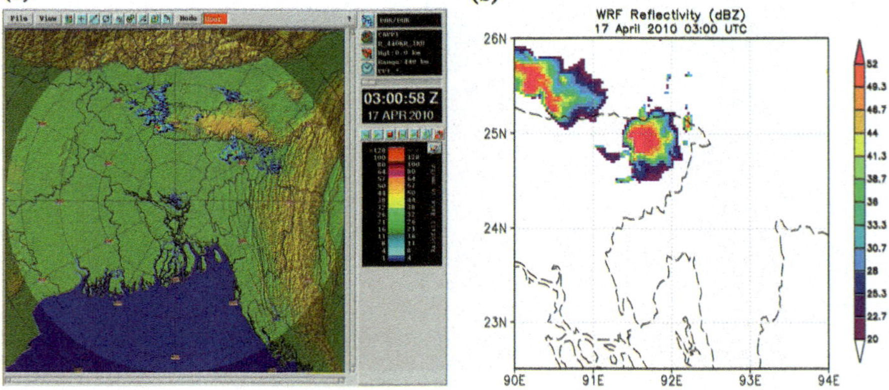

Fig. 3 Moulvibazar DWR derived (**a**) rainfall rate on 17 April 2010 at 0301 UTC, (**b**) WRF model simulated Reflectivity on 17 April 2010 at 0300 UTC

3.3 Wind at 950 hPa and Relative Humidity at 2 m

The model simulated vector wind at 950 hPa and wind speed at 10 m level are shown in Fig. 4a. The wind speed is 6–10 ms^{-1} at 10 m level in the northeast region of Bangladesh at 0600 UTC and 1800 UTC on 17 April 2010 and 0600 UTC on 18 April 2010. There is a strong cyclonic circulation at 950 hPa level over West Bengal and adjoining Bangladesh with a prominent trough extending towards the northeast. The 950 hPa horizontal wind shows strong southerly flow through southern Bangladesh with a trough extending over the north and north-eastern part of Bangladesh (Fig. 4a). There exists a micro-circulation during 0000–1200 UTC on both 17 and 18 April 2010 over the north-northeast area of Bangladesh as can be found from Fig. 4a Such a micro-circulation has been responsible for the generation of a severe thunderstorm with higher rainfall. The model simulated feature is consistent with surrounding radiosonde observations. The strong southerly flow incurs the moisture in the lower levels. For triggering a convective activity, the stronger south-southwesterly wind flow is important which has helped deliver moisture convergence over the convective zone.

The distribution of relative humidity at 2 m simulated by the WRF model over Bangladesh is shown in Fig. 4b at every 6 h on 17 and 18 April 2010. The dry and moist lines are found to intersect at the place of occurrence of the storm as shown in Fig. 4b. This point of intersection shows the place of occurrence of the severe storm which caused the flash flood.

3.4 TRMM Retrieved and Model Simulated Rainfall

The spatial distributions of rain intensities retrieved from TRMM 3B42RT (Huffman 2016) for the flash flood event over Bangladesh that occurred on 17 April and 18 April 2010. The daily rainfall starting from 0130 UTC and ending at 0130 UTC of next day are shown in Fig. 5. The rainfall area covers almost the northern, northeastern and eastern parts of Bangladesh. In the morning, there are two rainfall areas over northern and northeastern Bangladesh (Fig. 5a). But from 1200 UTC on 17 April 2010, the small areas are aggregated into one intense and large rainfall area which continued up to 2100 UTC, the maximum intensity being found during 1500–1800 UTC. Rainfall amount of 32 mm day^{-1} and 64 mm day^{-1} are found from 1500 UTC of 17 April to 0000 UTC of 18 April 2010. The rainfall amount is found to decrease from 0300 to 0900 UTC with its position shifted slightly north-eastward. Later, rainfall areas are found over the northeastern region of Bangladesh and the Meghalaya region with an amount greater than 64 mm. On 18 April 2010, the whole country becomes rainfall free during 0900–1200 UTC as can be shown in Fig. 5b. From 1200 UTC, some systems are found to develop over Assam and adjoining areas and the system has intensified and moved to the south/southeastward when heavy rainfall might have occurred over the Meghalaya Plateau and adjoining

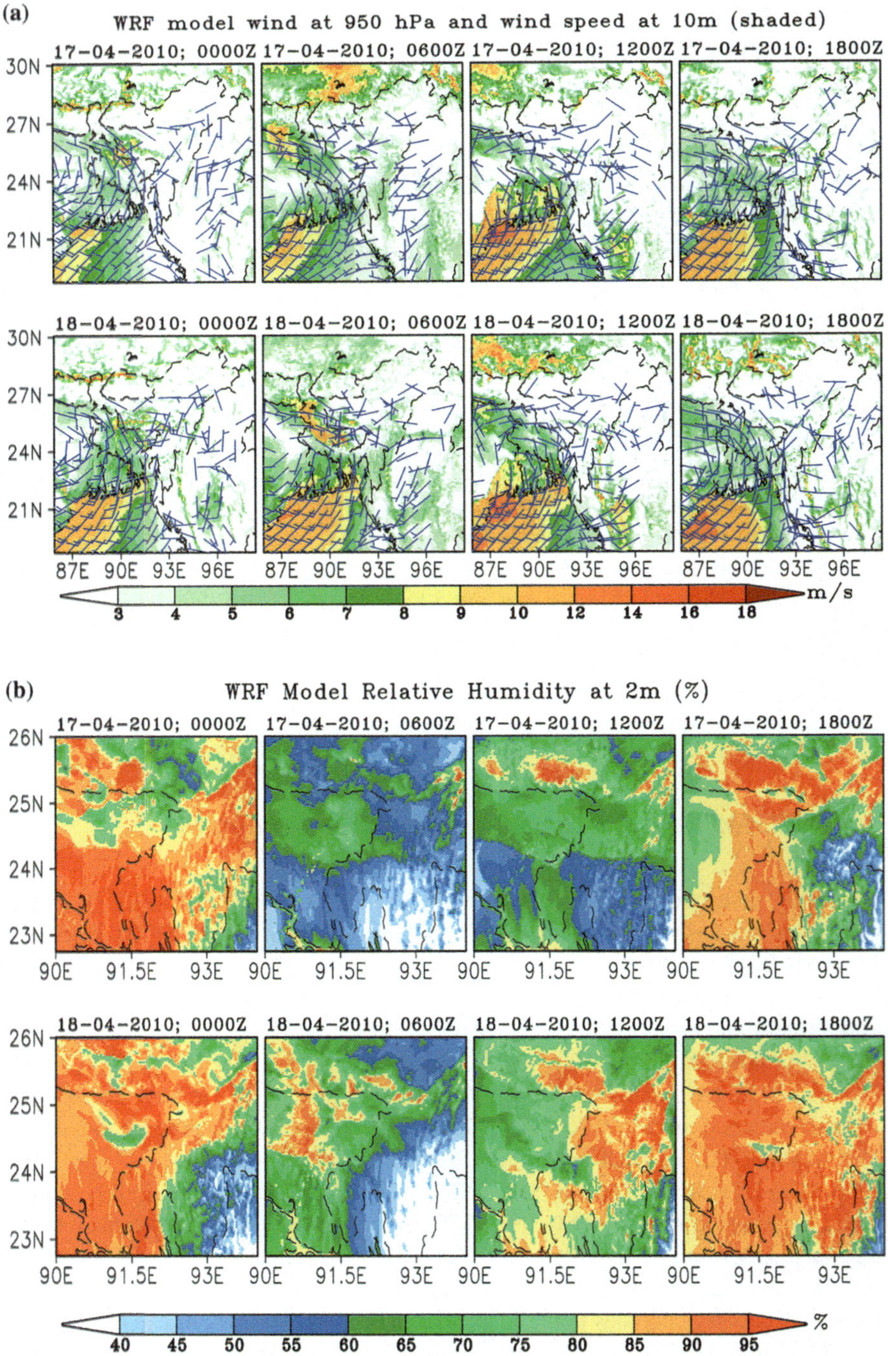

Fig. 4 WRF model simulated (**a**) wind at 950 hPa and wind speed at 10 m (shaded) and (**b**) spatial distribution of relative humidity at 2 m (%)

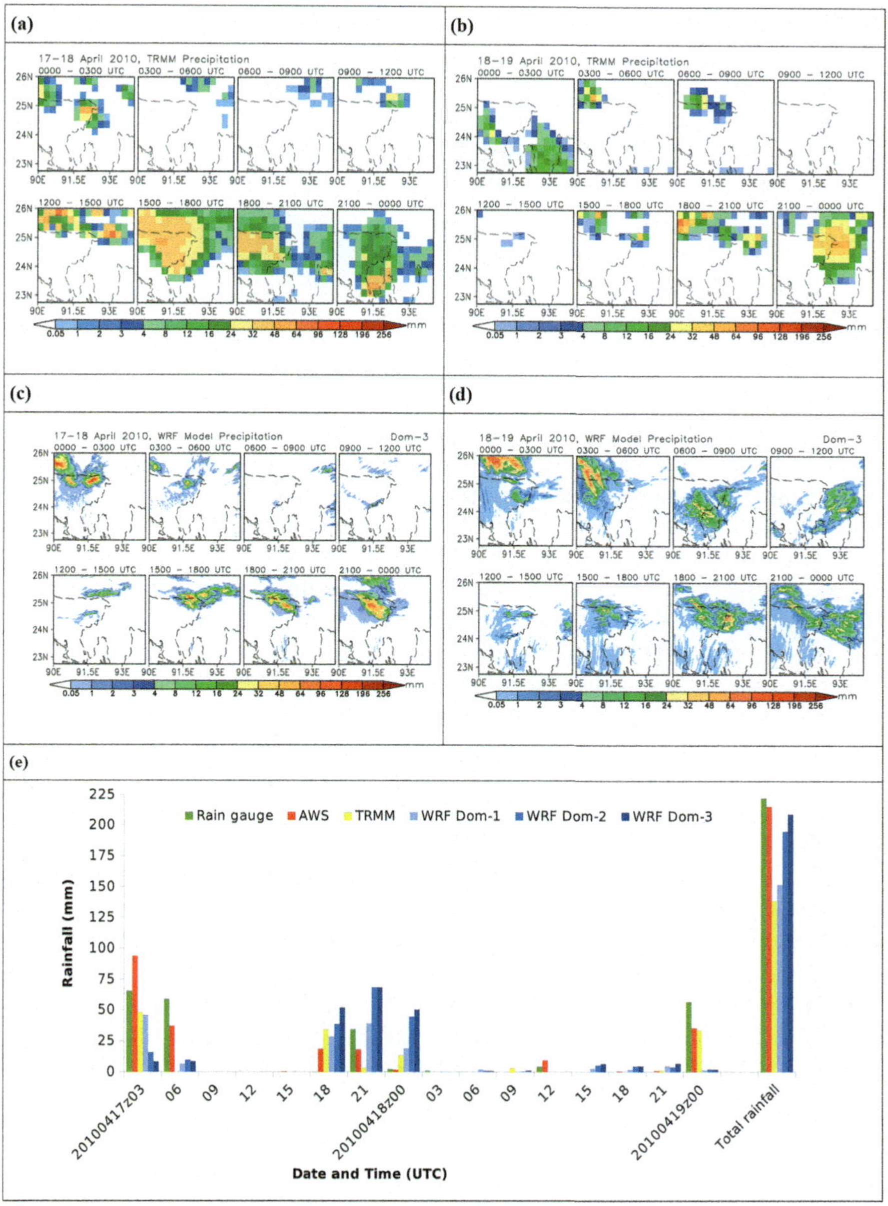

Fig. 5 Accumulated 3 hourly rainfall retrieved from TRMM on (**a**) 17 April 2010 and (**b**) 18 April 2010. WRF model simulated accumulated 3 hourly rainfall on (**c**) 17 April 2010, (**d**) 18 April 2010 and (**e**) 3 hourly rainfall calculated over Sylhet (24.9 °N and 91.88 °E) station

Table 2 Statistical analysis of three hourly station observed, TRMM derived and model simulated rainfall

	Rainfall (mm)					
				WRF Model		
	Rain gauge	AWS	TRMM 3B42RT	Dom-1	Dom-2	Dom-3
Total	222.6	215.9	138.8	152.1	195.3	209.4
% of underestimation		3.0%	37.6%	31.7%	12.3%	**5.9%**
RMSE		11.9	16.7	22.3	29.8	31.6

northeastern Bangladesh over the Haor areas. The rainfall area of the first day is more expansive compared to the second day which also indicates more rainfall had occurred on 17 April 2010. The model simulated three-hourly rainfall data, which is comparable with the spatial distribution of rainfall derived from TRMM (Fig. 5a, b) is shown in Fig. 5c. While the model results show the shifting of the areas of rainfall both in time and locations, the intensities of rainfall rates are simulated very well.

The model underestimated rainfall over the region during the flash flood event. Recorded 48-h rainfall accumulated during 17–19 April 2010 at the synoptic observatory of Sylhet was 222.6 mm. For the domain 1–3, 48-h rainfall volume differs with observations and are 152 mm, 195 mm and 209 mm respectively over the Sylhet region. TRMM retrieved 48-h rain amount is 139 mm, which also underestimates the rain gauge recordings. There is also 7 mm difference in rainfall between BMD rain gauge at Sylhet and special observation taken by Kyoto University of Japan team by using Automatic Weather station (AWS) at Sylhet International Airport which is about 5 km north of BMD rain gauge.

The statistical analysis shows that the root mean square error (RMSE) compared to BMD rain gauge station with TRMM is 17 mm and with the model simulated rainfall is about 22, 29 and 31 mm respectively for domain 1, 2 and 3. Although the RMSE for domain 3 is higher, it simulates heavy rainfall events quite well. The model domain 3 has underestimated 5.9% whereas domain 2 has underestimated 12.2% and domain 3 has underestimated 31.7%. Three-hourly rainfalls from the surface observatory, TRMM 3B42RT derived, model-simulated and their statistical analysis is shown in Table 2. It is also noted that from 0000 to 0600 UTC of 17 April 2010, the recorded rainfall is 124 mm in Sylhet station. Such a heavy rainfall plays a significant role in producing a flash flood event.

4 Conclusion

A number of conclusions can be drawn from the study which are presented below.

- The model simulated results provide a basis to study the microphysical and dynamical characteristics of the flash flood producing rainstorms, which are not generally available from the data produced by meteorological stations.

- The large-scale processes are found to be the significantly favorable conditions for producing rain-bearing convective storm over the steep topography of the Meghalaya region. This study reveals that the WRF model is able to simulate precipitation and associated parameters well. However, there is a spatial shift of model-simulated cloud and precipitation compared to the actual observations and TRMM derived precipitation.
- DWR derived rain rate provide an indication of the meso-convective event. The parallel bow-shaped convective lines had an elongated length of approximately 300 km. The value of reflectivity is found to be more than 52 dBZ in the model simulated reflectivity, indicating severe convection and heavy precipitation.
- The WRF model with a 3-km resolution has simulated the formation of the convective storm, and the initiation of the storm and cloud cluster nearly at the time of occurrence. The model has captured the rainfall distribution reasonably although the total rainfall amount is underestimated compared to the observations.
- The WSM6 scheme as microphysics schemes, the Kain-Fritsch with the Noah land surface model as cumulus parameterization scheme, and the planetary boundary layer scheme as the YSU PBL scheme have produced the best results for heavy rainfall prediction over this region. The model has well captured the rain and reflectivity of hydrometeor close to the place of occurrence of the event.
- The low-level wind field shows that there is a southwesterly flow from the Bay of Bengal towards the northeast Bangladesh, which later converges over the north-northeastern region of Bangladesh. The strong southwesterly flow helps to transport a high amount of moisture from the Bay of Bengal across the southern, the southeastern region of Bangladesh and neighboring areas. The strong low-level convergence has carried the moisture up to 200 hPa level.
- The model has underestimated in capturing the strength of the flash flood in general due to the weakness in producing heavy precipitation. The 48-h simulated rainfall is about 152 mm for outer domain-1, 195 mm for inner domain-2 and 209 mm for the innermost domain-3 but actual rainfall is 222.6 mm as observed by BMD rain gauge at Sylhet in Bangladesh. Model domain-3 underestimated precipitation by only 5.9%.

References

Abhilash, S., Das, S., Kalsi, S. R., Das Gupta, M., Mohankumar, K., George, J. P., Banerjee, S. K., Thampi, S. B., & Pradhan, D. (2007). Impact of doppler radar wind in simulating the intensity and propagation of rainbands associated with mesoscale convective complexes using MM5-3DVAR system. *Pure and Applied Geophysics, 164*(8–9), 1491–1509. https://doi.org/10.1007/978-3-7643-8493-7_4.

Basher, M., Stiller-Reeve, M., Islam, A. S., & Scott, B. (2017). Assessing extreme rainfall trends over the northeast regions of Bangladesh. *Theoretical and Applied Climatology*, 1–12. https://doi.org/10.1007/s00704-017-2285-4.

CEGIS. (2012). *Master plan of the Haor area, Bangladesh Haor and Wetland Development Board (BHWDB)*, Summary Report, Volume II, Ministry of Water Resources, Government of Peoples Republic of Bangladesh.

Chatterjee, P., Pradhan, D., & De, U. K. (2008). Simulation of hailstorm event using mesoscale model MM5 with modified cloud microphysics scheme. *Annales de Geophysique, 26*, 3545–3555.

Das, M. K., & Debsarma, S. K. (2012). Cyclone Aila moisture effects on heavy rain and flooding in Bangladesh, Bhutan, NE-India and Nepal. *International conference on ensemble methods in geophysical sciences*, 12–16 November 2012, Toulouse, France. http://www.meteo.fr/cic/meetings/2012/ensemble.conference/

Das, S., Mohanty, U. C., Tyagi, A., Sikka, D. R., Joseph, P. V., Rathore, L. S., Habib, A., Baidya, S. K., Sonam, K., & Sarkar, A. (2014). The SAARC STORM: A coordinated field experiment on severe thunderstorm observations and regional modeling over the South Asian region. *Bulletin of the American Meteorological Society, 95*, 603–617.

Das, M. K., Chowdhury, M. A. M., Das, S., Debsarma, S. K., & Karmakar, S. (2015). Assimilation of doppler weather radar data and their impacts on the simulation of squall events during pre-monsoon season. *Natural Hazards, 77*(2), 901–931. https://doi.org/10.1007/s11069-015-1634-9.

Das, M. K., Saiful, I. A. K. M., Jamal Uddin, K. M., & Samarendra, K. (2017). Numerical simulation of flash-flood-producing heavy rainfall of 16 April 2016 in NE regions of Bangladesh. *Vayu Mandal, 43*(2), 97–108.

Dimri, A. P., Chevuturi, A., Niyogi, D., Thayyen, R. J., Ray, K., Tripathi, S. N., Pandey, A. K., & Mohanty, U. C. (2017). Cloudbursts in Indian Himalayas: A review. *Earth-Science Reviews, 168*, 1–23. https://doi.org/10.1016/j.earscirev.2017.03.006.

Dudhia, J. (1989). Numerical study of convection observed during the winter monsoon experiment using a mesoscale two-dimensional model. *Journal of the Atmospheric Sciences, 46*, 3077–3107.

Hasan, M. A., & Islam, A. S. (2018). Evaluation of microphysics and cumulus schemes of WRF for forecasting of heavy monsoon rainfall over the southeastern hilly region of Bangladesh. *Pure and Applied Geophysics*, 1–30.

Hong, S.-Y., & Lim, J.-O. J. (2006). The WRF single-moment 6-class microphysics scheme (WSM6). *J Korean Meteorological Society, 42*, 129–151.

Houze, R. A., Jr., Rasmussen, K. L., McMurdie, L., Chaplin, M. M., & Kumar, A. (2017). Synoptic and mesoscale aspects of the June 2013 flooding in Uttarakhand, India. *Monthly Weather Review, 145*, 4447–4466. https://doi.org/10.1175/MWR-D-17-0004.1.

Huffman, G. (2016). TRMM (TMPA-RT) near real-time precipitation L3 3 hour 0.25 degree $\times$ 0.25 degree V7, Greenbelt, MD, Goddard Earth Sciences Data and Information Services Center (GES DISC). Available at: https://disc.gsfc.nasa.gov/datacollection/TRMM_3B42RT_7.html. Last access: 2 November 2017.

IMD (India Meteorological Department). (1944). *Nor'wester of Bengal*. IMD technical note no. 10.

Islam, M. N., Hayashi, T., Terao, T., Uyeda, H., & Kikuchi, K. (2005). Characteristics of rainfall systems analyzed from radar data over Bangladesh. *Journal of Natural Disaster Science, 27*(1), 17–23.

Kain John, S. (2004). The Kain–Fritsch convective parameterization: An update. *Journal of Applied Meteorology, 43*, 170–181.

Karmakar, S., Quadir, D. A., & Das, M. K. (2017). Numerical simulation of physical and dynamical characteristics associated with the severe thunderstorm on 5 April 2015 at Kushtia and Jhenaidah. *Natural Hazards, 86*(3), 1127–1146. https://doi.org/10.1007/s11069-016-2733-y.

NCEP (National Centers for Environmental Prediction): National Weather Service/NOAA/U.S. Department of Commerce. (2000). *Updated daily, NCEP FNL Operational Model Global Tropospheric Analyses, continuing from July 1999, Research Data Archive at the National Center for Atmospheric Research, Computational and Information Systems Laboratory*. Available at: https://doi.org/10.5065/D6M043C6. Last access: 2 November 2017.

Nowreen, S., Murshed, S. B., Islam, A. K. M. S., Bhaskaran, B., & Hasan, M. A. (2014). Changes of rainfall extremes around the haor basin areas of Bangladesh using multi-member ensemble RCM. *Theoretical and Applied Climatology, 119*, 363–377.

Pradhan, D., & Sinha, V. (2005). Thunderstorm genesis over Kolkata- a case study. *Vatavaran, 29* (1), 20–27.

Rajeevan, M., Kesarkar, A., Thampi, S. B., Rao, T. N., Radhakrishna, B., & Rajasekhar, M. (2010). Sensitivity of WRF cloud microphysics to simulations of a severe thunderstorm event over Southeast India. *Annales de Geophysique, 28*, 603–619. https://doi.org/10.5194/angeo-28-603-2010.

Rasmussen, K. L., & Houze, R. A., Jr. (2012). A flash flooding storm at the steep edge of high terrain: Disaster in the Himalayas. *Bulletin of the American Meteorological Society, 93*, 1713–1724.

Roy, B., Islam, A. S., Islam, G. M. T., Bhattacharya, B., Ali, M. H., Khan, A. S., Hossain, M. S., Sarker, J. C., & Khan, M. U. (2019). Determining flash flood danger level at gauge stations of the North East Haor regions of Bangladesh. *Journal of Hydrological Engineering, 24*(4), 05019004.

Skamarock, W. C., Klemp, J. B., Dudhia, J., Gill, D. O., Barker, D. M., Duda, M. G., Huang, X.-Y., Wang, W., & Powers, J. G. (2008). *A description of the advanced research WRF Version 3 (NCAR Tech. Note NCAR/TN-475+STR,* 113 pp). https://doi.org/10.5065/D68S4MVH

Skamarock, W. C., Klemp, J. B., Dudhia, J., D. Gill, O., Liu, Z., Berner, J., Wang, W., Powers, J. G., Duda, M. G., Barker, D. M., & Huang, X.-Y.. (2019). *A description of the advanced research WRF Version 4* (NCAR Tech. Note NCAR/TN-556+STR, 145 pp). https://doi.org/10.5065/1dfh-6p97.

Webster, P. J., Jian, J., Hopson, T. M., Hoyos, C. D., Agudelo, P., Chang, H.-R., Curry, J. A., Grossman, R. L., Palmer, T. N., & Subbiah, A. R. (2010). Extended-range probabilistic forecasts of Ganges and Brahmaputra floods in Bangladesh. *Bulletin of the American Meteorological Society, 91*(11), 1493–1514.

Assessment of Rainwater Quality from Harvested Rainwater in the Coastal Region of Bangladesh

Al Imran Mostafa, Zaman Samina, Chakraborty Tapos Kumar, and Ghosh Gopal Chandra

1 Introduction

As the population is continually growing, the demand for safe drinking water is also increasing day by day (Fewkes 2012), and it is seen that the demand for water doubles every 21 years (Li et al. 2010). In Bangladesh, around 98% of the people have access to technologically improved water sources and 96% have access to improved water sources that are within 30 min of collection time. It is also of concern that only 39% of the Bangladeshi population has access to safe drinking water within household premises and around 74 million people are drinking water from unimproved water sources (World Bank 2018). The SDG drinking water goal 6.1 aims to 'achieve universal and equitable access to safe and affordable drinking water for all' (WHO 2016). In the coastal region of Bangladesh, there is an abundance of water but scarcity of safe drinking water, mainly because of salinity intrusion and arsenic contamination, tidal surge, flooding and so on. Due to unavailability of safe drinking water, around 15 million people are forced to drink saline water and around 30 million people are unable to collect potable drinking water in the coastal regions of the country (Hoque 2009). The available contaminated water contributes to a high incidence of waterborne diseases (Hunter et al. 2010).

The main sources of drinking water in the coastal area are Harvested Rainwater (HRW), Pond Sand Filter (PSF), pond water and ground water. The PSF water and pond water is microbiologically contaminated, which makes it unsafe for drinking purposes (Kabir et al. 2016; Moniruzzaman and Rahman 2017). Rainwater Harvesting System (RWHS) has received a significant amount of attention as a potential alternative and sustainable option for drinking water supply in the coastal

A. I. Mostafa (✉) · Z. Samina · C. Tapos Kumar · G. Gopal Chandra
Department of Environmental Science and Technology, Jashore University of Science and Technology, Jashore, Bangladesh

A. Haque, A. I. A. Chowdhury (eds.), *Water, Flood Management and Water Security Under a Changing Climate*, https://doi.org/10.1007/978-3-030-47786-8_12

areas of Bangladesh. Bangladesh is a tropical country, receives high seasonal rainfall, where the average annual rainfall is about 2400 mm in the majority part of the country. Therefore, people usually have access to rainwater for about 6–-8 months on average (Ghosh et al. 2015). Several studies in Bangladesh have found that the harvested rainwater meet the physical and chemical drinking water quality standards but not the microbial quality standard (Howard et al. 2007; Karim 2010; Ghosh et al. 2015). However, coastal population use harvested rainwater as drinking water without any further treatment, which can be unsafe for them.

The objective of this study is to assess the physical, chemical and microbial quality of harvested rainwater, and their relation to sanitary inspection of the rainwater collection and storage in the south-west coastal region of Bangladesh.

2 Methodology

2.1 Study Area

This study was carried out in the south-west coastal region of Bangladesh. The Morrelganj upazila (the upazilas are the second lowest tier of regional administration in Bangladesh) of Bagerhat district (the districts are the first tier of the administrative units of local government in Bangladesh) was selected for this study (Fig. 1 a and b) due to its geographical location, and sensitive to climate change issues. The area of Morrelganj upazila is 460.91 sq. km, and located between 22°20'N and 22°37'N latitude and between 89°42′E and 89°54′E longitude (Banglapedia 2018). It is populated by 267,446 people and 69,482 households. The sources of drinking water in the upazila are: tap water (2.7%), tube-well water (28.7%), and others (68.6%) – mainly harvested rainwater (BBS 2012).

2.2 Sampling Technique

The samples for this study were collected during the rainy season, and there were no incidents of rainfall for four days prior to the sampling in the study area. For this study, a total of fifteen RWHS were selected in the study area (Fig. 1c) and a total of fifteen harvested rainwater samples were collected from the RWHS point of use i.e. outlet of the storage tank during 10–29 August 2018. From each sampling point two samples were collected, one for physical and chemical parameter analysis, and the other for microbial analysis. All samples were collected in 500 mL sterilized plastic sampling bottles. The samples were kept in icebox (~4 °C) just after collection and transported to the laboratory for analysis. The samples were analyzed within 24h of collection.

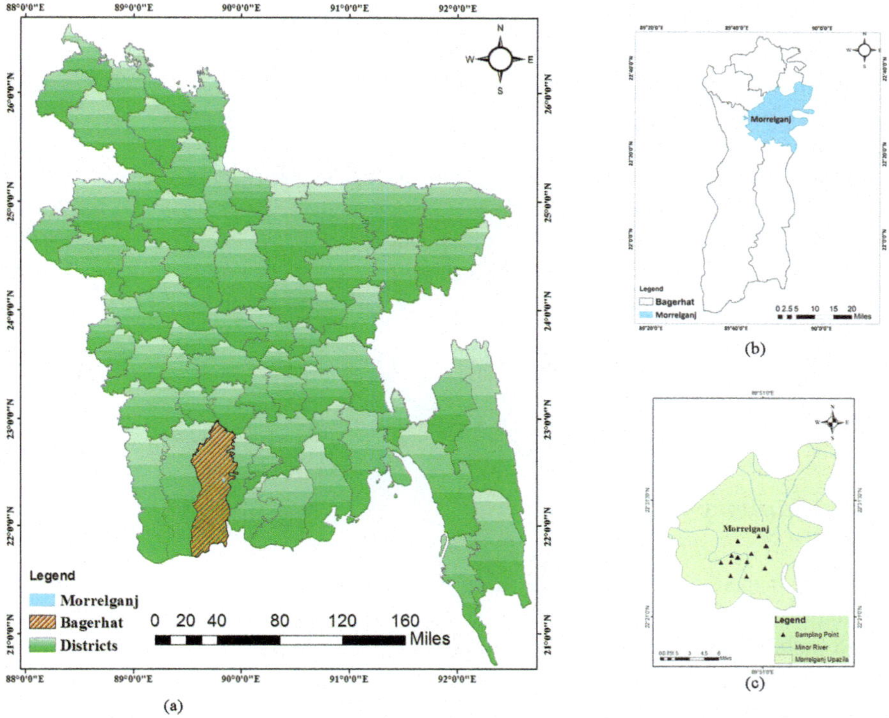

Fig. 1 Map of the study area (**a**) Bangladesh, (**b**) Morrelganj Upazila, and (**c**) Sampling location

2.3 Physical, Chemical and Microbial Analysis

The physical and chemical parameters for water quality analysis was measured in terms of turbidity, pH, electrical conductivity (EC), total dissolved solids (TDS), ammonia, nitrate, phosphate, magnesium, total hardness (TH) and chloride. pH was recorded using MARTINI instrument, PH 56 PHWP. Electrical Conductivity (EC) and total dissolved solids (TDS) was recorded using HACH Sension 156 multi–parameter meter. Turbidity, ammonia, nitrate, phosphate, chloride, magnesium, and total hardness were measured using HACH DR/3900 Spectrophotometer. Total coliforms (TC) and faecal coliforms (FC) were tested by the membrane filtration technique (APHA 1995).

Table 1 Sanitary inspection form

1. Is there any visible contamination of the roof catchment area (plants, dirt, or excreta)?	Yes/No
2. Are the guttering channels that collect water dirty?	Yes/No
3. Is there any deficiency in the filter box at the tank inlet (e.g. lacks fine gravel)?	Yes/No
4. Is there any other point of entry to the tank that is not properly covered?	Yes/No
5. Is there any defect in the walls or top of the tank (e.g. cracks) that could let water in?	Yes/No
6. Is the tap leaking or otherwise defective?	Yes/No
7. Is the concrete floor under the tap defective or dirty?	Yes/No
8. Is the water collection area inadequately drained?	Yes/No
9. Is there any source of pollution around the tank or water collection area (e.g. excreta)?	Yes/No
10. Is a bucket in use and left in a place where it may become contaminated?	Yes/No
Total risk score: .... Out of 10	
Risk score: 0–2 = low; 3–5 = medium; 6–8 = high; 9–10 = very high	

2.4 *Sanitary Inspection*

Sanitary inspection (SI) form was used to evaluate actual and potential sources of contamination for *rainwater* collection and storage. The sanitary inspection forms used included ten risk factors on potential sources of pollution. Sanitary risk factors were in the form of binomial categorical data. To examine the relationship between the risk factors and faecal contamination, faecal coliforms data was transformed into categorical variables. In this study, SI was performed during sample collection and details about the SI form are presented in Table 1.

3 Results

The HRW quality data is presented in Table 2. The turbidity value ranges from 0.52 NTU to 1.65 NTU and the mean value was 1.028 NTU. The mean pH value was 7.63 (ranging from 6.5 to 8.5). The highest EC value was 440 μS/cm and the lowest was 10 μS/cm with 68 μS/cm as the mean value. The TDS values ranged from 5 mg/L to 220 mg/L with a mean value of 31.33 mg/L. The ammonia value ranged from 0.01 mg/L to 0.18 mg/L with a mean value of 0.05 mg/L. The nitrate value ranged from 0.5 mg/L to 1.2 mg/L with an average value of 0.91 mg/L. Phosphate value ranged from 0.02 mg/L to 0.65 mg/L with an average of 0.11 mg/L. The magnesium value ranged from 1.46 mg/L to 30.88 mg/L with an average value of 6.55 mg/L. TH value ranged from 5 mg/L to 77.5 mg/L with an average value of 18 mg/L. Among the fifteen samples, only seven rainwater samples contained chloride, which ranged from 2 mg/L to 85.61 mg/L with a mean value of 12.33 mg/L. The results of the analysis of biological parameters of the water samples are presented in Fig. 2. This reveals that the harvested rainwater is not free from microbial contamination. All harvested rainwater samples were contaminated by TC (range: 175–470 CFU/

Table 2 Physical, chemical and microbiological analysis result of harvested rainwater

Parameters	Unit	Mean ± SD	Minimum	Maximum	WHO [15]	ECR [16]
Turbidity	NTU	1.028 ± 0.33	0.52	1.65	10	10
pH	–	7.63 ± 0.6	6.5	8.5	6.5–8.5	6.5–8.5
EC	µS/cm	68 ± 104.13	10	440	2000	2000
TDS	mg/L	31.33 ± 53.02	5	220	1500	1000
Ammonia	mg/L	0.05 ± 0.04	0.01	0.18	0.5	0.5
Nitrate	mg/L	0.913 ± 0.28	0.5	1.2	45	10
Phosphate	mg/L	0.114 ± 0.15	0.02	0.65	5–6	6
Magnesium	mg/L	6.55 ± 7.22	1.46	30.88	–	30–35
TH	mg/L	18 ± 17.68	5	77.5	500	200–500
Chloride	mg/L	12.23 ± 21.17	2	85.61	250	150–600
FC	CFU/ 100 mL	56 ± 79	0	255	0	0
TC	CFU/ 100 mL	326 ± 102 175 470	0	0		

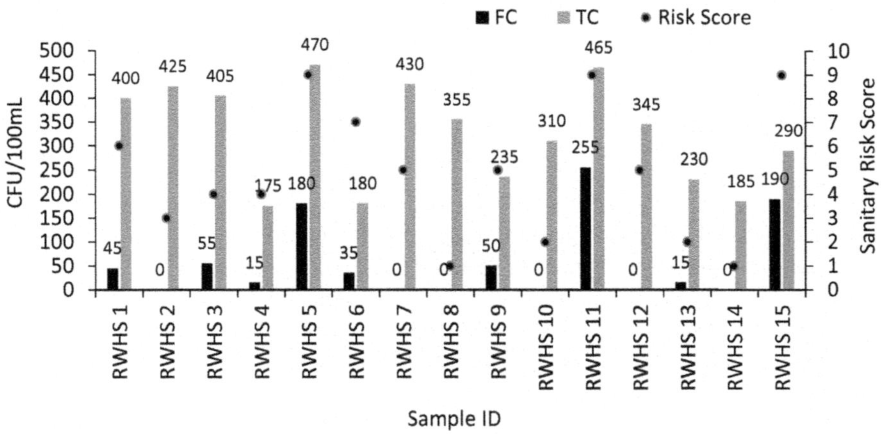

Fig. 2 Risk score of the RWH systems and microbiological concentration (TC and FC) in the RWHS

100 mL; mean: 326 CFU/100 mL) whereas FC was found in nine samples out of fifteen (range: 0–255 CFU/100 mL; mean: 56 CFU/100 mL) (Table 2).

Figure 3 shows the distribution of sanitary risk scores and microbial analysis results of the RWHS and HRW, respectively. According to SI risk score based on WHO risk categories it was found that four samples (RWHS-8, 10, 13 and 14) in low risk, six samples (RWHS-2, 3, 4, 7, 9 and 12) in moderate risk, two samples (RWHS-1 and 6) in high risk and three samples (RWHS-5, 11 and 15) in very high risk categories. On the other hand, six samples (RWHS-2, 7, 8, 10, 12 and 14) in safe level, five samples (RWHS-1, 4, 6, 9 and 13) in moderate risk, one sample in high

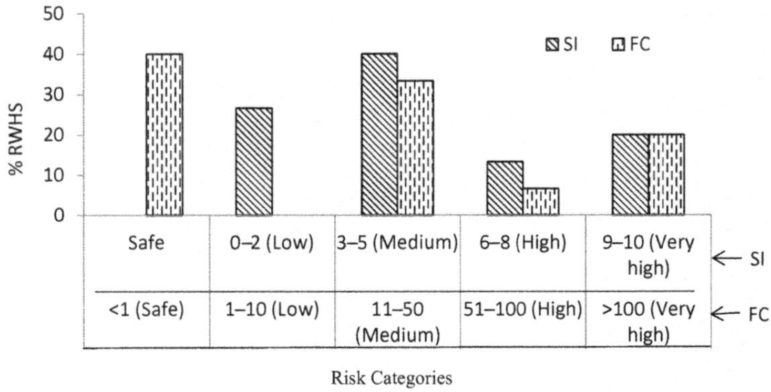

Fig. 3 Distribution of sanitary risk scores and WHO drinking water quality (microbial risk) of the RWHS for water sampling

risk (RWHS-3) and three samples (RWHS-5, 11 and 15) in very high risk categories (Fig. 3). However, eight samples out of fifteen samples fall in the same categories of SI and WHO risk categories: RWHS-4, 5, 8, 9, 10, 11, 14 and 15.

4 Discussion

The physical and chemical quality of harvested and stored rainwater depends on the characteristics of the individual area, such as topography, weather conditions and proximity to pollution sources (Evans et al. 2006). From this study results, it revealed that that all the physical and chemical parameters meet with Bangladesh and WHO standards, which are mostly similar to other studies (Ghosh et al. 2015; Biswas and Mandal 2014; Ayog et al. 2016; Lina 2016). The variation of conductivity (10–440 μS/cm) demonstrates the influence of the sea environment and storage tank itself (Chang et al. 2004; Zhu et al. 2004). Relatively low concentration of ammonia, nitrate and phosphate were detected in the HRW as the coastal area of Bangladesh is relatively pure from traffic and industrial emissions.

The microbiological quality of the HRW depends on the various types of storage tanks (Dillaha and Zolan 1985; Evison and Sunna 2001), the type of catchment area (Chang et al. 2004; Zhu et al. 2004), and the handling and management of the water (Evison and Sunna 2001; Pinfold et al. 1993). The health risks associated with suspended sediments remain poorly understood but should be noted as a possible risk during sanitary surveys. Thus, local knowledge of possible sources and environmental pathways through which animal pathogens can reach humans should form part of the sanitary inspection (Bartram and Fewtrell 2001).

According to the SI score, it was seen that among fifteen households, 27% were in low risk category, 40% in medium risk category, 13% in high risk category and 20%

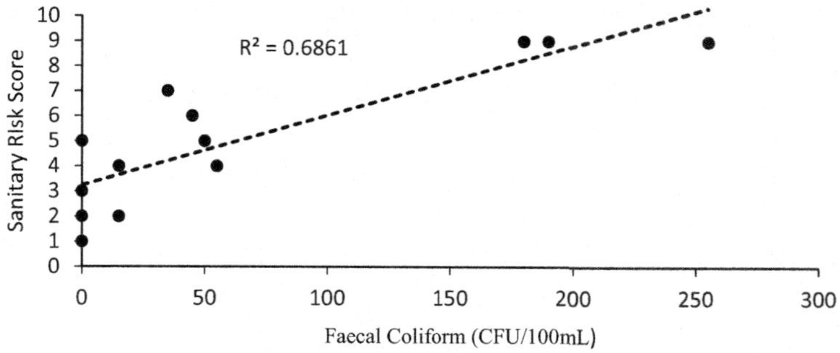

Fig. 4 Correlation between faecal coliforms and sanitary risk score

in very high category (Fig. 3). From the investigation it was seen that a majority of the RWHS systems have low to medium risks of contamination, and some also fall into high to very high risk category. From this result it can be said that rainwater harvesting systems are overall good. On the other hand, a positive correlation ($R^2 = 0.6861$) was found between the FC and risk score (Fig. 4) and eight samples fall into the same category for both SI score and WHO risk categories. Microbial contamination in the HRW may be due to the presence of bird or animal excreta, plants or dusts on roof catchments, dirty or blocked gutter, leaking or defective tap, defective floor under the tap, presence of pollution source around the storage tank or water collection area (Karim 2010). The physical and chemical characteristics of harvested rainwater were reasonably satisfactory but unsuitable for drinking due to microbial contamination.

5 Conclusion

As climate change threatens water security in the coastal regions of Bangladesh, rainwater harvesting can prove to be one of the most promising alternatives for supplying drinking water in the coastal parts of the country. In this study, all the physical and chemical parameters for harvested rainwater are within WHO and Bangladesh drinking water quality standard. The sanitary conditions of major percentage of rainwater harvesting systems are within low to medium contamination risk. However, microbial contamination was found in harvested rainwater in medium to very high level of risk. A positive correlation was found between the faecal coliforms and sanitary inspection risk score, and routine sanitary inspection is recommended to reduce microbial contamination. Finally, a low-cost water treatment system for the harvested rainwater is required in order to deliver the safely managed water in the coastal regions of Bangladesh.

Acknowledgement This study was partially supported by Jashore University of Science and Technology, Bangladesh and the Ministry of Science and Technology of Bangladesh. The authors would like to extend thanks to The World Academy of Science (TWAS) for instrumental facility under the COMSTECH-TWAS Joint Research Grants Programme (TWAS Ref: 13-371 RG/ENG/AS_C; UNISCO FR: 3240279207).

References

APHA. (1995). *Standard methods for the examination of water and wastewater* (19th ed.). Washington, DC: American Public Health Association, AWWA, WEF.

Ayog, J., Dullah, S., & Ramli, R. (2016). Harvested rainwater quality assessment on the effects of roof materials to the first flush runoff. *Transaction on Science and Technology, 3*(1–2), 271–276.

Banglapedia. (2018). Title: Morrelganj Upazila. http://en. banglapedia. org/index. php? Title = Morrelganj Upazila.

Bartram, J., & Fewtrell, L. (Eds.). (2001). *Water Quality: Guidelines, Standards and Health: Assessment of Risk and Risk Management for Water-related Infectious Disease* (p. 424). London: International Water Association (IWA) publication. ISBN: 1900222280-924154533X.

BBS. (2012). *Population and housing census – 2011*. Bangladesh Bureau of Statistics: Statistics and Informatics Division, Ministry of Planning, Government of the People's Republic of Bangladesh.

Biswas, B. K., & Mandal, B. H. (2014). *Construction and evaluation of rainwater harvesting system for domestic use in a remote and rural area of Khulna*. Bangladesh, International Scholarly Research Notices, 2014.

Chang, M., Mcbroom, M. W., & Beasley, R. S. (2004). Roofing as a source of nonpoint water pollution. *Journal of Environmental Management, 73*(4), 307–315.

Dillaha, T. A., III, & Zolan, W. J. (1985). Rainwater catchment water quality in Micronesia. *Water Research, 19*(6), 741–746.

Evans, C. A., Coombes, P. J., & Dunstan, R. H. (2006). Wind, rain and Bacteria: The effect of weather on the microbial composition of roof-harvested rainwater. *Water Research, 40*(1), 37–44.

Evison, L., & Sunna, N. (2001). Microbial regrowth in household water storage tanks. *Journal-American Water Works Association, 93*(9), 85–94.

Fewkes, A. (2012). A review of rainwater harvesting in the UK. *Structural Survey, 30*(2), 174–194.

Ghosh, G. C., Jaha, S., Chakraborty, B., & Akter, A. (2015). Potential of household rainwater harvesting for drinking water supply in Hazard prone coastal area of Bangladesh. *Nature Environment and Pollution Technology, 14*(4), 937.

Hoque, M. R. (2009). *Access to safe drinking water in rural Bangladesh: Water governance by DPHE*. Doctoral dissertation, Brac University.

Howard, G., Ahmed, M. F., Teunis, P., Mahmud, S. G., Davision, A., & Deere, D. (2007). Disease burden estimation to support policy decision-making and research prioritization for arsenic mitigation. *Journal of water and health, 5*(1), 67–81.

Hunter, P. R., Macdonald, A. M., & Carter, R. C. (2010). Water supply and health. *PLoS Medicine, 7*(11), e1000361.

Kabir, A. E., Chakraborty, T. K., & Ghosh, G. C. (2016). Bio-Sand Filter (BSF): A simple water treatment device for safe drinking water supply and to promote health in Hazard prone hard-to-reach coastal areas of Bangladesh. *American Journal of Environmental Protection, 5*(5), 107–112.

Karim, M. R. (2010). Microbial contamination and associated health burden of rainwater harvesting in Bangladesh. *Water Science and Technology, 61*(8), 2129–2135.

Li, Z., Boyle, F., & Reynolds, A. (2010). Rainwater harvesting and greywater treatment systems for domestic application in Ireland. *Desalination, 260*(1–3), 1–8.

Lina, M. A. T. (2016). *Rainwater harvesting system: Quality and impacts on public health*. Doctoral dissertation, Birzeit University, West Bank, Palestine.

Moniruzzaman, M., & Rahman, M. A. (2017). Examine the water quality of pond sand filter (PSF): A study on Khontakata Union of Sarankhola Upazila, Bangladesh. *Journal of the Bangladesh National Geographical Association, 39*(1–2), 97–108.

Pinfold, J. V., Horan, N. J., Wirojanagud, W., & Mara, D. (1993). The bacteriological quality of Rainjar water in rural Northeast Thailand. *Water Research, 27*(2), 297–302.

WHO. (2016). *World health statistics 2016: Monitoring health for the SDGs Sustainable Development Goals*. World Health Organization.

World Bank. (2018). *Promising progress: A diagnostic of water supply, sanitation, hygiene, and poverty in Bangladesh*.

Zhu, K., Zhang, L., Hart, W., Liu, M., & Chen, H. (2004). Quality issues in harvested rainwater in arid and semi-arid loess, plateau of northern China. *Journal of Arid Environments, 57*(4), 487–505.

Assessment of Water Availability for Agriculture and Other Uses – Development of Indices for Deduru Oya River Basin in Sri Lanka

D. L. K. Gunawardena and N. D. K. Dayawansa

1 Introduction

In the twenty-first century, water scarcity is a serious problem that is faced by different communities in the world. Water scarcity is generally defined as a situation where water availability in a country or in a region is below 1000 m^3 per person per year (Pereira et al. 2002). However, many regions in the world experience severe scarcity, with an availability of less than 500 m^3 of water per person per year. According to Rijsberman (2006) it is difficult to determine whether water is truly scarce at global scale or whether it is a problem of supply and demand. The threshold value of 2000 m^3 of water per person per year indicates that an area is water stressed, as under these circumstances the population faces large problems when a drought occurs or when man made shortages are created (Pereira et al. 2002) and Winpenny (1997) defines water scarcity as an imbalance of supply and demand under prevailing institutional arrangements and/or prices; an excess of demand over available supply; a high rate of utilization compared with available supply. According to the United Nations, water scarcity already affects every continent (www.un.org). Around 1.2 billion people, or almost one-fifth of the world's population, live in areas of physical scarcity, and 500 million people are approaching this situation. Another 1.6 billion people, or almost one quarter of the world's population, face economic water shortage where countries lack the necessary infrastructure to take water from the rivers and aquifers (www.un.org).

It is important to have an idea about the overall water availability in a country and or a region, when making decisions regarding agriculture, development and other

D. L. K. Gunawardena · N. D. K. Dayawansa (✉)
Department of Agricultural Engineering, Faculty of Agriculture, University of Peradeniya, Peradeniya, Sri Lanka
e-mail: dammid@pdn.ac.lk

A. Haque, A. I. A. Chowdhury (eds.), *Water, Flood Management and Water Security Under a Changing Climate*, https://doi.org/10.1007/978-3-030-47786-8_13

socio economic activities. According to Shiklomanov (2000), reliable assessment of water storage on earth is a complicated problem, mainly due to its dynamic nature. Many indices have been developed to quantitatively assess water resources vulnerability (e.g. water scarcity or water stress) in the past 20 years. There are many equally important facets to water use, supply and scarcity. If the amount of water that is necessary to meet human demands is known, then the water that is available to each person can serve as a measure of scarcity (Rijsberman 2006). The Falkenmark indicator is mostly used for assessing water stress and it identifies the amount of total annual runoff available for human use (Falkenmark et al. 1989). According to Vörösmarty et al. (2005) it is possible to identify climate and human induced water scarcity through the Falkenmark water stress index, because the basis of the index is individual water usage. The study by Ohlsson (2000) developed a Social Resource Water Stress/Scarcity Index (SWSI) combining traditional hydrological indices and the UNDP Human Development Index to capture the social impacts of water scarcity more accurately than earlier developed indices. Mapping of water availability will facilitate the identification of water scarce areas that can be given priority in water resource development and management. There are many past efforts in identifying and mapping water availability in the basin scale as well as the global scale (Falkenmark et al. 1989; Alcamo et al. 2003; Brown and Matlock 2011; Roy et al. 2012; Tidwell et al. 2014; Xu and Wu 2017).

Sri Lanka receives rainfall as a result of multiple activities and seasonal variabilities, such as monsoons, convectional activity and cyclonic activity. Mean annual rainfall varies from 900 mm (in the dry parts) to over 5000 mm (in the wet parts). Average annual rainfall is about 1900 mm at 75% probability. There are 103 perennial rivers and about 34,000 large and small tanks available in the country. This is a considerable amount of water when compared to the surface area of the island. Even though the country possesses a large amount of surface water resources and adequate rainfall, spatial and temporal variations of the water availability is a major challenge for agricultural and other socio-economic activities (Perera and De Silva 2009).

Within the 103 major river basins in Sri Lanka, many basins provide water for multiple uses while facing water scarcity during some periods of the year. There is a huge spatial and temporal variability of water availability in some river basins. Mapping of water availability will facilitate the identification of water scarce areas quickly and this will enable the decision makers to identify areas that should be given priority in water resources development, planning and management. Hence this study attempted to assess the water availability in Deduru Oya basin which has reported water scarcity issues in the recent past (Katupotha 2009; Saumyarathne et al. 2016).

Under the above circumstances, this study was carried out with the objectives of developing Water Availability Indices (WAI) for Deduru Oya basin in Sri Lanka and mapping spatial and temporal variability of water availability using the developed indices.

2 Methodology

2.1 Study Area

This study was conducted in Deduru Oya basin which is located in the North Western province of Sri Lanka, centering the Kurunegala District. Figure 1 presents the location of the Deduru Oya basin and the Divisional Secretariat (DS) divisions (administrative divisions) within it. There are 28 DS divisions in the basin. Three DS divisions namely; Ukuwela, Yatawatta, and Poojapitiya were not considered for this study since their land area contribution to the total basin is not very significant (Fig. 1).

Though water does not follow administrative boundaries, most of the data used in this study have been collected or available at the administrative division level. In addition, the management interventions are also mostly applied at the scale of administrative divisions. Hence, it was decided to use the administrative boundary (Divisional Secretariat-DS) as the spatial unit for mapping in this study. Though it is appropriate to use the smallest administrative unit (Grama Niladhari Division) to depict the spatial variability more effectively, the next higher administrative unit of DS had to be used considering the data availability.

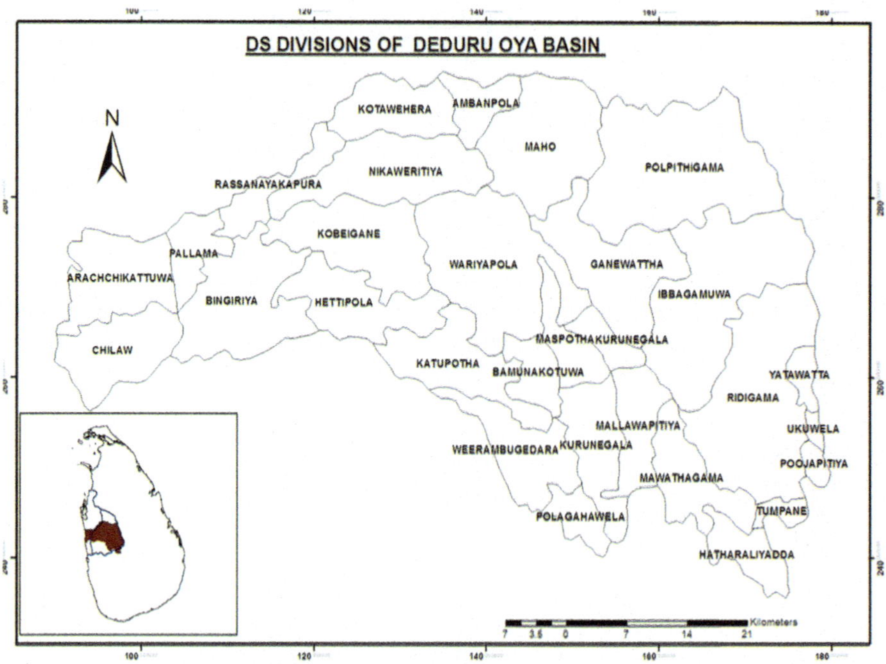

Fig. 1 The location of Deduru Oya basin and the Divisional Secretariat boundaries available within the river basin

2.2 Criteria in Developing Water Availability Index

Three components were identified as affecting the water availability of an area, which includes available water resources, ecosystems present in the area and the water consumption. Under these components, eight parameters were selected to develop the Water Availability Index (WAI) as given in Table 1.

Most of the data were collected by using secondary data sources available at government departments/Institutions. Drainage density (total length of streams per unit area of DS division) was calculated using the drainage network of the basin and the DS division boundary data layer with the help of Geographical Information Systems. Water quality data were collected using existing literature. Collected parameter values at each DS level are presented in Table 2.

2.3 Relative Contribution of Each Parameter to Water Availability

Multi criteria analysis was used for assigning weight to the factors affecting water availability to identify their relative importance. In this study, Analytical Hierarchical Process (AHP) (Saaty 1980) was used in assigning weightages. It is one of the commonly used methods in multi criteria decision making. It has the ability to measure both qualitative and quantitative characteristics (Saaty 1990). Saaty (1990) has demonstrated mathematically that the eigenvector solution was the best approach in the AHP analysis. In the process, a pairwise comparison matrix was created by using the eight parameters used in assessing water availability. Then expert knowledge was sought in identifying the relative importance of each parameter for water availability comparing two parameters at a time. According to the relative importance, a numerical value was assigned ranging from 1 (equally important) to 10 highly important. The given values were normalized by dividing with column totals of each parameter. Criteria weights were calculated by dividing summation of normalized row totals by total number of parameters. Accordingly, the calculated weightages are shown in Table 3.

Table 1 Parameters affecting water availability

Component	Parameter
Availability of water resources	Rainfall
	Surface water resources
	Drainage density
	Pipe borne water network
	Agro well density
Ecosystem impacts	Water quality
	Land use
Water consumption	Population density

Table 2 Collected parameters and their values at each DS

DS Division	Rainfall (mm)	Agro well density (wells/ km^2)	Surface water resources (% of total area)	Drainage density (km/km^2)	pipe water coverage (%)	Population density (persons/ km^2)	paddy land coverage (% of total area)	Water quality
Ambanpola	1100	1.13	16.35	0.31	73.37	164.61	17	Poor
Arachchikattuwa	1100	0.92	2.01	0.31	29.09	264.5	18	Moderate
Bamunakotuwa	1250	0.39	6.62	0.35	0	360.78	19	Good
Bingiriya	1100	0.41	2.69	0.7	0	300.46	12	Moderate
Chilaw	1175	0.46	4.17	0.38	38.63	637.5	18	Moderate
Ganewattha	1156	0.5	15.37	1.08	10.95	276.4	20	Poor
Hatharaliyadda	1500	0	0	2.28	18	481.7	2	Good
Hettipola	1135	0.94	21.9	0.66	0	395.02	26	Good
Ibbagamuwa	1200	1.04	9.29	0.96	22.47	402.58	21	Good
Katupotha	1250	0.58	9.06	0.59	17.25	321.8	13	Good
Kobeigane	1100	2.14	12.91	0.76	62.32	282.87	15	Poor
Kotawehera	1100	0.99	13.6	0.13	51.81	115.22	16	Poor
Kurunegala	1315	0.31	6.06	0.8	13.65	796.45	19	Good
Maho	1100	0.23	13.16	0.64	36.06	221.66	21	Poor
Mallawapitiya	1309	0.84	1.46	1.14	28.38	702.98	14	Good
Maspotha	1240	0	2.86	1.11	74.6	589.28	16	Good
Mawathagama	1350	0.17	0.86	1	22.43	613.11	15	Good
Nikaweritiya	1100	1.09	21.92	0.7	86.27	254.86	32	Poor
Pallama	1100	0.65	1.08	1.46	22.82	204.9	13	Moderate
Polagahawela	1400	0.26	2.08	1.12	41.36	683.6	22	Good
Polpithigama	1100	3.02	17.22	1.11	36.25	194.79	23	Good
Poojapitiya	1500	0	14.09	1.56	0	981	0	Good
Rassanayakapura	1100	1.33	5.51	1.02	82.31	178.32	17	Poor
Ridigama	1308	0.08	0	1.04	53.25	416.53	14	Good

(continued)

Table 2 (continued)

DS Division	Rainfall (mm)	Agro well density (wells/ km^2)	Surface water resources (% of total area)	Drainage density (km/km^2)	pipe water coverage (%)	Population density (persons/ km^2)	paddy land coverage (% of total area)	Water quality
Tumpane	1500	0	13.29	2.13	12	691.8	4	Good
Ukuwela	1235	0	2.74	0.93	15	877.2	4	Good
Wariyapola	1143	1.23	0	0.92	67.52	339.75	19	Poor
Weerambugedara	1250	0.18	0	1.03	29.24	362.12	17	Good
Yatawatta	1240	0	0	0.98	10	478	5	Good

Table 3 Weights calculated for each parameter contribute to water availability

Parameter	Weightage (%)
Rainfall	19
Surface water resources	15
Land use	14
Drainage density	12
Agro well density	10
Pipe water coverage	10
Water quality	10
Population density	10

2.4 Calculation of WAI

Each parameter was stored in the GIS database at DS division scale as individual attributes. Since the units of measurement of each parameter were different, all the parameters were brought into a single scale between 0 and 20 by normalization. In normalization, the maximum and minimum values (based on their importance to water availability) of each parameter were identified among the DS divisions. The score for each parameter value in the 0–20 scale was calculated using Eq. 1.

$$Score = \left[\frac{Value - Min}{Max - MIn}\right] * 20 \qquad (1)$$

Subsequently, each DS was given a score between 0 and 20 for each parameter, based on actual parameter values.

Water availability of an area is dependent on multiple parameters and they can affect the water availability negatively or positively. As an example, rainfall has a positive impact since it is a source of water whereas population density has a negative impact as it leads to high water usage. At the same time, some parameters can contribute positively or negatively to the water availability depending on the context of their use. The agro well density can be considered as positive as the availability of agro wells can improve water accessibility for agriculture (Agro wells are used to abstract water for agriculture). From the point of view of consumption, the agro wells can contribute to high groundwater abstraction thus reducing the availability of groundwater. Hence two arguments were tested in developing the WAI for the study area. In general water availability (WAIg), all the parameters except population density and poor water quality were taken as positive and in consumption related water availability (WAIc) land use (paddy) and agro well density were taken as negative. The seasonal rainfall (*Yala – minor cultivation season* and *Maha – Major cultivation season*) is the factor which determines the temporal water availability. Under these two indices, temporal variation of the water availability was also calculated considering *Yala* and *Maha* rainfall. Equation 2 was used to calculate the WAI for both conditions.

Table 4 Classification of Indices

Index value	Class
<0.5	Low water availability
0.5–0.65	Moderate water availability
>0.65	High water availability

$$WAI = \sum \frac{Sc * Wc}{20} \quad (2)$$

Where,

S – Score value of each criteria.
W – Weightage of each criteria.

2.5 Mapping of Water Availability

Arc GIS 10.2 software was used for the mapping purpose. WAI was calculated as an attribute in each DS division. Then the water availability index was classified into different classes using the natural breaks classification option in ArcGIS to develop the spatial and temporal variation of the WAIg and WAIc (Table 4). Maps were prepared to show the spatial variation of water availability using two indices for two rainfall seasons; *Yala* and *Maha*.

3 Results

3.1 Water Availability Index Based on Consumption (WAIc)

Table 5 presents the calculated WAIc values for each DS while Fig. 2 presents the spatial distribution of WAIc where paddy and agro well densities were considered as negative contributors to water availability. According to Fig. 2, Bingiriya, Maho, Ambanpola, Kotawehera and Arachchikattuwa DS divisions are under low water availability. While majority of the DS divisions are in moderate water availability, Hathraliyadda and Ridigama DS divisions show high water availability.

3.2 Temporal Variation of WAIc

Figure 3 (a and b) shows the temporal variation of water availability in Deduru Oya basin based on WAIc. Accordingly, in *Yala* season, many DS divisions are affected by low water availability. Hence, there is a possibility for water shortages for domestic and agricultural purposes during this period of the year. However, due to

Table 5 Calculated WAIc and WAIg values for each DS division

DS division	WAIg	WAIg Yala	WAIg Maha	WAIc	WAIc Yala	WAIc Maha
Ambanpola	0.579	0.550	0.655	0.489	0.460	0.565
Arachchikattuwa	0.374	0.346	0.450	0.451	0.423	0.527
Bamunakotuwa	0.568	0.464	0.511	0.533	0.429	0.476
Bingiriya	0.358	0.330	0.434	0.435	0.407	0.511
Chilaw	0.423	0.347	0.394	0.570	0.494	0.541
Ganewattha	0.576	0.500	0.547	0.541	0.465	0.512
Hatharaliyadda	0.546	0.451	0.546	0.748	0.653	0.748
Hettipola	0.724	0.619	0.695	0.584	0.479	0.555
Ibbagamuwa	0.731	0.655	0.702	0.536	0.460	0.507
Katupotha	0.593	0.489	0.536	0.586	0.482	0.529
Kobeigane	0.564	0.535	0.640	0.502	0.473	0.578
Kotawehera	0.490	0.461	0.566	0.483	0.454	0.559
Kurunegala	0.521	0.398	0.474	0.556	0.433	0.509
Maho	0.638	0.609	0.714	0.443	0.414	0.519
Mallawapitiya	0.532	0.408	0.456	0.650	0.526	0.574
Maspotha	0.628	0.523	0.571	0.621	0.516	0.564
Mawathagama	0.547	0.452	0.528	0.610	0.515	0.591
Nikaweritiya	0.718	0.689	0.794	0.523	0.494	0.599
Pallama	0.418	0.390	0.494	0.536	0.508	0.612
Polagahawela	0.661	0.566	0.642	0.591	0.496	0.572
Polpithigama	0.755	0.726	0.831	0.560	0.531	0.636
Poojapitiya	0.551	0.456	0.551	0.753	0.658	0.753
Rassanayakapura	0.600	0.571	0.676	0.510	0.481	0.586
Ridigama	0.597	0.473	0.549	0.715	0.591	0.667
Tumpane	0.613	0.518	0.613	0.815	0.720	0.815
Ukuwela	0.450	0.288	0.421	0.652	0.490	0.623
Wariyapola	0.613	0.509	0.585	0.523	0.419	0.495
Weerambugedara	0.577	0.472	0.520	0.612	0.507	0.555
Yatawatta	0.470	0.308	0.441	0.672	0.510	0.643

Maha rains, water availability in these areas improve up to moderate water availability. Wariyapola and Bamunakotuwa DS divisions still show low water availability in the *Maha* season. It could be due to high water consumption (high agro well density) in this area, especially in Wariyapola. Bamunakotuwa area has a low drainage density, which could be the reason behind low water availability due to lack of surface water resources.

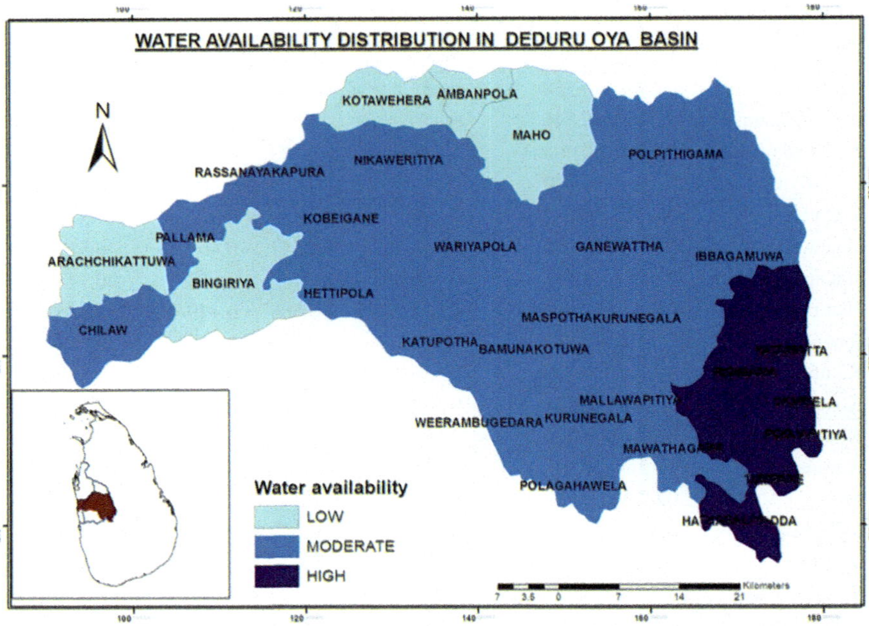

Fig. 2 Spatial distribution of water availability (WAIc)

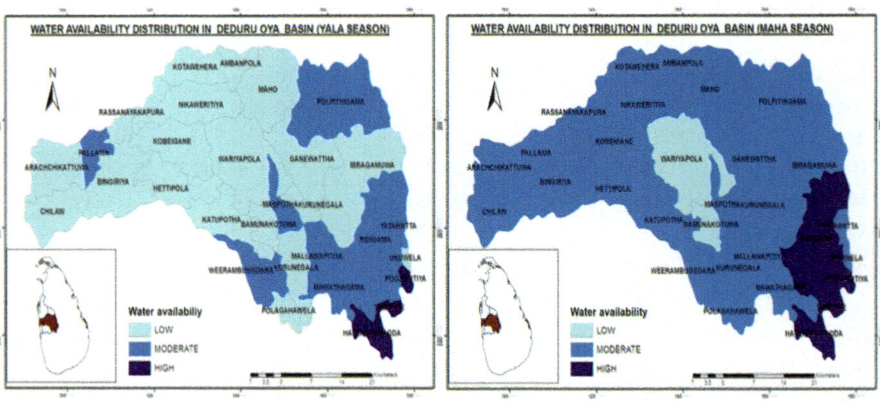

(a) (b)

Fig. 3 Temporal variation of water availability (WAIc) (**a**) Water availability in the *Yala* season (**b**) Water availability in the *Maha* season

3.3 General Water Availability Index (WAIg)

Table 5 presents the calculated WAIg values for each DS and Fig. 4 shows the general water availability in Deduru Oya basin. In this index (WAIg), high agro well density and presence of paddy lands were considered positively since they are

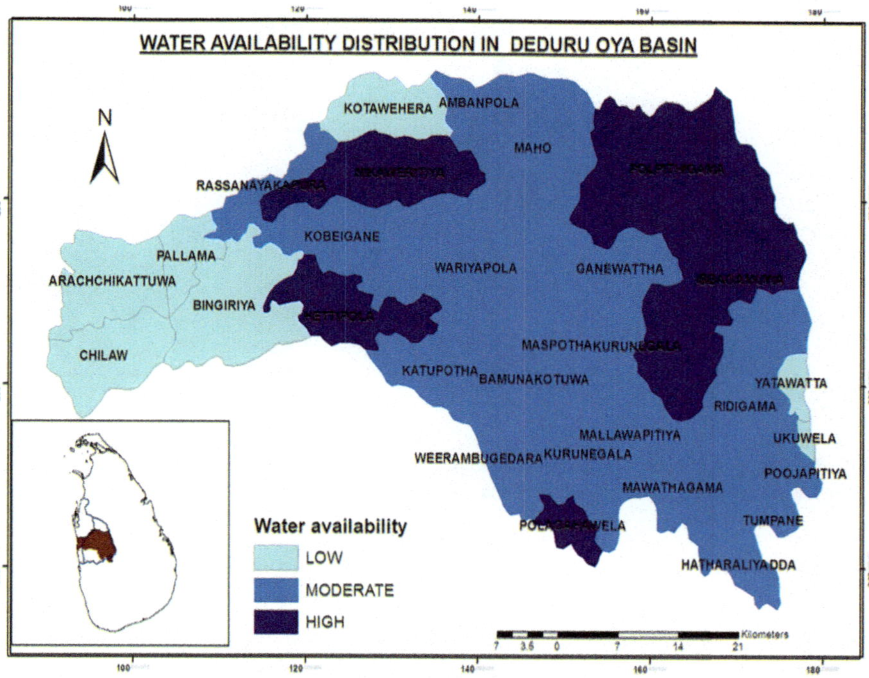

Fig. 4 Water availability in Deduru Oya basin (WAIg) (based on accessibility)

indicators of water availability. According to Fig. 4, Arachchikattuwa, Chilaw, Pallama, Bingiriya and Kotawehera DS divisions indicate low water availability. Most of the DS divisions are in the moderate level of water availability. Accessibility to water is high in Polpithigama area, due to high drainage density and agro well density. This has positively contributed to high WAIg in this area.

3.4 Temporal Variation of WAIg

According to Fig. 5, three DS divisions (Ibbagamuwa, Polpithigama, and Nikawaratiya) show high water availability in *Yala* season. Spatial variability of water availability is high in *Yala* season. In Maha season, the entire area gets rainfall and water availability increases as a result. Hence, majority of the DS divisions show either high or moderate water availability in *Maha*. Pallama, Bingirya, Kurunegala and Mallawapitiya DS divisions have low surface water resources, drainage density and agro well density; hence reduces accessibility to water. These factors have also contributed negatively to water availability in the *Maha* season.

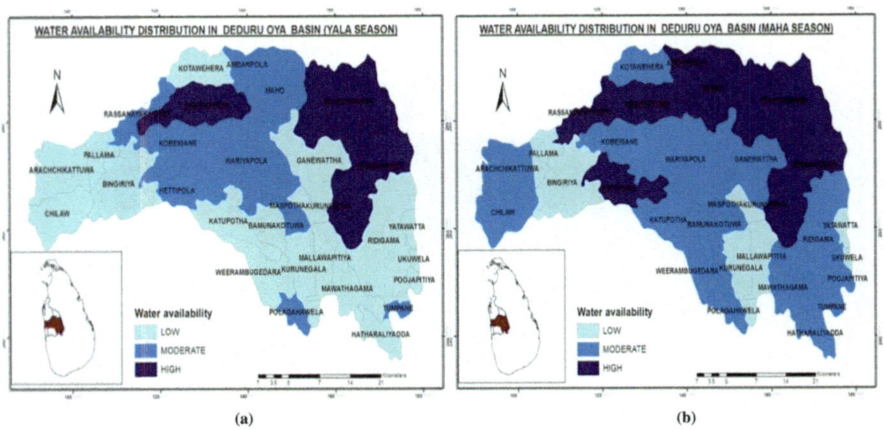

Fig. 5 Temporal variation of water availability (based on accessibility) (**a**) Water availability in Yala season (**b**) Water availability in Maha season

4 Discussion

Deduru Oya river basin shows a considerable spatial variation in availability of surface water resources, drainage density, population density, agro well density and other parameters which determine the water availability. It also consists of three agro ecological regions namely; IL (Intermediate Low Country), IM (Intermediate Mid Country) and WM (Mid Country Wet Zone), showing rainfall variability within the basin. Temporal variability in the indices was also seen between two major cultivation seasons namely *Yala* and *Maha*, which has significant difference in the amount of rainfall received. Relatively low consumption related water availability (WAIc) compared to general water availability (WAIg) in some DS divisions indicate high water use for different sectors. Field verification of these indices will help refine them further and selection of a small spatial unit for mapping (Eg. GN division) will further enhance the results by improving the ability to represent spatial variability more effectively. To get more accurate results, this method can be further improved by developing sub-indices under each parameter. This will also help incorporate both water accessibility and affordability issues.

5 Conclusion

WAIg and WAIc developed in this study can be considered as suitable indices to identify general water availability in the study area and to assess the water availability based on consumption. Maps of these indices show a considerable variation among DS divisions indicating inadequacy of water in some parts of the basin. These indices show temporal variability too, especially during *Yala* and *Maha* seasons.

There are some areas with relatively low water availability even in *Maha* season which brings high rains to the basin. Though rainfall is a critical factor to determine water availability in a given area, it cannot compensate for the issues created by other factors such as low surface water resources, drainage density etc.

Spatial variability of the developed indices is controlled by the size of the mapping unit. A smaller spatial unit such as Grama Niladhari Division will help visualize the spatial variability more effectively.

References

Alcamo, A., Doll, P., Henrichs, T., Kaspar, F., Lehner, B., Rosch, B., & Siebert, S. (2003). Development and testing of the WaterGAP 2 global model of water use and availability. *Hydrological Sciences Journal, 48*(3), 317–337.

Brown, A., & Matlock, M.D. (2011). A review of water scarcity indices and methodologies, food, beverage and agriculture, the sustainability consortium, White Paper 106, April, 2011.

Falkenmark, M., Lunduist, J., & Widstrand, C. (1989). Macro-scale water scarcity requires micro-scale approaches: Aspects of vulnerability in semi-arid development. *Natural Resources Forum, 13*, 258–267.

Katupotha, K. N. J. (2009). Water shortage in the lower Deduru Oya basin. Proceedings of the national conference on water, food security, climate change in Sri Lanka. In P. Weligamage, G. G. A. Godaliyadda, & K. Jinapala (Eds.), *National conference on water, food security and climate change in Sri Lanka* (pp. 1–21). Colombo: International Water Management Institute.

Ohlsson, L. (2000). Water conflicts and social resource scarcity. *Physics and Chemistry of the Earth Part B, hydrology oceans and atmosphere, 25*(3), 213–220.

Pereira, L., Cordery, I., & Iacovides, I. (2002). *Coping with water scarcity, technical documents in hydrology, 58* (269 p). Paris: UNESCO.

Perera, E. N. C., & De Silva, R. P. (2009). Development of district wise Water Availability Index (WAI) for Sri Lanka. In N. D. K. Dayawansa, S. Pathmarajah, & M. I. M. Mowjood (Eds.), *Proceedings of the water professionals' day symposium, water resources research in Sri Lanka* (pp. 139–146). Colombo: Geo-Informatics Society of Sri Lanka.

Rijsberman, R. F. (2006). Water scarcity: Fact or fiction? *Agricultural Water Management, 80* (1–3), 5–22.

Roy, S. B., Chen, L., Girvetz, E. H., Maurer, E. P., Mills, W. B., & Grieb, T. M. (2012). Projecting water withdrawal and supply for future decades in the US under climate change scenarios. *Environmental Science & Technology, 46*, 2545–2556.

Saaty, T. L. (1980). *The analytic hierarchy process*. New York: McGraw-Hill.

Saaty, T. L. (1990). How to make a decision: The analytic hierarchy process. *European Journal of Operational Research, 48*, 9–26.

Saumyarathne, N. G. R., Gunawardena, E. R. N., & Dayawansa, N. D. K. (2016). Water conflicts among different water users and uses in the Hakwatuna Oya watershed in the Deduru Oya basin, Sri Lanka. *Tropical Agricultural Research, 28*(1), 38–49.

Shiklomanov, I. A. (2000). Appraisal and assessment of world water resources. *Water International, 25*(1), 11–32.

Tidwell, V.C., Moreland, B.D., Zemlick, K.M., Roberts, B.L., Passell, H.D., Jensen, D., Forsgren, C., Sehlke, G., Cook, M.A., King, C.W. & Larsan, S. 2014. Mapping water availability, projected use and cost in the Western United States, Environmental Research Letters 9, 064009, 16p.

Vörösmarty, C. J., Douglas, E. M., Green, P. A., & Revenga, C. (2005). Geospatial indicators of emerging water stress: An application to Africa. *AMBIO Journal of Human Environment, 34*(3), 230–236.

Winpenny, J. T. (1997). *Managing water scarcity for water security*. A discussion paper prepared for the 1st FAO E-mail conference on managing water scarcity, www.fao.org.

www.un.org, http://www.un.org/waterforlifedecade/scarcity.shtml

Xu, H., & Wu, M. (2017). *Water availability indices – A literature review*. Lemont: Energy Systems Division, Argonne National Laboratory, ANL/ESD-17/5.

Flash Flood Forecasting of Jadukata River Basin at Laurergarh, Sunamganj from Real Time Satellite Precipitation Product by Using HEC-HMS

Nusrat Khan and Md. Mostafa Ali

1 Introduction

Flash flood refers to the rapid inundation of water in a certain area. All around the world, relatively flat areas located close to hilly regions are exposed to frequent flash floods. Bangladesh also faces frequent flash floods, particularly in the north-eastern region of the country (Haque and Islam 2017). The north-eastern region of Bangladesh is very prone to flash floods sourcing from the hills of Meghalaya and Tripura, attributing to its topographical features, geographical location and low lying lands. Figure 1 represents the study area where the west and the south-west Khasi hill contain the basin for the study river Jadukata.

For the purpose of the study, Jadukata river basin located within the West and South-West Khasi Hills districts, has been selected. Figure 1 represents the study area. Jadukata river has its watershed in the West Khasi Hills and South West Khasi Hills districts. Heavy rainfall from the Meghalaya hills in India causes flood inundation in the low-lying north-eastern zone (Hashem and Meer 2016). Strong surges of water from upstream and incessant rainfall in the north-eastern zone triggers flash floods in Tahirpur, inundating vast areas of farmlands. Tahirpur is a sub-district in Sunamganj which has 22 haors. Jadukata-Rakti river contributes largely to the flow of water to the low lying areas of Tahirpur, rendering it to be one of the worst sufferers of flash floods.

N. Khan (✉) · M. Mostafa Ali
Bangladesh University of Engineering and Technology, Dhaka, Bangladesh

A. Haque, A. I. A. Chowdhury (eds.), *Water, Flood Management and Water Security Under a Changing Climate*, https://doi.org/10.1007/978-3-030-47786-8_14

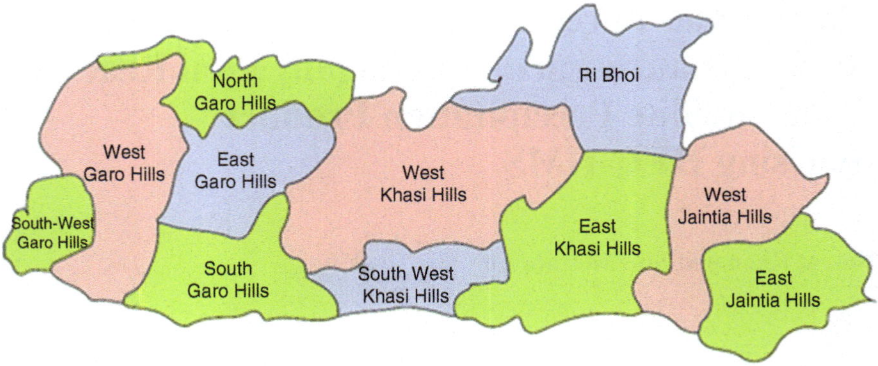

Fig. 1 Study area

2 Methodology

2.1 Collection of Data

For the purpose of the study, data collected includes the Digital Elevation Model (DEM), land use pattern, river network shapefiles, soil distribution data, and necessary maps. DEM was collected from USGS (https://earthexplorer.usgs.gov/). The USGS EarthExplorer (EE) tool provides users the ability to query, search, and order satellite images, aerial photographs, and cartographic products from several sources. The land cover maps were developed at Boston University, using data from the Moderate Resolution Imaging Spectroradiometer (MODIS) instrument on NASA's Terra satellite. Soil map was downloaded from World Digital Soil Map. Precipitation data was collected from CHRS data portal. PERSIANN-Cloud Classification System (PERSIANN-CCS) is a real-time global high resolution ($0.04 \times 0.04°$ or $4\ \text{km} \times 4\ \text{km}$) satellite precipitation product developed by the Centre for Hydrometeorology and Remote Sensing (CHRS) at the University of California, Irvin (UCI). PERSIANN-CCS system enables the categorization of cloud-patch features based on cloud height, aerial extent, and variability of texture estimated from satellite imagery.

2.2 Data Processing

The data which was collected from different websites were prepared for different processes. The DEM was prepared to provide watershed and other information like drainage area, slop, stream network and topographical features of the particular area of concern. For this purpose, HEC-GeoHMS was used as an extension of Arc GIS. Soil data and land use data were prepared to get the surface storage, imperviousness, and canopy storage information. Precipitation data was downloaded as NETCDF format and then converted to CSV format.

2.3 Model Execution

The project setup is mainly dependent on a successful project generation through terrain pre-processing. Using GEO-HMS fill, flow direction, flow accumulation, stream definition, stream segmentation, catchment processing and drainage line processing were done. The basin model and meteorological model were prepared by Geo-HMS. The catchment was divided into 49 sub-basins with a defined stream network from a 30S resolution DEM. The developed terrain was exported in HEC-HMS for further detailing of the basin. Figure 2 indicates the basin Meteorology prepared from HEC Geo HMS with the help of terrain processing. Soil Moisture Accounting, Recession, and Muskingum were adopted as Loss method, Base flow method and Routing method respectively. Observed precipitation data from CHRS Data Portal was assigned for 49 sub-basins. The obtained discharge as the outcome of the model was calibrated with the observed data of BWDB at Laurergarh for the years 2003–2009 and validated for 2010. These were directly exported to HEC HMS. Following this, control specifications were specified and time series data were added.

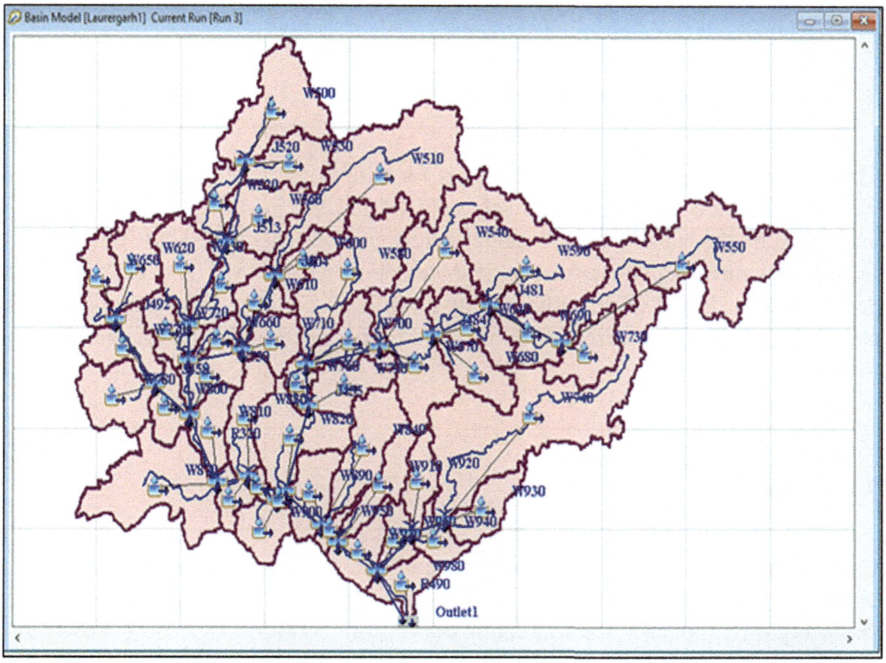

Fig. 2 Meteorological Model for the basin

2.4 Calibration

Calibration was done by optimizing parameters that have more sensitivity for the years 2003–2009. Goodness-of-fit calibration can be evaluated through visual comparison and statistical measures. In this model, the only discharge station in Bangladesh is Laurergarh station. This station is selected as the calibration point. And calibration period is from 1 January 2003 to 31 December 2009.

2.5 Validation

Verification and validation of computer simulation models were conducted during the development of a simulation model with the ultimate goal of producing an accurate and credible model.

2.6 Application of Model in Forecasting

Using daily rainfall data from 22 March to 10 April the model was run to determine the magnitude of flash flood that actually occurred in Sunamganj.

2.7 Goodness of Fit Estimators

There are many goodness of fit estimators for hydrologic modeling to evaluate how well the simulated and observed values fit (Rahat 2016). Here goodness of fit is calculated by the Nash–Sutcliffe model efficiency coefficient (NSE), PBias (Percent bias), Root mean square error-standard deviation ratio (RSR) and Coefficient of determination (R^2). Table 1 indicates the formulas used to find out Goodness of Fit estimators. Table 2 represents the values of Goodness of Fit Estimators for this model.

3 Results

Calibration of parameters was performed until the simulated discharge matched the observed discharge. Calibration was done from 1st January 2003 to 31st December 2009. Figure 3 indicates the simulation curve from 2003 to 2009. Here Tables 3, 4 and 5 present the optimized values of sensitive parameters after calibration. The validation was performed for 2010 and Fig. 4 shows the validation.

Table 1 Formulas of Goodness of Fit

Formula	Value	Rating
$NSE = 1 - \left[\frac{\sum_{i}^{n}(xobs(i)-ymod(i))^2}{\sum_{i}^{n}\left(xobs(i)-\overline{xobs}\right)^2}\right]$	>0.65	Very good
	0.54–0.65	Adequate
	>0.50	Satisfactory
$PBIAS = \left[\frac{\sum_{i}^{n}(xobs(i)-ymod(i))}{\sum_{i}^{n}xobs(i)}\right]$	< ± 20%	Good
	±20 % to ± 40%	Satisfactory
	> ± 40%	Unsatisfactory
$RSR = \left[\frac{\sqrt{\sum_{i}^{n}(xobs(i)-ymod(i))^2}}{\sqrt{\sum_{i}^{n}\left(xobs(i)-\overline{xobs}\right)^2}}\right]$	0.0 ≤ RSR ≤ 0.5	Very good
	0.0 ≤ RSR ≤ 0.6	Good
	0.0 ≤ RSR ≤ 0.7	Satisfactory
	RSR ≥ 0.70	Unsatisfactory
$R^2 = \left[\frac{\left[\sum_{i}^{n}\left(xobs(i)-\overline{xobs}\right)\left(ymod(i)-\overline{ymod}\right)\right]^2}{\left[\sum_{i}^{n}\left(xobs(i)-\overline{xobs}\right)^2\sum_{i}^{n}\left(ymod(i)-\overline{ymod}\right)^2\right]}\right]$		Satisfactory

Note: xobs = observed flow, ymod = model/simulated flow

Table 2 Summary of Goodness of Fit

	Year	NSE	PBIAS	RSR	R^2
Calibration	2003–2009	0.53	27.83%	0.66	0.81
Validation	2010	0.51	37.24%	0.68	0.68

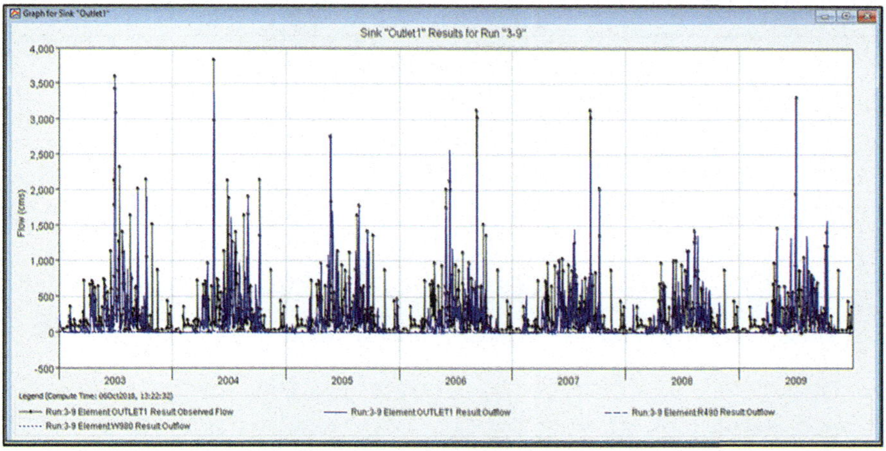

Fig. 3 Calibrated discharge for Laurergarh (2003–2009)

Among different types of estimators NSE, PBias, RSR and R^2 have been used to estimate this model's accuracy. The values were mostly adequate and satisfactory. Figure 5 depicts the simulated value for 22 March to 10 April. It indicates an abrupt rise in discharge. The highest value showed here does not match the exact value of

Table 3 Calibrated values for soil moisture accounting parameters

Sub basin	All
Soil (%)	45
GW1 (%)	25
GW2 (%)	15
Soil maximum infiltration rate (mm/hr)	0.0001
Imperviousness (%)	0
Soil storage capacity (mm)	20
Soil tension storage capacity (mm)	0.01
Soil maximum percolation rate (mm/hr)	0.0001
Groundwater 1 storage capacity (mm)	100
Groundwater 1 max. percolation rate (mm/hr)	0.001
Groundwater 1 storage coefficient (hr)	36
Groundwater 2 storage capacity (mm)	100
Groundwater 2 max. percolation rate (mm/hr)	0
Groundwater 2 storage coefficient (hr)	480

Table 4 Calibrated values for clark unit hydrograph

Sub basin	Time of concentration (hr)	Storage coefficient (hr)
All	12	12

Table 5 Calibrated values for recessions

Initial discharge (m^3/s)	Recession constant	Ratio to peak
5	0.8	0.1

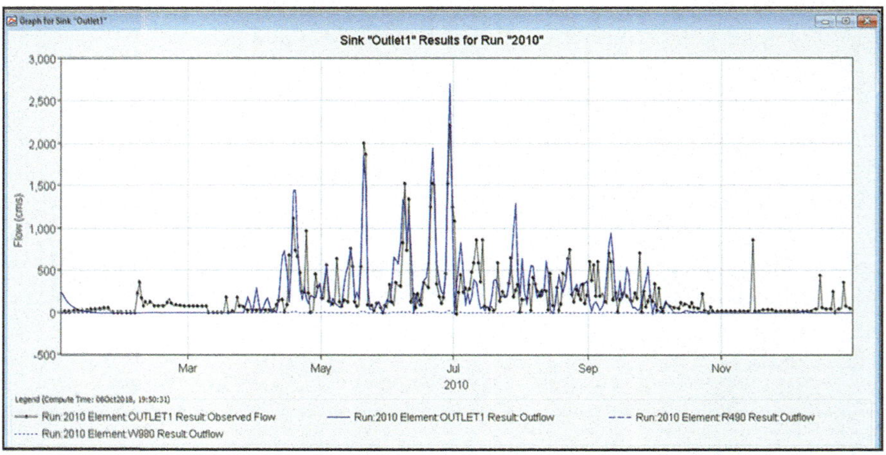

Fig. 4 Validation at Laurergarh for 2010

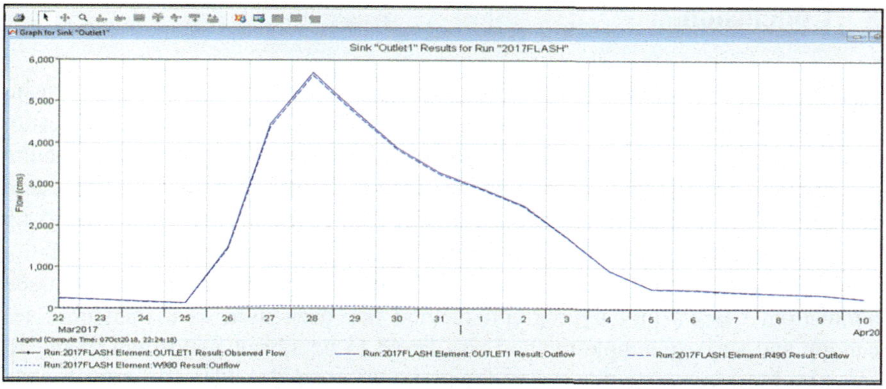

Fig. 5 Discharge for 22nd March to 10th April

the observed one, but it shows the exact indication of a flash flood in a qualitative way.

4 Discussion

Adjusting several parameters of the model setup is required to perform calibration and validation, which ensures quality of the model. The estimation of hydrological model parameters is a challenging task. The reasons for this are the highly non-linear nature of hydrological processes and the fact that different parameter vectors driving models describing the physical processes might have the same effect on the discharge. This means that changes in some parameters might be compensated by others. There are numerous parameters in hydrological models that can be classified as physical parameters. The iterative parameter estimation procedure in HEC-HMS is known as optimization. "Univariate gradient method" was used in to estimate all parameters for sub-basin as an initial trial. Other parameters estimated for calibration are Lag time, Peaking Coefficient, Initial Discharge, Muskingum K. Percent impervious area is kept 0%. Around 19 parameters e.g. (Canopy Storage, Capacity Surface Storage, Capacity Soil Storage Capacity, Soil Tension Storage Capacity, Soil Maximum Infiltration Rate, Groundwater 1 Storage Capacity, Groundwater 1 Max. Percolation Rate, Initial GW2 Storage, Initial Canopy Storage, etc.) have been outnumbered in order to find the most sensitive one. Along with flash flood forecasting, this hydrological model can be utilized for further study of climate change, flood inundation mapping, morphological studies, tracing effects of changes in land-use etc.

5 Conclusion

In this study, a flash flood forecasting model is prepared for Laurergarh by using HEC- HMS (3.5.0) with the collaboration of PERSIANN-Cloud Classification System (PERSIANN-CCS) which is a real-time global high-resolution satellite precipitation product that successfully assessed the flash flood in 2017. The model has been developed using the latest available DEM (Digital Elevation Model) file resolution of 30 m × 30 m to assess the statistical flow for the designated calibration points. For loss method, Soil Moisture Accounting method has been used for model simulation. Clark's unit hydrograph method has been used as a transform. For canopy and surface, simple method was found to be advantageous and finally for routing, Muskingum method was found to be most feasible. By detecting the sensible parameters for the model results in a satisfactory calibration for long term (2003–2009) simulation for Laurergarh point. The NSE values are 0.53 and 0.51 for calibration and validation period which suggests a satisfactory to an adequate range. The correlation coefficient of determination (R^2) values are 0.81 for the calibration and 0.68 for the validation period. The PBIAS and RSR values are found to be 27.83 and 0.66 in the calibration stage and 37.24 and 0.68 in the validation stage. Subsequently, a satisfactory calibration and validation model was simulated for 22 March to 10 April, so that it can show the flashflood which occurred on the 28 March. From this project, it is visible that a high rise of discharge within only 2 days of the 27 and 28 March was present. This sudden rise of discharge occurred which created flash flood in that area on 28 March. So it can be said that the model can detect a high discharge pattern properly even if the magnitude can be not exactly equal to the observed discharge on the field.

Acknowledgement The work presented in this paper entitled "Flash Flood Forecasting of Jadukata River Basin at Laurergarh, Sunamganj from Real-Time Satellite Precipitation by Using HEC-HMS", had been submitted to Department of Water Resources Engineering, BUET as the thesis for degree of Bachelor of Science. This is the outcome of an original work done by the author under the supervision of Dr. Md. Mostafa Ali, Professor, Department of Water Resources Engineering, BUET.

References

Haque, S. & Islam, A. K. M. S. (2017). *Event based flash flood simulation at Sunamganj using HEC-HMS*. 6th International Conference on Water & Flood Management (ICWFM-2017), ISBN: 978-984-34-2207-1.

Hashem, K., & Meer, A. (2016). *South Asia Flash Flood Guidance (SAsiaFFG) project*. Delhi: South Asian Flood Mitigation.

Rahat, S. H. (2016). *Impact assessment of future climate scenario on the upper Meghna river basin using HEC-HMS*. Under graduate Thesis, Department of Water Resouces Engineering, BUET.

Awareness and Attitudes of Sri Lankans About the Value of Potable Water and Water Security

Pavithra Abhayawardana

1 Introduction

1.1 Importance of Water

Water is a vital resource for human beings. However, with rapid population growth and fast-paced development across the world, acquiring safe, potable water is emerging as a major issue for citizens globally. Currently, over one billion people in the world do not have access to safe drinking water (Jayawardana 2012). Thus, water security poses one of the biggest global challenges of the twenty-first century. As populations grow, humanity faces the prospect of uncertain future water supplies both due to increasing demands and adverse changes in climate (Srinivasan et al. 2017). According to Dublin Principle No.1, fresh water is a finite and vulnerable resource, essential to sustain life, development, and the environment (Solanes and Gonzalez-Villarreal). In addition to being a basis for survival, it is also at the core of sustainable development. This further indicates water as being critical for social as well as economic development considering its influence on health, welfare as well as production sectors. Though essential for human life, access to improved water represents a day-to-day struggle for millions, particularly in developing countries. Over the last few decades, identifying the factors accounting for household demand for water quality is evident by the growing body of literature on household water behavior (Jalan et al. 2009). Compared to most countries in the world, Sri Lanka is a water rich country with annual rainfall exceeding 1500 mm per year in average. In a place where water scarcity is not a major threat, availability of the resource is not an issue but rather the way in which the resource is used and managed that poses challenges (Falkenmark 2001).

P. Abhayawardana (✉)
National Water Supply and Drainage Board, Colombo, Sri Lanka

A. Haque, A. I. A. Chowdhury (eds.), *Water, Flood Management and Water Security Under a Changing Climate*, https://doi.org/10.1007/978-3-030-47786-8_15

1.2 Potable Water

Drinking water, also known as potable water or improved drinking water is water safe enough for human consumption by means of drinking and food preparation (Wikipedia 2019). Water supplied for public use must thus be potable. This implies it must meet certain satisfactory standards for drinking purposes from the standpoint of its chemical, physical and biological characteristics. Drinking water should preferably be obtained from a source free from pollution. Raw water directly available from surface water sources is however not generally suitable for drinking purposes. Countries across the globe are doing everything at their disposal for keeping water clean and converting raw water to potable water. Sri Lanka is not an exception and spends over 20,000 million rupees a year for the provision of potable water.

1.3 Water Pollution

Contaminated drinking water is a major health hazard in developing countries and diarrhea is the most common disease associated with it. This is particularly true for the poor and vulnerable with low access to healthcare facilities and services. Wells tend to be polluted with coliforms, algae, heavy metal and other contaminant chemicals at times. When the extraction rate is more, there is a tendency that the contaminants in the neighboring soil are transported into the well with water.

1.4 Water Treatment

Thus, the objective of water treatment is to produce safe and potable drinking water. While Sri Lanka is endowed with ample sources of surface and groundwater, with increasing population growth and environmental pollution, it is not feasible to utilize it as drinking water without any type of purification (Abhayawardana 2013).

1.5 Awareness About Safe Water

Economics literature suggests a positive correlation between households' awareness about health effects arising from availing poor water quality, and the consumption of improved water. Several studies have found that education plays a strong role in determining individual and household demand for water purification (Jalan et al. 2009). As Sri Lankans exhibit higher levels of literacy and access to education compared to other countries in South Asia, there is a demand for better sanitation and the national government has been doing its part to provide the necessary infrastructure by initiating major water supply development projects all over the country. The attitudes and culture of Sri Lankans also play a role towards the country being a nation with better health standards. Household awareness about health and hygiene can be raised through formal education and through education imparted (directly and indirectly) by the media (Jalan et al. 2009). When compared with illiterate headed-households, the more educated headed-households tend to show a higher probability of choosing improved sources of drinking water. Households owning a TV are also likely to adopt improved sources of water. Although, the possession of a radio positively affects household decisions on water usage. The effect is not significant in the equation of demand for improved drinking water sources (Jalan et al. 2009).

A household survey from urban India was used to estimate the effects of awareness proxies (education of adult household members, their exposure to mass media like newspapers, television or radio, and their occupations) on home water purification (Jalan et al. 2009). Also, several studies on the determinants of households' choices of water sources in developing countries (DC) have proved that household education has a strong impact on choice. According to literature, the influence of household education and radio possession on choosing a water source choice was tested (Abebaw et al. 2010). This study suggests that access to education has a strong correlation with a household's water choices. Correlation with access to radio was however found to be negative or statistically insignificant. Awareness is widely proxied by schooling and exposure to mass media in previous studies (Abebaw et al. 2010). The questionnaire yielded several information on household access to water. Households were asked to identify their primary sources of drinking water. Very limited study has been carried out in order to analyse the impact of increased awareness regarding the detrimental effects of using unsafe water on household's demand for water quality. Hence, the aim of this study was to examine the impact of household awareness on the household's choice of adopting improved water sources (Fotue 2013).

1.6 Safe Water Coverage in Sri Lanka

Though the country has achieved 79% coverage of improved water supply according to United Nations Development Programme (UNDP), there are persisting inequalities between rural and urban areas when it comes to water availability and access. In Sri Lanka, domestic water supply is charged according to a two part tariff system where there is a fixed rate of Rs.50 per month and an increasing block rate for the variable part (Thrikawala et al. 2008). According to the data given by Department of Census & Statistics, 40% of the Sri Lankan population have organized water supply facilities and 59.4% is depending on other sources such as wells, tube wells, streams and rivers etc., including 10% on unprotected sources. Sri Lankan Government targets to provide safe drinking water supply for all by 2025 with 60% piped born water supply coverage by 2020 (Sarath Gamini).

1.7 National Water Supply and Drainage Board (NWSDB)

National Water Supply and Drainage Board (NW & DB) is the national institution for provision of potable water in Sri Lanka. The NW & DB has so far been able to provide 43.4% of population with pipe borne water. There are other forms of water supply through rural water schemes run by local governments as well as consumer societies and hand pumps that account for additional 25% of population. As such the total safe drinking water supply coverage in the country is closer to 65% of population.

NW & DB is the semi-government agency responsible for provision of pipe borne water supply for the population and it has a consumer database exceeding 1.5 million at the moment. Every year more than 100,000 new consumers are added to the database and the demand for pipe borne water is ever increasing. This is due to the pollution of surface water sources and also the increase in population who are getting used to living in smaller plots of lands, barring from going for dug wells of their own (Sri Lanka Socio-Economic Data 2014). NW & DB collects raw water from the sources, produces clean water, transmits it to the demand centres, stores and distributes to domestic and various other users. NW & DB is the supplier; and the consumer is the buyer, in business terms. Any business, to be sustainable, needs to have a positive balance sheet and the NW & DB had been providing water free of charge until 1982. The organization now continues to suffer from a negative balance sheet even after the introduction of billing and there is a wide gap between production and selling prices. The production costs presently is over Rs. 150/=, while the average selling price is less than Rs. 25/= for majority of the customers (Sri Lanka Socio-Economic Data 2014). The quality of water produced by the NW & DB meets Sri Lanka Standards (SLS) for drinking water. However the uses of water do not stop there, as a lot of the time, clean water is also used for purposes that can easily be either avoided or satisfied by other water sources, resulting in inefficient use of water resources (Sri Lanka Socio-Economic Data 2014). This mirrors the situation in several other countries in the world where water supply remains a service that is significantly under-priced, especially for residential consumers (Nauges and Berg 2009).

2 Objective of the Study

Sri Lanka is an island nation, characterized by an abundance of water sources with 103 rivers and multiple origins of rainfall. Monsoonal, convectional and depression rain accounts for a major share of the annual rainfall. The mean annual rainfall varies from under 900 mm in the driest parts (south eastern and north-western) to over 5000 mm in the wettest parts (western slopes of the central highlands) (Climate Data 2019). The country has achieved 79% coverage of improved water supply according to UNDP. Therefore, water security issues pertaining to water scarce areas of the world are not common in national context.

However, it is evident that the general public do not possess adequate awareness regarding water security issues currently facing the world. No research has been conducted to assess public awareness on water security and water conservation. No data are available on the level of treatment given to water acquired from sources other than piped water prior to consumption and domestic rain water harvesting. Additionally, with the emerging impacts of climate change and rapid growth of population, water security issues is expected to become increasingly pressing for Sri Lankan citizens in the near future.

Therefore, the objective of this study was to assess the awareness and attitudes of Sri Lankans on the value of potable water and water security. The results would be used to identify the prevalent gaps in public knowledge and awareness on the issue, so that appropriate strategies and interventions can be recommended to the NW & DB for taking necessary actions.

3 Materials and Methods

3.1 Primary Data Collection

Cross sectional primary data were collected in 2018. The primary data utilized in the descriptive and empirical analyses of this study were collected using a structured questionnaire. The collected data included information on the socio-economic characteristics as well as existing water supply situation of the sample households (Mezbo and Ewnetu 2015).

3.2 Selection of Kegalle District for the Study

According to statistical data for the country, obtained through Socio-Economic Data 2014, of Sri Lanka the average economic, educational and social data of Sabaragamuwa Province is the most similar data to average Sri Lankan results (Table 1). Thus, it was assumed that selecting interviewees from Sabaragamuwa Province would give a fair representation of the island nation. Thus, Kegalle district, one of two districts belonging to Sabaragamuwa Province was selected for data collection.

3.3 Questionnaire Survey

A questionnaire survey was conducted among 100 randomly selected households in Kegalle district. The questionnaire was aimed at household heads assuming that they are responsible for paying water bills and communicating their water needs to

Table 1 Socio Economic Data of Sabaragamuwa Province and Sri Lanka (2014)

Item	Sri Lanka	Sabaragamuwa province
Average income, Rs. per month		
Per household	36,451	36,173
Per person	9104	9132
Per income receiver	20,427	19,418
Population distribution by gender, %		
Male	47.4	48.1
Female	52.6	52.6
By educational attainment, %		
No schooling	4.2	5.1
Up to grade 5	25.1	26.1
Grade 6–10	44.6	47.5
Passed G.C.E (O/L)	14.7	11.3
Passed G.C.E. (A/L) and above	11.2	9.9

outside parties. To account for people belonging to different ethnic groups and nationalities in Kegalle district, the questionnaire was prepared both in Sinhalese and English languages. Either of the two forms were administered based on the respondents' preference. The questionnaire was also divided into five sub sections.

The first section (questions 1–15) aimed to solicit general information about the household, to obtain demographic information about the household head and its members, as well as their education and income levels. The second section (questions 16–20) aimed at gathering information about the respondents' awareness regarding piped water (whether they are utilizing piped water from NW & DB). Third section (questions 21–22) inquired about the level and degree of treatment given to the water used by the household. Fourth section (questions 23–25) inquired about the attitude of household members on the use of safe drinking water in case of emergencies. In the fifth and final section (question 26), respondents were asked about their behaviours and habits regarding water usage, which can be used to interpret their values and attitudes towards safe drinking water.

As the proverbial creatures of habit, people tend to repeat the same behaviours in recurring contexts (Thrikawala et al. 2008). Their habits form their lives and the attitudes they have towards water is evident from these practices. Ultimately, whether they save or waste the water can be determined through these habits.

4 Results and Discussion

4.1 General Information About the Households

100 Sample households were included for the analysis of this study. Of these sample households, 46% were headed by women whereas the remaining 54% were male. The age of the surveyed household heads ranged from the minimum of 25 to a

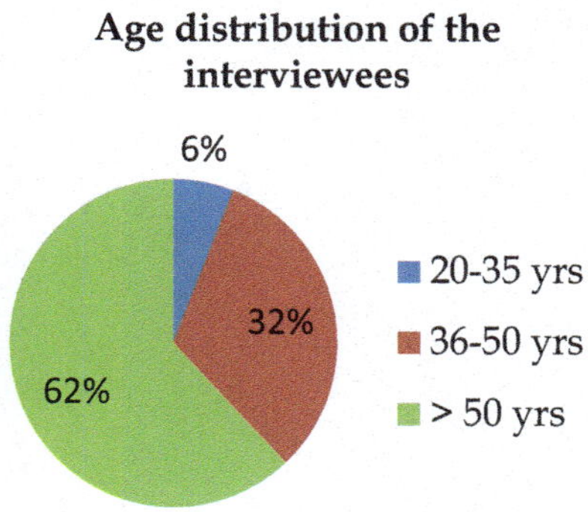

Fig. 1 Age distribution of the interviewees

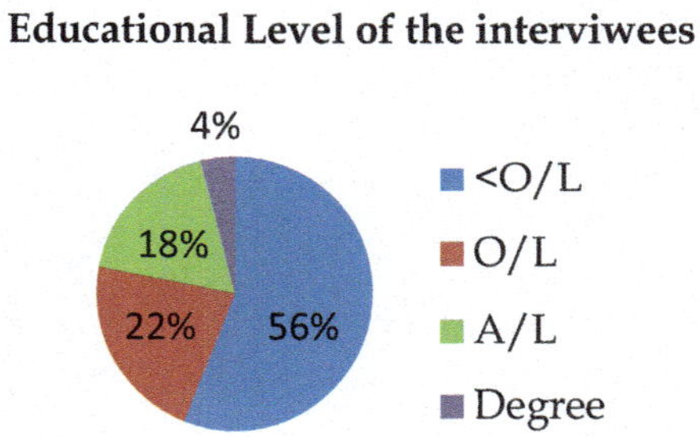

Fig. 2 Variation of educational level

maximum age of 80, with an average age of 38. The age distribution is provided in Fig. 1.

Data about the educational attainment of the surveyed household revealed that 56% of the surveyed households fall under educational category of less than Ordinary Level (O/L) and 22% of the respondents fall under the educational category of Ordinary Level (O/L). 18% have passed Advanced Level (A/L) and 4% have university degrees as indicated in Fig. 2.

Data from the survey indicated that out of the total households, 16% earned monthly income ranging from Rs.10,000 to 20,000, 74% from Rs. 20,000 to 50,000 and 10% above Rs. 50,000 per month. None of the households had a monthly income lower than Rs. 10,000.00. The income level distribution is given in Fig. 3.

34% of the group consumed piped water from the NW & DB, 14% received piped water from community-based organizations (CBOs), 52% acquired drinking water from wells and the remaining 4% used water from rivers or streams as given in Fig. 4.

In recent times, television and radio channels increasingly telecast various programmes to raise awareness among the public on important environmental and social issues, such as scarcity of water in global and local context, the need to preserve and conserve fresh water, climate change and the threats pertaining to it etc. Therefore, it is generally assumed that access to a television or a radio, or both, in the household, helped increase knowledge and understanding among its members regarding water quality, treatment procedures, tariff charges etc. 98% of the survey respondents were found to own televisions, while 92% had both a television and a radio. 98% stated that they used these on a daily basis. People were asked about their

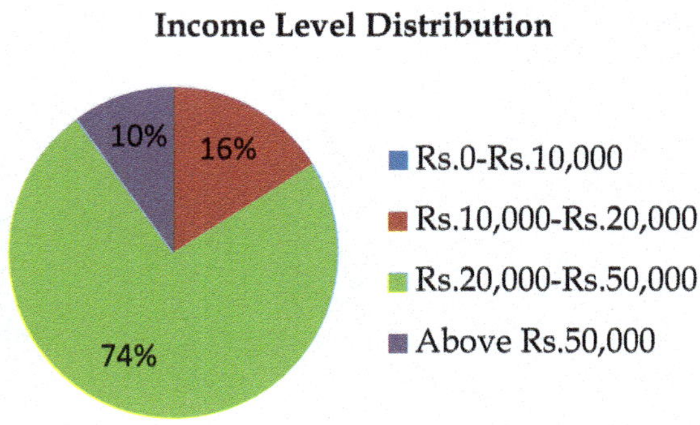

Fig. 3 Income level distribution

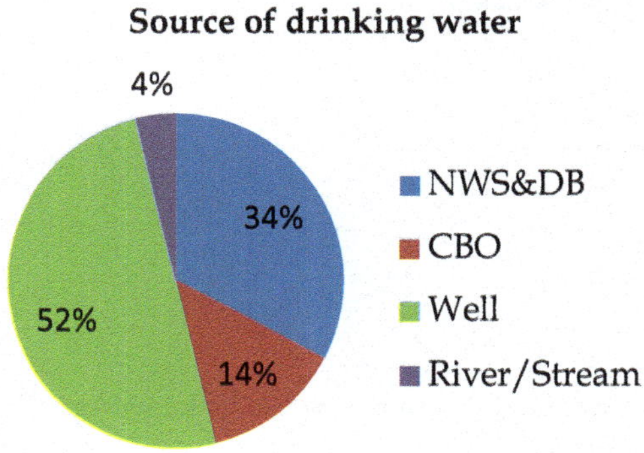

Fig. 4 Source of drinking water of the surveyed households

education and income levels, assuming that higher these factors were, the higher the awareness and ability to consume safe water would be.

4.2 *Awareness of the NW & DB Consumers*

Figure 5 shows the monthly tariff variation of the households who are consuming drinking water from NW & DB. Majority of the consumers (72%) have monthly tariffs varying from Rs.100–1000.

According to the survey, only 1 out of 34 people who consumed piped water from NWSDB was aware of the tariff structure. That implies that almost all the consumers (97%) did not know about the charges which underscore relative ignorance among the public about the utility. There were 10 respondents (28%) who pay a monthly tariff higher than Rs. 1000 per month but appeared to still be unaware of the tariff structure. To be precise, the respondents spend a considerable percentage of their monthly income for utilising potable water, without proper knowledge on how to calculate or identify the surcharges within the payable amount. With no knowledge of the tariff structure, people cannot be expected to come up with ways to reduce or

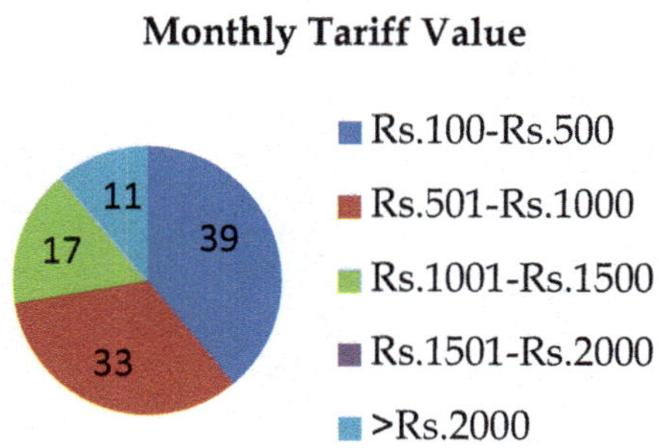

Fig. 5 Monthly tariff variation

minimized their tariffs, which can eventually lead to reduction of wastage of water. If they are made aware of the tariff structure, people will be more inclined to reduce the consumption, and in turn reduce their due monthly tariff.

Only 2% of the respondents were found to be aware of the amount borne by NW & DB to purify the water for drinking quality. Out of the others, 66% of them assumed it to be around Rs.1–20, 24% assumed it to be around Rs.21–50 and the other 10% assumed it to be around Rs.51–100 as indicated in Fig. 6. This indicates the low value people give to the commodity as well as their lack of knowledge. The cost to NW & DB for a unit (1 meter cube) of water is around Rs. 160 and it is provided to the public at a lesser rate.

If people were aware of the money spent on water, it can be assumed that they will pay more attention to curb the wastage of water. 15% of the people who consumed water from NW & DB were not aware of the original source of their water. 39% of the respondents were not aware of what is meant by a "unit" of water. As the unit used by NW & DB is a meter cube, 41% misinterpret to be one litre, while another 20% consider it to be 1000 gallons. This further highlights their lack of knowledge about the utility.

4.3 Using of Potable Water for Other Activities

89% of the people with piped water from NWSDB used it for washing purposes, while 100% used it for bathing, 61% used it to water gardens and 11% even used it to bathe pets as presented in Fig. 7.

4.4 Treatment of Water

Traditionally, Sri Lankans inherit the habit of treating their drinking water and this is still evident among the population regardless of educational, economical and spatial disparities. Among the people who consumed water from wells, streams and rivers,

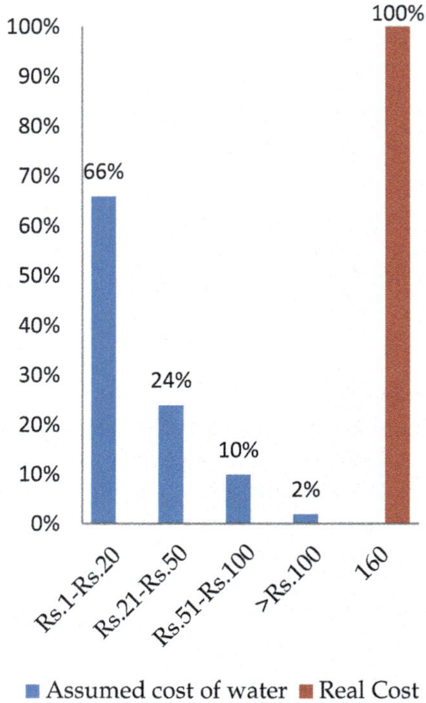

Fig. 6 Reals cost Vs. Assumed cost of water

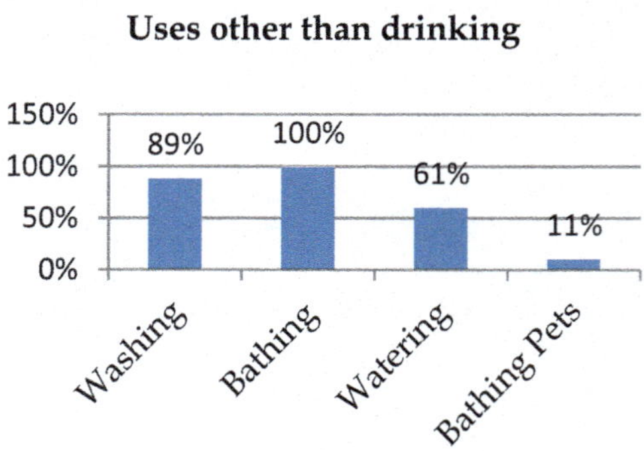

Fig. 7 Uses other than drinking water

79% treated the water before consuming it. Only 21% consumed the water as it is without subjecting it to some form of treatment. Figure 8 shows the levels of treatment of water before consuming.

However, ignorance of people about present issues regarding deterioration of water quality by addition of various pollutants is evident from the survey. 97% of the people who use water from wells, streams and rivers have not tested the water from a laboratory to assess its suitability for drinking. Though people are aware that the water must be purified, their knowledge seems to be out of date. The water pollutants include heavy metals, nitrates, phosphates etc. which can be harmful for humans.

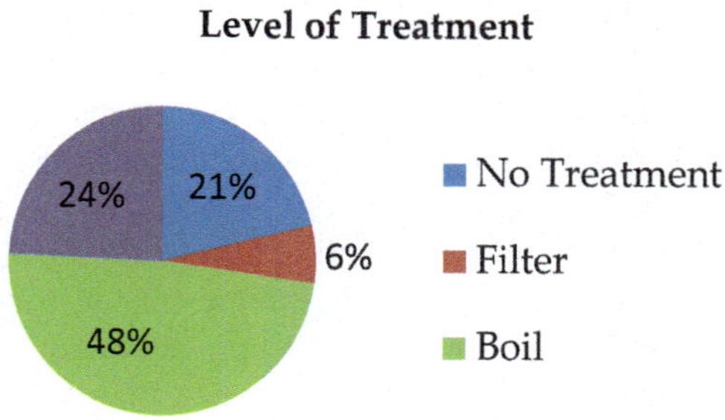

Fig. 8 Levels of treatment of water

4.5 *Habits of People Regarding Water*

Only 56% of the surveyed population took a clean bottle of water when they were going out of their houses. However, this habit appeared to be more common among females, as 88% of the females were found to carry water while only 17% of the men did this. This implies that women are more concerned about consuming safe drinking water. According to the survey, 12% of the population continuously keep the water flowing (tap on) while they wash their faces and brush their teeth. 66% of the people bathe using a shower which results in a lot of wasted water, a practice that can be replaced by using filled basins or tubs to bathe. 42% of the surveyed group stated that they filled a cup with water when they were thirsty, drank some and threw away the rest, resulting in further wastage. Only 58% said that they informed the NW & DB when they observed a leakage on the road. Even among the NW & DB consumers, only 64% would do this regularly. 20% stated that it was normal to have leakages in their homes. Only 6% reported that they practiced storing and harvesting rainwater for various uses. While Kegalle receives rainfall all throughout the year with average precipitation of 2493 mm, it is surprising to see that people do not capitalize on this abundance by more frequently practising rainwater harvesting (Climate Date 2019). Rainwater harvesting describes the small-scale concentration, collection, storage and use of rainwater runoff for production purposes. Domestic rainwater harvesting is one of the broad categories of rainwater harvesting where water is collected from rooftops, courtyards and similar compacted or treated surfaces, stored in underground tanks or aboveground tanks and used for domestic purposes, garden watering and small-scale productive activities (Kahinda et al. 2007).

4.6 *Availability of Water in Case of an Emergency*

The survey also inquired about the availability of water in case of an occurrence of a disaster. In the recent floods, many people were trapped in their houses without safe drinking water as the wells were inundated and the water supply was abandoned. Kegalle district also recently experienced disasters in many places in the form of landslides and floods. The recent explosion which took place at the Salawa Army Camp was also a reminder for local people to carry water when they go to safety in case of a sudden disaster. However, even in the aftermath of such disastrous events, people do not seem to be aware of the importance of keeping sufficient potable water ready for emergency or crisis situations.

Only 48% of the surveyed population had drinking water stored at their houses. Unlike congested urban settlements in districts like Colombo, most of the people in Kegalle district have a sufficient plot of land around their houses. However, they do not seem to utilize the space to store a barrel or can of water. Many families had children, who are vulnerable to water borne illnesses due to lack of safe water. However, even the adults of these households did not recognize storing water as a necessity. Only 62% stated that they included a water bottle in the things they would take in case of an emergency.

4.7 Awareness About Water

From the results of the questionnaire survey, it is very clear that there is highly unsatisfactory levels of awareness among the people regarding the different aspects of water use such as safe drinking water, the cost involved with water production, tariff structure etc. This is of concern, as lack of knowledge regarding the value of water, leads to ignorance regarding the need to conserve it or reduce the wastage. Majority of these people (98%) use the television or the radio on a daily basis, thus it can be stated that the media's contribution in making people aware about these issues is not sufficient. As the impacts of climate change takes precedence, Sri Lanka should also expect more occurrences of extreme scenarios like floods and droughts, which can incur both short and long-term impacts on water security. Therefore, it is essential that public are provided with basic knowledge on the value of water.

5 Conclusions

Issues relating to water security are emerging as a global threat. Lack of awareness regarding inefficient use of water resources and its consequences appeared to be particularly prevalent among Sri Lankan citizens.

In Kegalle district people obtain drinking water from wells, piped water from NW & DB and CBOs, and nearby rivers. However, their basic knowledge about the issues related to water is very poor and ill informed. Majority of the consumers of NW & DB do not have an idea about the costs involved, the tariff structure and the ways in which water could be used more efficiently and sustainably.

No disparities in terms of awareness and attitudes were found across gender, income levels or education levels. Most people practice habits which lead to wastage of water, and ones that can replaced with conservation practices. The school curriculums should also bear the responsibility for the lack of knowledge and for the ignorance about this vital resource. NW & DB must also be responsible for the lack of knowledge among their consumers and they should take steps to provide awareness considering it as a priority. The results represent the attitude of Sri Lankans about water and it shows that they still do not have adequate knowledge that water is an invaluable resource that should be not wasted. It can be concluded that the public has inadequate knowledge on water security and immediate actions should be taken to improve awareness. Spending a massive amount of national expenditure on providing water supply itself is not sufficient to eradicate water insecurity. With the rapid population growth and emerging impacts of climate change, water security issues will be major challenges for all nations in the coming future and practical initiatives should be taken up by relevant authorities to raise awareness among public to in order to successfully address these challenges.

6 Suggestions

Drawing from the above conclusions, the following recommendations can be considered to improve the awareness among the public about water use.

- Awareness should be improved among the public. Intensifying campaigns of sensitization of the population through mass media campaigns about the necessity of drinking potable water should be implemented (Handbook for Water Consumers).
- It is suggested that NW & DB distribute a leaflet including the tariff structure; the amount spent to produce one a meter cube of water etc. in general terms which will give the public an idea about the process of conversion of raw water to potable water and will allow them to alter their attitude towards the value of water. Alternatively, this data can be included in the monthly bill, which is certain to reach the consumers once a month.
- Water conservation measures are to be introduced to the school curriculum in order to promote and foster a culture of water conservation culture in the country (Nauges and Berg 2009).
- A range of visual aids/posters/items should be introduced to remind people to save water. (E.g. the Australian government has distributed free sand hour glasses for the bathrooms to help people count the minimum time needed for a shower, which lead to conserving a huge amount of water)
- Reducing leakages of water should be promoted. It is mainly due to poor quality of water fittings, and substandard plumbing systems. Regulatory measures are required in order to procure quality water fittings and installation services (Nauges and Berg 2009).
- Rainwater harvesting must be promoted as an alternative source of water supply. Although rainwater is abundantly available for a large area of the island for several months of the year, storage of rainwater is not practiced. Technical and financial assistance (in form of loans or installments) should be granted through government agencies to people for construction of rainwater collection tanks.

- Incentives such as reduction of bills or tariffs by a certain percentage should be applied to NW & DB consumers who utilize less water than the standard or average monthly amount.
- Media should also be reminded of their duty to make the nation more aware of important social and environmental facts and issues.

References

Abebaw, D., Tadesse, T., & Mogues, T. (2010). *Access to improved water sources and satisfaction with services evidence from rural Ethiopia*, International Food Policy Research Institute, Sustainable Solutions for Hunger and Poverty.

Abhayawardana, G. P. R. (2013). Suggested improvements on water treatment plants for sustainable development. *Engineer, XXXXVI*(03), 69–73.

Falkenmark, M. (2001). The greatest water problem: The inability to link environmental security, water security and food security. *Water Resources Development, 17*(4), 539–554.

Fotue, L.A.T. (2013). Awareness and the demand for improved drinking water source in Cameroon. *International Journal of Economic Practices and Theories*, 3(1), e-ISSN 2247–7225.

Handbook for Water Consumers (n.d.). First edition, Ministry of Water Supply and Drainage, National Water Supply and Drainage Board.

Jalan, J., Somanathan, E., & Chaudhuri, S. (2009). Awareness and the demand for environmental quality: Survey evidence on drinking water in urban India. *Environment and Development Economics, 14*(06), 665–692.

Jayawardana, A.W. (2012). *Challenges for sustainable water management*, ACESP 2012, Faculty of Engineering, University of Ruhuna, Sri Lanka.

Kahinda, J. M., Taigbenu, A. E., & Boroto, J. R. (2007). Domestic rainwater harvesting to improve water supply in rural South Africa. *Physics and Chemistry of the Earth, 32*, 1050–1057.

Mezbo, G. K., & Ewnetu, Z. (2015). Households willingness to pay for improved water services in urban areas: A case study from Neblet town, Ethiopia. *Journal of Development and Agricultural Economics, 7*(1), 12–19.

Nauges, C., & Berg, C. V. D. (2009). Demand for piped and non-piped water supply services: Evidence from Southwest Sri Lanka. *Environmental and Resource Economics, 42*, 535–549.

Sarath Gamini, P.H. (n.d.). *Challenges in the water sector and waste water sector.*

Solanes, M., Gonzalez-Villarreal, F. (n.d.). *The Dublin principles for water as reflected in a comparative assessment of institutional and legal arrangements for integrated water resources management*, TAC Background Papers No. 3.

Sri Lanka Socio-Economic Data. (2014), Central Bank of Sri Lanka, Vol. XXXVII.

Srinivasan, V., Konar, M., & Sivapalan, M. (2017). A dynamic framework for water security. *Water Security, 1*, 12–20.

Thrikawala, S., Gunaratne, L. H. P., & Gunawardena, E. R. N. (2008). Impact of different tariff structure on residential water demand: A case study from Kandy, Sri Lanka. *Tropical Agricultural Research, 20*, 60–72.

Distribution, Access and Gendered Roles of Common Property Water Resources in Bhotechaur, Nepal

Poonam Pandey, Mafruha Akter, M. M. G. S. Dilini, K. Deepika Laxme, Carolin Arul, Robert Dongol, and Sunil Tiwari

1 Introduction

Common Property Resources (CPR) are natural resources owned and managed collectively by a community or society (OECD 2001). CPR includes natural resources such as water, forests, pasture land, irrigation system, fishing grounds and the atmosphere. Distribution and access of these resources varies both spatially and temporally. In addition, these resources have been exploited, polluted and converted into other uses, unsustainably, in most countries. Increasing population, social construction; rapid development has created stresses on these resources. Water is one of the most essential resource for human existence. Naturally, the area with higher water distribution is occupied by the largest percentage of the world's population triggering serious issues for human survival.

The sustainability of freshwater supply is threatened because of the widespread depletion of groundwater, surface water pollution, and climate change impacts (IPCC 2007; Gleeson et al. 2012; WWAP 2012). Due to the declining level of freshwater in many parts of the world, a global water crisis would possibly occur in

P. Pandey (✉) · R. Dongol · S. Tiwari
Nepal Engineering College, Bhaktapur, Nepal
e-mail: robert@nec.edu.np

M. Akter
Institute of Water and Flood Management (IWFM), Bangladesh University of Engineering and Technology (BUET), Dhaka, Bangladesh

M. M. G. S. Dilini
Board of Study of Agricultural Engineering, Postgraduate Institute of Agriculture, University of Peradeniya, Peradeniya, Sri Lanka

K. Deepika Laxme · C. Arul
Centre for Water Resources, College of Engineering, Chennai, Tamil Nadu, India

A. Haque, A. I. A. Chowdhury (eds.), *Water, Flood Management and Water Security Under a Changing Climate*, https://doi.org/10.1007/978-3-030-47786-8_16

the near future if appropriate water conservation and adaptation measures are not in place. Springs are the primary source of domestic water supply for rural communities in the Hindu Kush-Himalayan (HKH) region; when they dry up or decline the resulting water shortages become major environmental threats (Negi and Josh 2002; Merz et al. 2003; Vaidya 2015). Tamb et al. (2012) reported that over the past decade there was a perceived decline in dry-period spring discharge by 48% in drought-prone areas and 35% in other areas of the Sikkim Himalaya. This apparently resulted in a serious water shortage during the dry period. Agarwal et al. (2012) found that the hill community in the Danda watershed of Uttarakhand (India) was getting just one third of its total domestic water needs. This widespread water scarcity has adversely affected public and ecological health, agricultural production, and livestock populations in the entire Himalayan region (ICIMOD 2015).

Almost 80% of the 13 million people residing in the hills and mountain areas of Nepal rely on springs as their primary source of water (CBS 2012; Tamb et al. 2012; Sharma et al. 2016). In 2015/16, less than half of the population used piped water supply (46.8%). In urban areas, 48.9% were using piped water supply compared to 45.4% in rural centers. Over double the households of the richest quintile use piped water compared to the households of the poorest quintile (63.4% richest and 28.5% poorest). More than one third (38.2%) are using hand pump/ tube wells that decline with increasing quintiles. Spring water is used by 6.4%, open well by 4.5%, covered well by 2.2% and other sources by 1.7% as a source of drinking water.

Competition for access and control of water increases the water conflicts, which indirectly affects livelihood of the people in the region. Existing water inequities are further aggravated by persisting disparities in gender, class, caste, location and power politics (Goodrich et al. 2017). The socio-economic characteristics, community's capability, incentive and sense of ownership are important components of understanding the resource utilization and management. In addition to these natural jurisdictions, in a country like Nepal, these are even more restricted by the local customs, values and norms. Access, distribution and gender roles are major factors in managing water resources for social harmony. These aforementioned attributes are shaped by social, cultural and economic conditions of a society. As such, this study tried to both analyze the accessibility of water in terms of its distribution, collection, use, and understand the gendered roles in the current socio-economic context.

2 Study Area and Methodology

2.1 Study Area Selection

Bhotechaur area of Melamchi Municipality lies in Sindhupalchowk District of Province 3 of the Federal Republic of Nepal (Fig. 1). Geographically, it is located at 27.80°N and 85.51°E. According to the municipality, Bhotechaur area is occupied

Fig. 1 Study area

by 800 households, out of which 60 households were chosen as surveyed households. Most of people in the area are Hindu in religion. The entire area is dominated by the households of Chaulagain caste and one household belongs to Tamang caste.

Based on the water management issues and for ease of study, Bhotechaur area was divided into two belts, the upper and lower belt. Out of the 60 selected households, the lower belt consists of 12 households, while the other 48 households were from the upper belt.

2.2 *Data Collection*

Various data about CPR for instance, water collection; use and access, were assembled through the individual household's interview. Apart from insight on water managerial issues, general information about the village were collected from the ward office.

3 Methods

The study has adopted different research tools based on qualitative and quantitative approaches. The various Participatory Rural Appraisal (PRA) tools like Key Person Interviews (KPIs), Timeline surveys; Resource Mapping and Time Use Survey were adopted to generate the required qualitative and quantitative data.

Respondents were identified purposively to represent the varying contexts of society, in terms of gender, caste and economic strata focusing on water accessibility, distribution and management. Time use survey was categorized as economic and non-economic activities and was analyzed using Statistical Package for Social Sciences (SPSS) software through frequency analysis. Further, economic activities were divided into System of National Account (SNA) and Non System of National Account (Non-SNA) activities.

Key Personnel Interviews (KPIs) were conducted with four persons, including one male representative from the ward and one male representative from the local police station and two female representatives. Timeline survey was conducted with one male and two female participants, who are different from those interviewed during KPIs. Resource mapping was done with two male and two female representatives. Quantitative data requirements were facilitated through the time use survey which has been conducted with 17 individual households. During which, work was divided into major and secondary activities. The people who accompanied during those work were also included in the time use survey. The hourly activities of a day were noted during the survey. SPSS software has worked for analyzing the time use survey data. Each and every activity was coded and re-coded in the following way (Fig. 2).

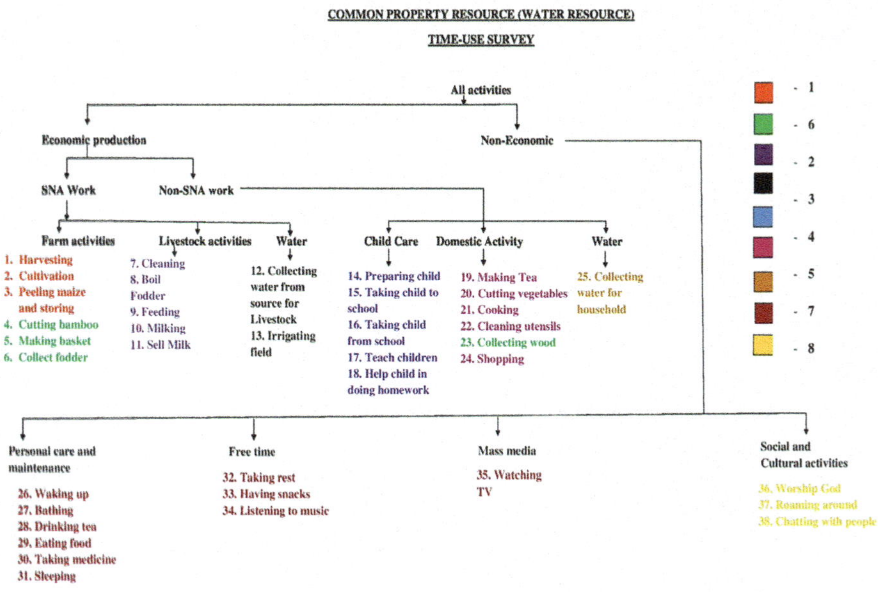

Fig. 2 Time use survey

Table 1 Timeline survey

Year	Temporal changes of water accessibility, collect-use and management
1940	Used farmer managed canal, because there were adequate numbers of springs. Used to walk half an hour to collect water for domestic purpose. Less numbers of houses available as many lived in joint families. Had alternative options for water supply including springs and streams for extreme conditions like drought, drying out of springs etc.
1965	Water collected from the canal. Bamboo pipes were connected to canal to tap water. High water consumption because of being a joint family, larger agricultural field and huge number of livestock. Water accessibility issues due to block and breach in the canal. Social conflicts existed when sharing water with lower caste people.
1980	Pipe system including reservoir tanks came into existence. Formation of nuclear families. Conversion of agricultural lands into settlements. Some of the springs were damaged due to settlement expansion.
1988	Pipeline system came into existence with the support of an INGO-Action Aid.
1993	The project of laying pipelines was handed over to the community.
2018	Direct access to water near home reduced burden for females but they lack structures/ systems to store them. Water consumption reduced as more people migrate in search of better lifestyle, education, job etc. In some places, tap installed lacked the pipeline connections. Political influence was blamed for the issue. Water users committee was formed but it was not active.

The first legend includes agricultural activities which directly yields economic benefits while the sixth legend represents non-water related benefit yielding works. All the livestock activities under SNA work is indicated by the second legend. Economic activities involving water resources are catered by the third legend. The activities which partially yield economic benefits are grouped under Non-SNA work which is inclusive of child care, domestic activity and collecting water for the household. Similarly, the activities which do not contribute to economic benefits are categorized under Non-economic activities (Table 1).

The timeline survey tool was used to figure out the gradual changes that occurred over years, in terms of water usage in the study area. The collective information was obtained from two females and one male respondent. Among the two females, one belonged to the Tamang community, which is considered to be of a lower caste in that area. It was found that, they had a canal system earlier; only during 1980s people got pipeline systems. Though it reduced the physical burden of fetching water from a longer distance, it does not impart any changes in social indifferences. Presence of the water user committee did not make an efficient effort to resolve water issues. This is due to the fact that the water user committee was controlled by upper caste and more influential people.

3.1 *Water (CPR) Resources of the Study Area*

The study area has plenty of springs and streams for water requirements. The area was divided as upper belt and lower belt based on the water management. The upper belt population depends on three major springs from where the water is being diverted and collected in a 10,000 liters storage tank. The collected water is distributed to the household through pipeline system. While, the lower belt population depends on spring source during wet season and stream source during dry season which is located far from the area referring to one of the key interviewer. There is no arrangement for storage tank to store water. The lower belt has been facing severe water scarcity as the pipeline system become dysfunctional because of the frequent landslides and clogging issue. In dry season, the sources get dried which aggravates the difficulties of the lower belt people.

4 Results and Discussions

Springs and streams are common water sources in the study area. In the past, local people used perennial spring sources for their water, which was transported to their settlements through a farmer-managed canal. Nowadays, the installation of pipes and taps are laid down to divert the water from a reservoir, which had made their quest for water easier.

The statement "No water issue", which was derived from most of the interviewed households from the upper belt was contradicted by the people from the lower belt of the area. A lady of age 50–55 was expressing her views on water availability through the following quote "We don't get water when we need it, but when we don't need water we have more than necessary". With this statement, it was clear that there are issues related to water resources in the area, therefore more in-depth enquiries were made with the people from the lower and upper belts of the area.

4.1 *Water Distribution and Access*

4.1.1 Socio-Economic Characteristics

Bhotechaur area is dominated by the hill Brahmins (Upper Caste) (51.7%) and remaining by Chettri and Tamang caste (Lower Caste) group. The upper belt area is inhabited by 48 households including two houses from lower caste (Tamang family) and the lower belt is inhabited by 12 houses (Upper caste). The livelihood of the people living in this area is mostly based on agricultural production (63.28%), in addition to this many of the people migrated to the nearest city Kathmandu for salaried job (13.57%), business (12.10 %) and others as (11.05%) (Census 2011).

The upper belt is economically more stable than the lower belt, except two lower caste households most of the families in the upper belt are dependent on both agricultural production as well as other activities including business and services for economic generation. Lower caste households have a small amount of land for agricultural production, they are mostly working in other peoples' land as labors and their access to water is controlled by the upper caste people. However, the lower belt people are mostly dependent on agricultural production.

The existence of discrimination between the upper caste and lower caste has been obtained from the timeline survey in sharing water resources. The survey was done with one upper caste women, one lower caste man and one lower caste woman. The upper caste women had difficulty in sharing water from the common canal before the pipeline came into existence. Now, due to the pipe system, they do not have to share the water from the common medium with the lower caste people, which according to the upper caste woman 'is like heaven now'. Similarly, referring to the conversation with lower caste respondents, they defined the water issues differently. Initially, they had to travel longer distance to collect the water but nowadays, the pipe system has reduced their burden. But they still have restrictions for water accessibility from the upper caste people.

4.1.2 Management (Access and Distribution)

The upper belt people are using water from four spring sources and collecting it into a reservoir and the lower belt people are using water from a spring and a stream. Water is delivered to the households through community water taps. One tap serves 6–7 houses and the taps are strategically located at various places. Though the piped water supply has brought water easily at their doorsteps, the study found disparities in water distribution in the study area. The disparities were observed between the rich and the poor, influential and non-influential, upper and lower cast and upper and lower settlement people within the community on water access and distribution.

The study found that the rich upper caste and influential local people are enjoying more privileges and have been using water from the common water taps in addition to private tanks, where excess rainwater is harvested during monsoon when the area receives maximum amount of rainfall. The rich people can transport water directly from the reservoir through a private piped connection for their household. The rich upper caste people revealed that there are no disparities between the local people but when triangulated with the lower caste people it was found that the water sources, accessibility and water distribution are still controlled by those people in the community.

The disparity was also observed between upper and lower settlements. Each household of the lower belt had paid NRs. 1200 at the time of pipeline installation. Unfortunately, the inhabitants of the upper settlements control the flow of water to the people of the lower settlements. The people from the lower settlements drive water from a river that flows through the forests through a temporary diversion. The temporary diversion is also a problem during monsoon where the discharge at the river is too huge, which will either displace or clog the pipeline. In such times, the local male from each of the house in the lower belt go to the source and remove or keep the pipeline intact. In the absence of male members from the house, female counterpart participates in the activity. There are few households in the lower belt headed by female members, for such households, it is very difficult for them to manage this but they do not have any other than going. These people have also approached the ward office for the construction of a reservoir tank at the water source, but their requests were frequently put down by the influential local people from the study area.

4.2 Gendered Roles (Access, Use and Management)

The gendered roles on water use and management was plotted based on economic activities as SNA and Non SNA and non-economic activities with time in a day versus frequency/number of persons. The results show that women frequently access the water source over a day's time in comparison to men. As indicated in Fig. 3, it is clear that contribution of men is more during the morning time due to women's engagement in household work.

According to Fig. 4, it is apparent that household work and water collection is predominantly done by women. Women are distinctively contributing both to household and farming activities; though they are not officially addressed as direct contributors to the System of National Account (SNA) work.

Similarly, under non-economic activities in Fig. 5, the involvement of men is more when compared to women as men have more spare time to engage in non-economic activities in comparison to women in a daily basis.

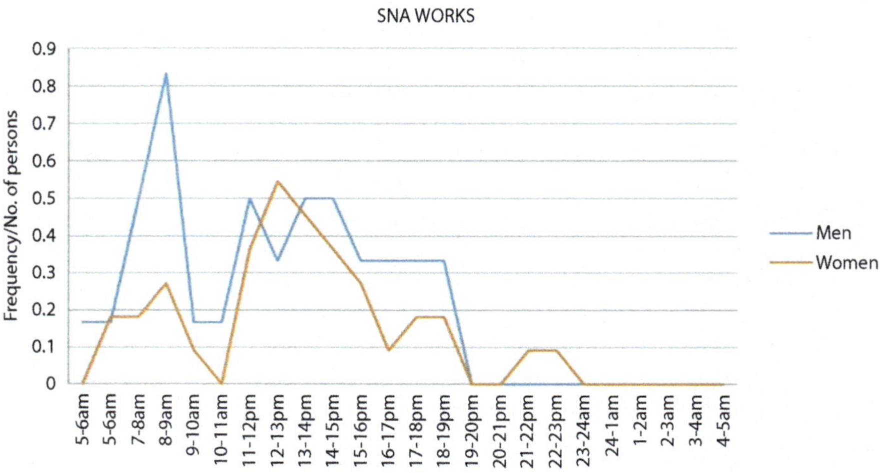

Fig. 3 SNA works

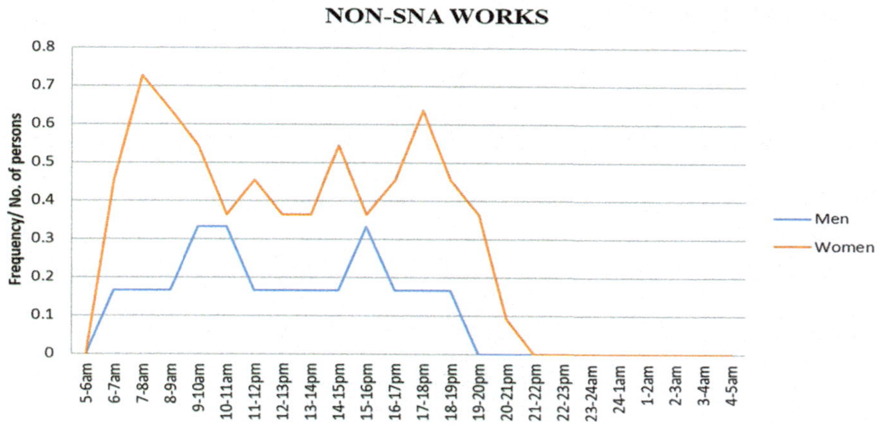

Fig. 4 Non-SNA works

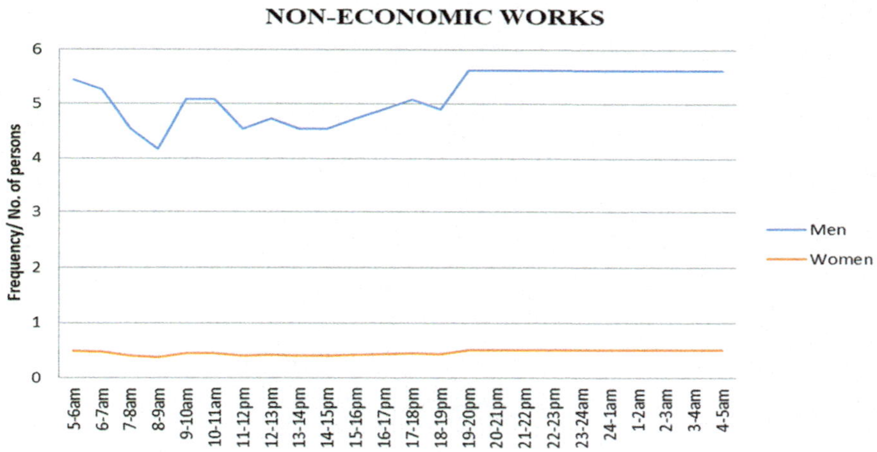

Fig. 5 Non-economic works

Similarly, the area has established a Water User Committee (WUA) to manage the water supply system. WUA comprises of two women members out of seven members. The involvement of women in decision making through WUA is minimal. Most of the women send their husbands or sons or father in law to attend the meetings time. One of the women hold the secretary position in the committee. While some are specially informed about the decisions made by the committee, others are called for simply signing the papers. The small maintenance work is carried out by the WUA members especially men.

4.2.1 Resource Mapping

A resource map was made with participation of the local people to understand the overall information on water use activities. Female respondents addressed their problems related to water when drawing the maps, where they mentioned time to fetch the water. While the men gave the overall ideas/perception about the village.

The number of respondents contributing to the survey was limited and only the present situation was considered. Therefore, it might be possible to imply that women are also involved in irrigation work during peak agriculture. Thus if possible, a study is needed in the future concerning all seasons and activities related to water to further figure out the real situation in this context (Fig. 6).

a

Drawn by a male respondent

b

Drawn by a female respondent

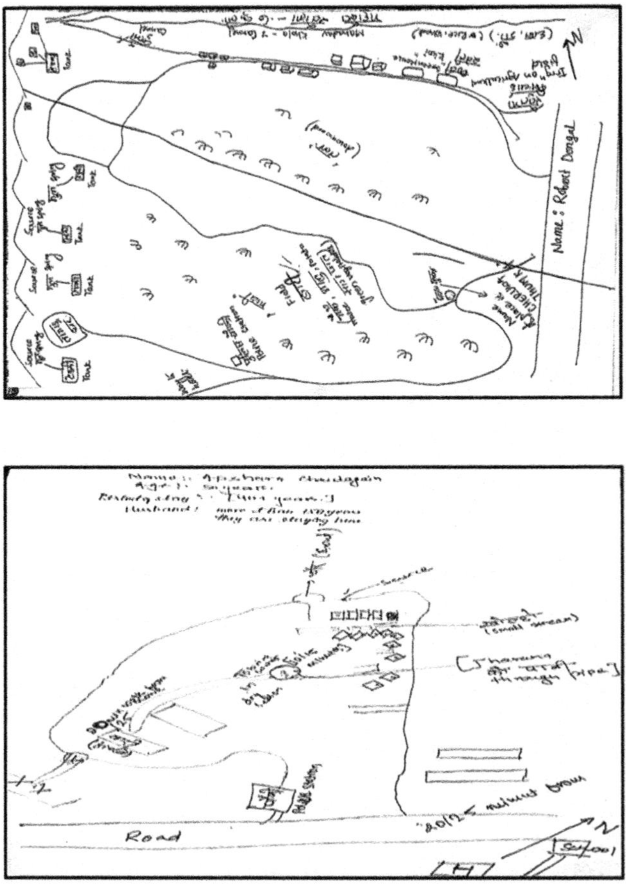

Fig. 6 Resource map

5 Conclusion

From the above study it is apparent that, although there are plenty of water resources surrounding the study area, local people still face various water related problems. This study gives a clear view on depletion of Common Property Resources (water), with the comparison of earlier years ranging back to the past 20–30 years. The upper belt is mostly dependent on pipeline systems while the lower belt uses the stream water to meet their needs. The pipeline system is governed by the male of the upper belt, which eventually creates disparity for people living in the lower portion. Water collection is more difficult for the lower belt women as they are fetching water in most cases. On the other hand, the women of the upper belt can have water in the house through the pipeline system. Within the study area, lower caste people and women of lower caste are severe victims of those water related issues. Women of both caste and belt are working more than the male in terms of working hours, but their input is not being appreciated in the GDP. However, the male members are working in the field more and are dealing with irrigation water which has great influence on SNA works and gives them more recognition for their involvement in economic growth. In this mountainous study area, while the women are the main collectors and users of water, it is still just the men who are considered for the water management committee. The Water Users Committee has reserved a few positions for the female members and the list contains some female names. But the interesting thing is that, some female members are totally unaware of this and some can only remember that they have only been summoned when the official papers needed to be signed. Consequently, it can be summarized that women are the ones who are majorly involved in collecting and using water but, their voices are still left unheard by the male dominated water management authority (Gleeson et al. 2012).

Acknowledgement We would like to acknowledge our indebtedness and render our warmest thanks to Dr. Sucharita Sen, former Executive Director, SaciWATERs and Dr. Vishal Narain, Professor, School of Public Policy who made this work possible. Their friendly guidance and expert advice have been invaluable throughout all stages of the work. We convey our sincere thanks to International Development Research Centre (IDRC), CANADA and the South Asia Consortium for Interdisciplinary Water Resources Studies (SaciWaters), Hyderabad for coordinating and nourishing us with valuable inputs to improve our research work.

References

Agarwal, A., Bhatnaga, N., Nema, R., & Agrawal, N. (2012). Rainfall dependence of springs in the Midwestern Himalayan Hills of Uttarakhand. *Mountain Research and Development, 32*, 446–455. https://doi.org/10.1659/MRD-JOURNAL-D-12-00054.1.

CBS. (2012). *National Population and housing census 2011*. Kathmandu, Nepal: Central Bureau of Statistics.

Gleeson, T., Wada, Y., Bierkens, M. F., & van Bee, L. P. (2012). *Water balance of global aquifers revealed by groundwater footprint*.

Goodrich, C. G., Udas, P. B., Gurung, M. B., Shrestha, N. T., & Shrestha, S. (2017). *Gender and social inclusion in local water planning: Lessons from water use master plan practices in Nepal. ICIMOD working paper 2017/16*. Kathmandu: ICIMOD.

ICIMOD. (2015). *Annual report*.

IPCC. (2007). In M. L. Parry, O. F. Canziani, J. P. Palutikof, P. J. van der Linden, & C. E. Hanson (Eds.), *Climate change: Impacts, adaptation and vulnerability. Contribution of working group II to the fourth assessment report of the intergovernmental panel on climate change* (p. 976). Cambridge, UK: Cambridge University Press.

Merz, J., Nakarmi, G., & Weingartner, R. (2003). Potential solutions to water scarcity in the rural watersheds of Nepal's Middle Mountains. *Mountain Research and Development, 23*(1), 14–18.

Negi, G. C., & Josh, V. (2002, February 1). Drinking water issues and development of spring sanctuaries in a mountain watershed in the Indian Himalaya. *Mountain Research and Development, 22*(1), 29–31.

OECD. (2001). *Common property resources (environmental)*. New York: Glossary of Environment Statistics.

Sharma B, Nepal S, Gyawali D, Pokharel GS, Wahid S, Mukherji A, et al. (2016). *Springs, storage towers, and water conservation in the Midhills of Nepal*. ICIMOD.

Tamb, S., Kharel, G., Arrawatia, M. L., Kulkarni, H., Mahamuni, K., & Ganeriwala, A. K. (2012). Reviving dying springs: Climate change adaptation experiments from the Sikkim Himalaya. *Mountain Research and Development, 32*(1), 62–72.

Vaidya, R. A. (2015). Governance and management of local water storage in the Hindu Kush Himalayas. *International Journal of Water Resources Development*, 253–268.

WWAP. (2012). *Managing water under uncertainty and risk*.

Gender Vulnerability Assessment due to Flood in Northern Part of Bangladesh (A Case Study on 2017 Flood)

Rabeya Sultana Leya, Debanjali Saha, Sujit Kumar Bala, and Hamidul Huq

1 Introduction

Bangladesh is a deltaic country situated within the Ganges-Brahmaputra-Meghna (GBM) basin. The basin is drained up by the three rivers, along with its 300 tributaries and distributaries which cover only 7% watershed area of about 1.75 million km^2 (Er-rashid 1978; Shahjahan 1983; Siddiqui 1983; Alexander 1989; Brammer 1990; Mukherjee et.al. 2009). An average of 844,000 million cubic meter of water flows into the country during the wet season (May–October) through these three key rivers (Dewan 2015). Bangladesh is highly vulnerable to different types of disasters due to a number of factors which include its climatic variability, high population density, high incidence of poverty, social inequality, poor institutional capacity, inadequate financial resources and poor infrastructure (Shaw et al. 2013). Bangladesh is considered to be the sixth most flood-prone country in the world (Azad et al. 2013). The unique geographical setting, physiographic features and its geographical location at the deltas of the three major rivers are responsible for frequent occurrence of floods, which have adverse impacts on the country's economy, society and natural environment.

Roughly once in every 10 years, one-third of the country is severely affected by floods. The catastrophic floods of 1987, 1988, 1998 and 2004 caused inundation of more than 60% of the country's area (DMB 2008). The floods in August 2017 were

R. S. Leya (✉)
Urban and Rural Planning Discipline, Khulna University, Khulna, Bangladesh

D. Saha · S. K. Bala
Institute of Water and Flood Management (IWFM), Bangladesh University of Engineering and Technology (BUET), Dhaka, Bangladesh

H. Huq
United International University (UIU), Dhaka, Bangladesh

A. Haque, A. I. A. Chowdhury (eds.), *Water, Flood Management and Water Security Under a Changing Climate*, https://doi.org/10.1007/978-3-030-47786-8_17

particularly impactful as they followed two earlier flooding episodes in late March and July that year, exacerbating the vulnerability of people, (Philip et al. 2019). The IPCC's latest 1.5 Degrees report pointed out the necessity of reducing the vulnerability of floods and to accelerate adaptive capacity (IPCC 2018). Vulnerability is the capacity of a person or group to anticipate, cope with, resist, and recover from the impact of a natural hazard, (Wisner and Blaikie 2004). The extent of vulnerability during disasters need to be described in the context of gender, whose components are sex, class, race, age and poverty and how it varies with these components (Morimoto 2019).

Gender refers to the social differences between women and men and have wide variations both within and between cultures. Gender relations are influenced by historical, religious, economic and cultural realities, which are changed and reshaped over time. Although it is a social construct, the position of women in society tend to be unfavorable to that of men in every aspect i.e. politically, economically, socially and culturally.

Vulnerability is considered to be composed of three main components – exposure, sensitivity and adaptive capacity, (Younus 2018). It is known as a set of conditions that affects the ability of countries, communities and individuals to prevent, mitigate, prepare for and respond to hazards (Briceno 2002; Mehta 2007). Both vulnerability and its antithesis, resilience, are determined by physical, environmental, social, economic, political, cultural and institutional factors. People living in vulnerable locations have less physical and economic capital to cope with disasters and as such suffer more – they are considered as gender in this study. Flood in particular is a highly devastating phenomenon for the marginalized and socio-economically disadvantaged cluster in Bangladesh.

During a flood, affected communities suffer majorly and their situations further worsen following the event. The most vulnerable groups in the event of a disaster are women, children, elderly, disabled and other disadvantaged people. Because of the social structure, women are more involved in household tasks, such as collecting water and fuel wood, cooking, managing food, and providing child care which have made them more vulnerable to natural disasters compared to men (Nasreen 2009). They are more affected by flood-induced vulnerabilities, experiencing deficits in food, clothing, communications, fuel wood, and increases in disease exposure, water quality problems, and sexual harassment Nasreen (2010).

Women suffer most because of their differential needs, roles and responsibilities and their vulnerability is also influenced by prevalent social norms, customs, religious beliefs and overall subordinate position within the society. Moreover, the elderly, children and the disable people also face crucial disadvantages as a group due to their dependency on other people. Furthermore, women and other distressed groups are often subjected to robbery, physical assault and violence during a flood disaster (Nasreen 2010; Azad et al. 2013). Poor and landless people, for whom their only asset is the roof over their heads, become destitute when their houses are damaged from flood inundation and their poverty does not allow them to rehabilitate timely.

Vulnerability of women and men to flood, lies in their socially constructed, gender differentiated roles and responsibilities. Access to information, mobility, education and ownership of resources by differentiated gender influence their resilience to natural disaster. As vulnerability is gender differentiated and context specific for this reason, the way women experience vulnerability is often largely different from men due to socially constructed gender roles and power relations (Schmidt et al. 2005).

The ability of a society to cope with and recover from disasters depends largely on the social, economic and cultural capacities, which vary among different groups of people. The local people of Bangladesh have developed few abilities to anticipate, prepare for, adapt with, and recover from natural disasters. The capacity developed by communities to face hazards, where capacity is understood as the set of knowledge, skills and resources, people resort to in dealing with natural hazards and disasters (Gaillard and Cadag 2013). To ensure that disaster management activities are more functional and effective, the approach to mitigate losses and damages from a disaster, need to be gender sensitive. The Ministry of Disaster Management and Relief (MoDMR) has developed a gender sensitive approach while a number of DRR plans and policies in Bangladesh have considered gender inclusion as a crucial aspect (Center for Excellence 2017).

People's individual capacities also help alleviate their vulnerability to any disasters. It is suggested that communities have traditions that enable them to cope with disasters (Campbell 2006). These traditions have been built on beliefs and behaviors over long periods and underpin the fabric of their societies (Campbell 2009).

The Capacity and Vulnerability Assessment (CVA) framework is one of the commonly used tools to measure the vulnerabilities and capacities of hazards affected people. CVA was first designed and tested around the 1980s by International Relief Development Project (IRDP), to make relief interventions more developmental. It is the most widely known and adopted format. It assesses vulnerability and capacity against physical/material, social/ organizational and motivational/ altitudinal context.

A similar tool called Vulnerability and Capacity Assessment (VCA) had been developed by the International Federation of Red Cross and Red Crescent Societies (IFRC). The tool was defined as a diagnostic tool, as well as a planning tool for response design in case of hazards and lacks mechanism to access multi-dimensional vulnerability.

The capacities and vulnerabilities approach (CVA) is one such gender analysis tool that provides a way to characterize gendered roles and responsibilities and power dynamics (March et al. 1999). Birks et al. (2016) stated that the CVA critically considers gender and its associated roles, responsibilities and power dynamics in a particular community and seeks to meet the social needs of that particular community. Although the original intent of the CVA was to guide humanitarian intervention and disaster preparedness, we adapted this framework to a different context, which focuses on identifying and addressing emerging problems and social issues in a particular community or area that affect their specific needs, such as an infectious disease outbreak or difficulty accessing health information and resources.

CVA of gender analysis has been used in this study to assess the vulnerability of 2017 flood and people's abilities to cope with the disaster. In order to reduce flood induced vulnerabilities this paper provides gender sensitive flood management solutions in disaster risk reduction, planning and management.

2 Study Area

In 2017, Bangladesh suffered one of the most devastating floods which affected 31 districts as per data from Ministry of Disaster Management and Relief. Three districts were selected in this study to assess the sufferings of gender. Two districts from north-west, Kurigram and Nilphamari, regularly experience riverine and rainfall flood while the third district – Netrokona is located in the haor region in the northeastern part of the country and is prone to flash floods (Fig. 1).

Kurigram is located in between 25°23′ and 26°14′ North latitudes and in between 89°27′ and 89°54′ East longitudes with an area of 2296.10 sq. km. Nilphamari covers an area of 1580.85 sq km and is located in between 25°44′ and 26°19′ North latitudes and in between 88°44′ and 89°12′ East longitudes. A number of major rivers such as Brahmaputra, Jamuna, Dharla, Dudhkumar, Teesta flow through the heart of the Kurigram district. Nilphamari district also contains many rivers of which Teesta, Jamuneshwari, Buri Teesta and Ghagat are some of the major ones.

Kurigram and Nilphamari districts are situated in the floodplains areas of Bramaputra-Jamuna. During 2017, intensity of both riverine and rainfall floods were higher than ever, causing unspeakable miseries to vulnerable communities inhabiting the riverbank and floodplain areas. Netrokona has also suffered massive damages due to early flash floods in 2017 from heavy rainfall in the northeastern hilly region.

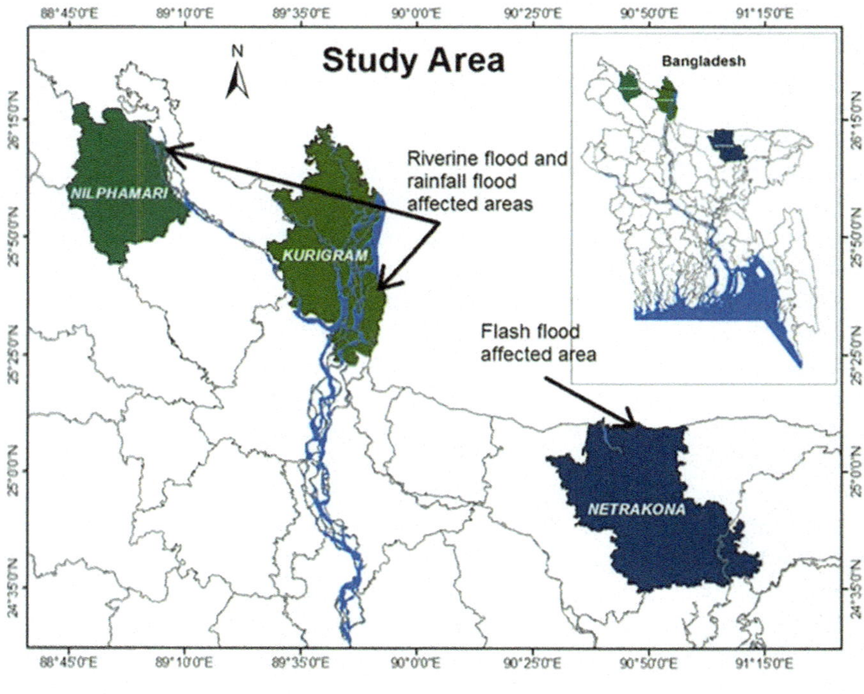

Fig. 1 Study area

The inhabitants of the study areas in Kurigram and Nilphamari have been associated with crop agriculture as their only livelihood means for several decades. During the monsoon (generally from June to September), these people cultivate Aman rice and during the dry season (generally from March to May) they mostly practice Boro rice cultivation and sometimes grow other crops like wheat, mustard, potato, etc. Some people, both men and women, work in road and embankment construction projects, some are small businessmen and wage earners, while in some families, male members travel to other districts as part-time workers. In Netrokona, apart from crop agriculture, people practice fish culture in household ponds, canals and nearby water bodies, during the monsoon and the dry season. However, all of these livelihood activities were severely affected in the wake of the 2017 flood event.

3 Methodology

Secondary data have been collected through literature survey for conceptualizing the overall damages and devastations of the 2017 flood. Published documents like reports, articles, journal papers and newspaper articles were reviewed to gather necessary information. Primary data were collected through two field surveys during November and December of 2017 in the three study districts. Participatory Rural Appraisal (PRA) tools were used to collect field level primary information. Twelve (12) Focus Group Discussions (FGDs) with flood affected women and marginalized people were conducted. Key Informant Interviews (KIIs) included local Union Parishad (UP) chairman, Upazila Nirbahi Officer (UNO), local elderly people, one parliament member who is also a Professor of a local college, founder of the local Non-Governmental Organization (NGO) 'Shabolombi', mid-level officials of other local NGOs namely Gono-Unnoyon Kotripokkho (GUK) and Unnoyon Sohojogi Team (UST) (Fig. 2).

Qualitative analysis of the collected data were performed using the CVA framework which helps to portray the vulnerabilities faced by gender during flood and their capacities to cope with it. The flood vulnerabilities faced by gender are categorized as physical, social and motivational in the CVA. To identify the reasons behind gender sensitivity during floods, existing flood management practices based on local peoples' capacities, local livelihood opportunities and access to information of different vulnerable groups of the study area were also discussed.

4 Findings and Discussion

When disasters like floods occur, people from all walks of life suffer, however those belonging to gender have to face additional challenges. The challenges and difficulties of gender were explored using questionnaire survey, FGDs and KIIs. Flood induced vulnerabilities faced by gender bring manifest in several ways, including physical, social, economic and environmental.

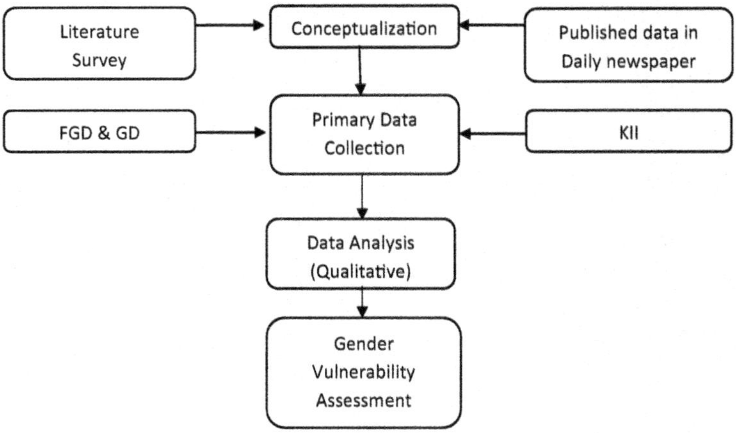

Fig. 2 Methodological framework

4.1 Losses and Damages of 2017 Flood

Different informants in all the upazilas of the study area stated that while they considered floods from the previous year i.e. 2016, to be majorly disastrous, the floods in 2017 flood were far worse. On a scale of 1 to 5, the respondents rated severity of floods in 2016 at 3 while floods in 2017 were rated at 5. 2016 flood afflicted about 1.5 times more damage than previous events, while the 2017 flood caused 3 times more damage than any other major flood in the history in terms of duration, inundation depth, loss of assets and livelihood (Table 1).

According to data collected, the riverbank area of Kurigram was the most affected among the different locations within the study area. This is the *Char* area of Kurigram, inhabited by communities who are habituated to regular flood occurrence. There is no embankment around the area, which increased their losses and damages during the flood. Sudden onrush of water meant that people were too afraid to save their crops, livestock and valuable assets. An inundation period of about 3 weeks, with a depth of 3–4 ft water in highlands, severely damaged their crop lands and leaving them with major challenges to tackle post flood, including the need to rebuild their houses without relief and support from GOs and NGOs. Floodplains area of Kurigram was less affected than the riverbank area. However, floods are fairly a new phenomenon for the area and most of the people reported never having faced any flood and thus have no experience in dealing with the disaster. As such, their vulnerability to the flood was somewhat high.

Nilphamari is home to many rivers and is therefore highly prone to floods. Floods are responsible for regular damages to crops, however compared to other areas, people's daily lives are less affected. This loss of crops and damages of crop lands however make it difficult for inhabitants to bounce back following a flood event.

Netrokona was hit by flash floods during 2017. Though the inundation period was long for Netrokona, they suffered less than other areas. Lower depth of water

Table 1 Losses and damages of 2017 flood in three districts

	Kurigram			
	Riverbank area	Floodplain area	Nilphamari	Netrokona
Depth of flood water	3–4 ft. on the road	3 ft. on the road	2–3 ft. on the road, up to 10 ft. in some other places	2–3 ft. on the road
Duration of inundation	15–20 days	7–8 days	7–8 days	2–3 months
Impacts of flood	Sudden water surge, crop land damaged, heavily damaged housing and other infrastructure, loss of livestock and assets	Crop land damaged, damaged housing and other infrastructure, damaged groundwater sources, loss of livestock and assets	Crop land damaged, loss of livestock and assets to some extent	Damaged crop land and fish resources, loss of livestock and assets to some extent
Place of stay during the flood	Left houses and stayed in shelters, embankments and roads	Left houses and stayed in shelters, nearby schools	Some left houses and stayed in shelters	Some left houses and stayed in shelters or relatives' houses
Specific gender impact during flood	Lack of privacy, Sanitation problem, Mental pressure, Malnutrition, Women's traditional attire, Reduction of shame, self-respect and dignity, Lack of security, Easily affected by diseases.	Mental pressure, Reduction of shame, self-respect and dignity, Easily affected by diseases.	Losses of mental strength, Reduction of shame, self-respect and dignity, Easily affected by diseases.	Lack of privacy, Sanitation problem, Easily affected by diseases, Malnutrition.

allowed them to continue their daily activities. Not too many people had to leave their houses for shelters. Due to yearly flood occurrence in the district, people tend to be more prepared for such a disaster. As a result loss of livestock and crops were minimum in this area (Fig. 3).

4.2 *Capacities and Vulnerabilities Framework for 2017 Flood*

Physical vulnerability is the most visible as the severity of a disaster can be measured by physical aspects. Poor people who have very little savings, low income levels and limited access to resources tend to suffer more in this aspect. Social and motivational vulnerability contribute to victimization. Internal social organizations and associated

Fig. 3 Some pictures of losses and damages of 2017 flood

conflicts and management techniques are also vital, and render people more vulnerable in the event of a disaster. Whereas, people's mental strength, ability and motivation during a disaster are constitute the motivational arena of vulnerability and capacity.

4.2.1 Physical

To understand physical vulnerabilities, data regarding loss of life, physical injuries, loss of crops, infrastructural damages etc. were collected.

In Kurigram, the lands were eroded deeply from few settlement areas as a result of the flood. Several trees were uprooted due to strong flows of water, which exacerbated the post flood environment. On the contrary, about 10 inches of sand was deposited on the agricultural land which made the land infertile for agricultural production. Infrastructure including roads, bridges and culverts in the study area were also severely damaged. There were no embankments or pucca road in Boldipara village, Jatrapur Union of Kurigram. Local people considered the lack of an embankment to be a major factor contributing to the severity of damages. Sudden onrush water during a flood meant people did not have access to roads to moving to an emergency shelter.

In the study area, most of the houses are made of earthen and semi-earthen materials, which are highly susceptible to flood inundation. Following the flood, most of the victims were unable to repair their houses and they had to leave their houses to live in schools, mosques, roads and culverts, converted to emergency shelters, for a long period of time. During the 2017 flood, Netrokona was less damaged compared to Kurigram and Nilphamari due to lower depth of inundation.

Flood victims of the study area had to suffer even after the flood as there is no flood shelter in vicinity. Schools are generally used as shelters in the event of a flood. However, these shelters cannot satisfy the minimum requirements of living because, the space available, cooking facilities, toilet facilities etc. are considerably inadequate for the number of victims involved. In these emergency shelters, men and women have to share the same toilet with no necessary amenities for women. Special facilities like care for children, elderly and the disabled are hardly considered. Embankments are generally higher in elevation compared to flood affected lands. Thus, affected people also take shelter on embankments along with their cattle and valuable assets. Very often, flow of flood water breaches these embankments within few days of flood water inundation and cause a loss of valuable assets of affected people. Taking shelters on highways, bridges and embankments also hinders the transportation system and creates obstacles in relief work.

Maximum tube-wells in the flood affected areas had been submerged by flood water, resulted in acute shortage of safe drinking water. The quality of water was not suitable for drinking and cooking purposes as flood water contaminated the ground and surface water. Affected people in Nilphamari attempted to boil flood water for drinking, however they were unable to do so due to lack of fuel wood. In Netrokona, groundwater sources suffered relatively less damage. During a flood, it is very difficult for women to go out for collecting safe water from distant sources. Men sometimes went to fetch safe water for cooking and drinking, however this was inadequate against the demand. Due to lack of safe water, women, children and elderly people were subjected to sanitation problems in the post-disaster period. An outbreak of water borne diseases like diarrhea, cholera, skin diseases, cold and fever had occurred as a result of stagnant and even after two months after the flood, the poor and the disadvantaged were found to be afflicted with these. Lack of sufficient medicine and medical support worsened their situation. Daily activities in the post-flood period were thus severely affected and children had miss 1–2 months of school due to illness.

Flood forecasting and warning system during the 2017 floods, failed to deliver due to lack of on-time information dissemination. According to the respondents, the damages and losses were much higher because of an ineffective warning system. Some people reported hearing about the flood from the union growth center, however they had no information regarding forecasted water level increase and intensity of the flood. Generally, since women tend to stay at home, they were not aware of the news regarding the flood circulating in local markets and tea-shops. Especially, people residing in char areas and along the river bank were completely unaware about the flood event. Thus local communities were largely unprepared for the expected floods and suffered huge losses and damage. People of Jatrapur union of Kurigram district reported that if they were informed about the flood in advance, they would have been able to save some of their valuable materials which had been washed away by flood water. It is a common belief among majority of the flood affected community, that early warning system is one of best and imperative solution to lessen the devastation from floods. It can help them to take necessary preparation for an incoming disaster including moving to safer places during the flood and saving their assets and livestock.

4.2.2 Social

People are more vulnerable to disasters when there is an existence of divisions among them. Data regarding religion, caste, class, and race were collected to understand the magnitude of social divisions. In most of the cases, good social relations among the flood victims were present in the time of flood. However, in time of relief distribution, people belonging to a relatively higher class, and families and friends of local powerful leaders received better treatment. In Sholmari union of Nilphamari District, fishers' communities, a gender group, were deprived from most of the relief. In Netrokona, many affected people informed that they took shelter at their relatives' houses in other districts, and were saved from the challenges. Social capability of such extent can make people less vulnerable to crises. In Nilphamari, local people reported that, male members of the community worked closely to shift female and children to safer places.

Many government and non-government organizations had played roles after the flood by providing relief, but had not performed well on flood preparedness, response and rehabilitation. Different Upazila and Union parishads have Disaster Management Committees (DMC) on paper, however these are poorly functioning. Local DMCs did not arrange monthly meetings, or provide necessary disaster preparedness training or worked for capacity building activities for the vulnerable communities. These DMCs were particularly inactive in rural areas. In the time of a disaster, DMC's presence and involvement could be somewhat noticeable, but gender group however remain highly neglected in these emergency situation. In Nilphamari, female members were not allowed to participate in the upazila level DMC, In Kurigram, though female were allowed to participate, but they did not have a lot of say in decision making. These female members were also not responsive to the issues of local women and other gender groups, either because they were not aware of such issues, or they did not find it important enough to be considered during an emergency.

4.2.3 Motivational

Collected data revealed that people believe floods to be inevitable due to the country's geographical setting. But proper precaution can however reduce the extent of losses and damages. In Kurigram, the local villagers of char areas stated that floods are highly common and local people are fairly habituated to it. Superstition, local myth and religious beliefs however limited the flood victims' ability to relocate themselves to safer places during the flood. In Kurigram, many people did not relocate themselves with the belief that water will rise slowly and they will get enough time to take safe shelter. Local respondents reported that flood water has increased from half to five feet within one day and did not even allow sufficient time to move valuable assets to a shelter. Flood affected people of Nilphamari demonstrate a fighting spirit against a disaster like flood. They believe they are capable of taking charge of a post-flood situation without needing to rely on support from the authority. People with common religious beliefs pray together in the time of a disaster which provides common faith and strength to deal with losses. Survey results revealed that people from all walks of life, devoted themselves to rebuild

Table 2 Capacities and vulnerabilities matrix

Sectors	Vulnerabilities	Capacities
Physical/material	Earthen and semi-earthen houses Community in char areas and river-bank Absence of embankments Poorly constructed infrastructures Absent of emergency food & medicine storage Lack of flood shelter Women's attire Poor & ineffective warning system	Fertile land Availability of labor Available indigenous housing materials
Social/ Organizational	Social conflict Absence of public awareness Poverty Lack of disaster planning and preparedness Limited access to power Lack of adequate access to information Vulnerable occupation Debt to microcredit organizations Relief or welfare dependency Mono-Crop agriculture	Inter & intra family relationship Group, community & area-wide organization Local leadership Local NGO's Adaptive strategies Multi-crop agriculture
Motivational/ Attitudinal	Traditional customs and religious beliefs Superstition Indecision to leave home for shelter Mental shock Lack of seriousness towards magnitude of disaster	Mentally prepared to face disaster Religion to provide common faith and strength Motivation to fight back Belief on own capacity Good relationship

their lands, houses and economic activities and also to build better and stronger cropping system in the aftermath of a disaster.

The capacities and vulnerabilities of gender in 2017 flood in northern region of Bangladesh are portrayed in this framework in three broad areas: physical, social and motivational category basis (Table 2).

It can be seen that physical vulnerabilities are high but capacities are too low to combat those vulnerabilities. On the other hand, motivational capacities are high which means people are committed to fight back after a flood, while their vulnerabilities still act as hindrances In the social context, social cohesion within flood affected communities generally exist, but socially constructed organizations are not as active as they should be. GOs and NGOs are however becoming increasingly active and the scenario is changing gradually, but there are still significant inadequacies considering the requirements. Moreover, prevalent social structures are still not gender friendly in the event of a disaster, because in emergency situations, social organizations tend to mainly cater to the needs of the masses.

5 Conclusion

The devastation of 2017 flood, enormously affected some of the most vulnerable groups of people including women, children, elderly, poor, disadvantaged and the disabled. Flood vulnerabilities like food deficits, communication gap, lack of privacy in the shelter, lack of fuel wood etc. were some of the major problems faced by differentiated gender. They also suffered from increased incidence of diseases such as cholera, dysentery, skin diseases, diarrhea, and jaundice. Disaster affected people were also subjected to mental health problems, with long-lasting trauma. Considering these issues of 2017 flood, some measures are recommended in this study to lessen the vulnerabilities and enhance the capacities of the community.

Emergency and quick response should be taken by the Ministry of Disaster Management and Relief in the event of a flood, where special attention should be given to the most affected and vulnerable groups of people. Rehabilitation, necessary food, safe water and medicine distribution, and proper information dissemination should be considered. Gaps between policies and practical implementation of policies need to be lessened in order to ensure comprehensive disaster management. A chain of accountability should be ensured in every step of the disaster management committee. Support provided by GOs and NGOs should be coordinated properly to ensure proper and full coverage of relief distribution in affected areas. NGOs and government organizations are also aware of the special needs of women, children and elderly but they would need to work in a coordinated way. For this, they provide gender-friendly relief products during flood disaster, however the quantity is insufficient.

Awareness and capacity building programs on disasters need to be propelled for local people. Capacity building and training for the vulnerable groups of the society can improve their conditions and support their disaster management initiatives in a progressive way. To reach vulnerable and affected groups of people, a committee of local people headed by local Union Parishad Members need to be formed. Flood resistant crops and the distribution of seeds in flood prone areas can be introduced. Diversified economic activities such as livestock rearing, small businesses, homestead kitchen gardening, sewing or making traditional products for commercial purposes, etc. can also be introduced among these flood affected people. Access to flood warning and relevant information of differentiated gender group before a disaster, can increase the efficiency of flood management immensely. Disaster management committees in the area need to be properly functioning and women's participation should be thoroughly ensured, so that the special needs of women and vulnerable communities are properly addressed during relief operations and other management activities.

Flood is the most common and frequently occurring natural disaster in Bangladesh. Differentiated genders are more affected by flood vulnerabilities. Flood vulnerabilities of gender in northern region of Bangladesh have been assessed against the capacities and vulnerabilities framework (CVA) in this study. It can be concluded that flood induced vulnerabilities can be minimized with the help of individual capacity, local adaptive strategies as well as government and non-government initiatives. While, there are gender-sensitive policies in place, they are not well followed or implemented in reality. . Lack of proper coordination among organizations, inefficient warning systems, lack of capacity to cope with flood vulnerability and gaps between documented policies and real practices, all combine to make the situation more disastrous. However, effective disaster management, capacity building interventions, access to diversified economic opportunities and practical implementation of gender sensitive policies can ensure effective and efficient disaster management in flood affected regions of the country (Fig. 4).

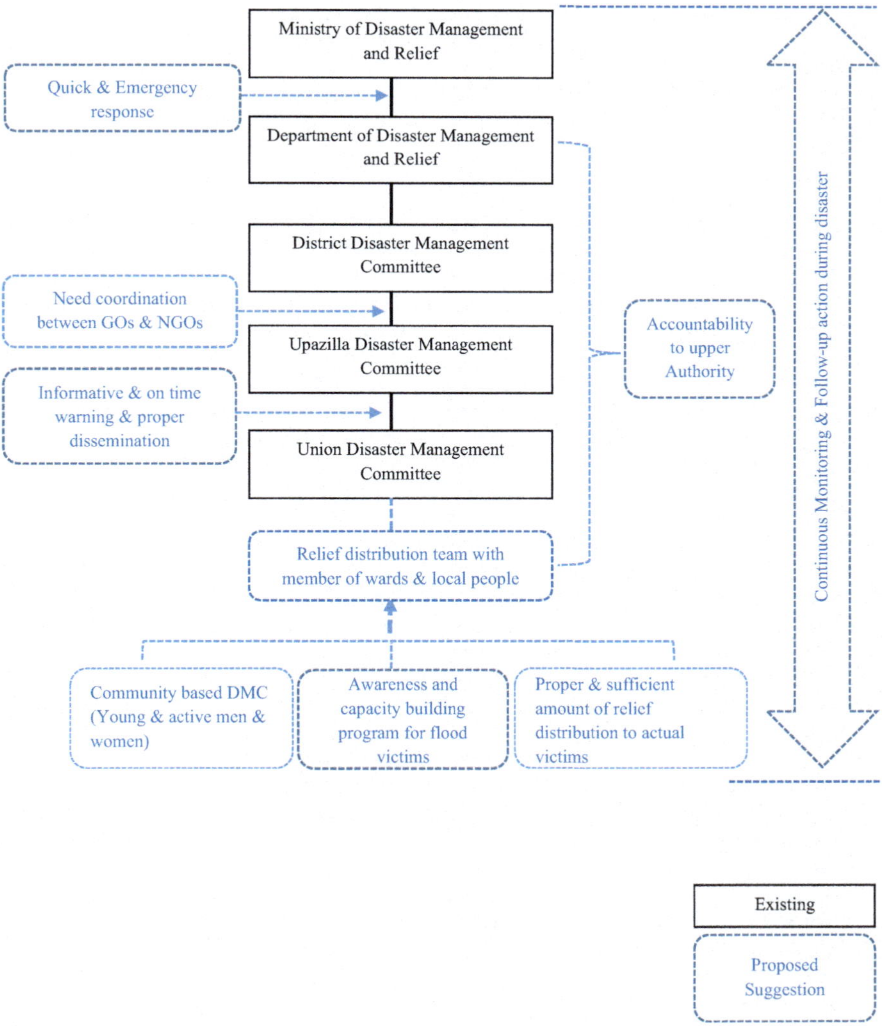

Fig. 4 Proposed action framework for reduction of gender vulnerability

Acknowledgments This study was performed under the "Research on Disaster Prevention/Mitigation Measures against Floods and Storm Surges in Bangladesh (SATREPS)" project funded by JICA-JST, Japan. The authors would like to express their gratitude to all the informants in the field for their help and support. The authors would also like to thank JICA-JST for providing financial support for this study.

References

Alexander, D. (1989). *The case of Bangladesh* (15 pp). Amherst: University of Massachusetts/Mimeo.

Azad, A. K., Hossain, M. K., & Nasreen, M. (2013). Flood-induced vulnerabilities and problems encountered by women in Northern Bangladesh. *International Journal of Disaster Risk Science, 4*(4), 190–199.

Birks, L., Powell, C., & Powell, J. (2016). Adapting the capacities and vulnerabilities approach: A gender analysis tool. *Health Promotion International, 32*, 930–941. https://doi.org/10.1093/heapro/daw032.

Brammer, H. (1990). Floods in Bangladesh: Geographical background to the 1987 and 1988 floods. *Geographical Journal, 156*(1), 12–22.

Briceno, S.. (2002, March 6). *Gender Mainstreaming in Disaster Reduction Panel. Commission on the Status of Women*. New York Secretariat of the International Strategy for Disaster Reduction. Geneva: UN/ISDR. www.unisdr.org

Campbell, J. R. (2006). *Traditional disaster reduction in Pacific island communities*. Avalon: GNS Science Report 2006/38.

Campbell, J. R. (2009). Islandness: Vulnerability and resilience in Oceania. *Shima: The International Journal of Research into Island Cultures, 3*(1), 85–97.

Center for Excellence. (2017, June). *Bangladesh disaster management reference handbook*. Hickam: Center for Excellence in Disaster Management and Humanitarian Assistance.

Dewan, T. H. (2015). Societal impacts and vulnerability to floods in Bangladesh and Nepal. *Weather and Climate Extremes, 7*, 36–42. https://doi.org/10.1016/j.wace.2014.11.001.

DMB (Disaster Management Bureau). (2008). *National plan for disaster management 2008–2015*. Dhaka: DMB (Disaster Management Bureau), Disaster Management and Relief Division, Government of the People's Republic of Bangladesh.

Er-Rashid, H. (1978). *Geography of Bangladesh* (579 pp). Boulder: Westview Press.

Gaillard, J., & Cadag, J. R. D. (2013). *Participatory 3-dimensional mapping for disaster risk reduction: A field manual for practitioners* (p. 178). London: CAFOD.

IPCC. (2018). *Global warming of 1.5°C. Summary for policy makers*. Geneva: IPCC.

March, C., Smyth, I., & Mukhopadhyay, M. (1999). *A guide to gender- Analysis frameworks*. London: Oxfam Great Britain.

Mehta, M. (2007). *Gender matters: Lessons for disaster risk reduction in South Asia*. International Center for Integrated Mountain Development, Kathmandu, Nepal.

Morimoto, T. (2019). Spatial analysis of social vulnerability to floods based on the MOVE framework and information entropy method: Case study of Katsushika Ward, Tokyo. *Sustainability, 11*(2), 529.

Mukherjee, A., Fryar, A. E., & Thomas, W. A. (2009). Geologic, geomorphic and hydrologic framework and evolution of the Bengal basin, India and Bangladesh. *Journal of Asian Earth Sciences, 34*(3), 227–244.

Nasreen, M. (2009). *Impact of climate change on agriculture and food security in Bangladesh*. Paper presented in special conference on Climate Change and Bangladesh Development Strategy: Domestic Tasks and International Cooperation on 2 January 2009. Dhaka: Bangladesh Poribesh Andolon (BAPA) and Bangladesh Environment Network New Zealand.

Nasreen, M. (2010). Rethinking disaster management: Violence against women during floods in Bangladesh. In S. Dasgupta, Siriner, I. and De, P. S. (eds.). *Women's encounter with disaster* (pp. 232–244). London, UK: Frontpage Publications Limited.

Philip, S., Sparrow, S., Kew, S., Van Der Weil, K., Wanders, N., Singh, R., & Otto, F. (2019). Attributing the 2017 Bangladesh floods from meteorological and hydrological perspectives. *Hydrology and Earth System Sciences Discussions, 23*(3).

Schmidt, A., Bloemertz, L., & Macamo, E. (2005). *Linking poverty reduction and disaster risk management* (p. 88). Bonn: GTZ.

Shahjahan, M. (1983). Regional cooperation in the utilization of water resources of the Himalayan rivers. In M. Zaman (Ed.), *River basin development* (pp. 114–130). Dublin: Tycooly International Publishing.

Shaw, R., Islam, A., & Mallick, F. (2013). National perspectives of disaster risk reduction in Bangladesh. In: *Disaster risk reduction (methods, approaches and practices).* Tokyo: Springer

Siddiqui, M. F. (1983). Management of river system in the Ganges and Brahmaputra Basin for development of water resources. In M. Zaman (Ed.), *River basin development* (pp. 137–149). Dublin: Tycooly International Publishing.

Wisner, B. P., & Blaikie, P. M. (2004). *People's vulnerability and disasters, At risk. Natural hazards* (2nd ed.). London: Routledge.

Younus, M. (2018). Climate change vulnerability assessment and adaptation of Bangladesh: Mechanisms, notions and solutions. *Sustainability, 11*(10), 1–20. https://doi.org/10.3390/su10114286.

Groundwater Fluctuation in Response to Annual Rainfall in North-West Region of Bangladesh

Rashed Uz Zzaman, Sara Nowreen, and Imran Hossain Newton

1 Introduction

Groundwater (GW) is one of the most significant sources for freshwater. Almost one-third of the world population depends primarily on groundwater (GW) for domestic (e.g., drinking, bathing), industrial, and agricultural purposes (Ahmadian, And Chavoshian 2012). Globally, GW is supplying approximately 50% of current potable water, 40% of the water demand of the self-supplied industry and 20% of the water used in irrigation (Hossain et al. 2015). Almost 95% of the earth's useable freshwater is stored as GW, among which almost 50% exists within the earth crust down to a depth of 800 m, and only 1.5% exists in rivers and lakes (Livingston 2009). As the surface water (SW) is not available uniformly over the year, using GW is often the only option for irrigation, industry, and safe drinking water for now and for tomorrow. With time the acceleration of the population, agriculture and industrial development is causing a continuous increase in the rate of GW withdrawal. However, the trend of abstraction and use of GW is not similar for all countries.

In Bangladesh, about 68.5% of irrigation water and almost 95% of drinking water comes from GW sources (Livingston 2009). However, this GW resource is gradually depleting due to aquifer over drafting and mismanagement (Mustafa et al. 2017). In the North-West (NW) region, GW use is much higher compared to the other parts of Bangladesh. Rivers and channels in this area become completely waterless during the dry season, making people entirely dependent on GW (Shahid and Behrawan 2008). To provide irrigation facility to these vast areas, BADC (Bangladesh agricultural Development Corporation) is now operating 25,559 DTWs, 12,871 LLP with about 7,97,907 STWs for irrigation purposes. In several studies, (Shamsudduha

R. U. Zzaman (✉) · S. Nowreen · I. H. Newton
Institute of Water and Flood Management (IWFM), Bangladesh University of Engineering and Technology (BUET), Dhaka, Bangladesh

A. Haque, A. I. A. Chowdhury (eds.), *Water, Flood Management and Water Security Under a Changing Climate*, https://doi.org/10.1007/978-3-030-47786-8_18

et al. 2009; Shahid 2011; Dey et al. 2012) it is found that Ground Water Level (GWL) has been declining substantially since the last decade and becoming a threat to the sustainability of GW affecting other GW dependent sectors. Further, due to the lack of appropriate knowledge and non-availability of modern technology, farmers are lifting water improperly without caring much about the declined GW table position. Some areas of the North-West region are already experiencing GW mining (Shamsudduha et al. 2009; Shahid 2011; Dey et al. 2012). An excessive GW abstraction can become a serious threat for further GW depletion.

In general, lowered GW table in Bangladesh is filled up by natural processes, i.e., direct diffused recharge by rainfall or indirectly focused recharge by the river flow. However, if these filling processes are hampered by any means it will create a permanent water table depletion, with a shortage of GW amount, frequent shortages of GW in this NW region already has social, economic and environmental (Sajjan 1998; Dey et al. 2012). This permanent depletion, defined as long-term (GWL) declination, is claimed to occur mainly by incessant groundwater pumping (Bartolino and Cunningham 2003), in particular, when pumping exceeds the recharge and continues for a longer period (Gleeson et al. 2010). During the dry season, GW abstraction becomes the topmost as rainfall is almost null. This abstracted volume is replenished directly by rainfall and indirectly by river flow during the wet season. Thereby, the difference in GWL between the dry season and the wet season, also known as Water Table Fluctuation (WTF), represents the actual recharge phenomenon. Due to various human-induced and climatic factors, the trend of this WTF is, in general, getting higher. The fluctuation, which was about 5–10 m in 1960, has now become almost 10–20 m, posing various threats like groundwater unavailability or need for going deeper into the aquifer to find the water table (Shahid and Hazarika 2010).

Several studies have been carried out, revealing severe declination of long-term trends in different parts of Bangladesh. For example, (Sarkar and Ali 2009) paid attention to trend analyses in Dhaka city, (Ali et al. 2012) observed non parametric GWL trends of 35 observation wells of NW region, (Adhikary et al. 2014) worked with GWL trends for few districts in NW part of Bangladesh, and (Jahan et al. 2010) studied parametric linear regression for trends in the Barind area. However, no comprehensive studies so far have been conducted on GW fluctuation with rainfall over the NW region of Bangladesh. Therefore, the objectives of the present research are to: (1) to explore the spatial variability and trend of (GWL) for the dry season, the wet season and the water table fluctuation (2) to find the relationship between (GWL) and annual rainfall.

2 Study Area

The North-west (NW) region of Bangladesh, stretches between $23°50'-26°30'$N latitude and $88°10'-89°45'$E longitude, covering an area of 32,000 km^2 (Fig. 1). The area is broadly categorized into six distinct landforms, viz., (1) Old Himalayan piedmont plain, (2) Tista alluvial fan, (3) Bhramaputra Jamuna floodplains, (4) Ganges river floodplains, and (5) Barind Tract. The old Himalayan piedmont plain is

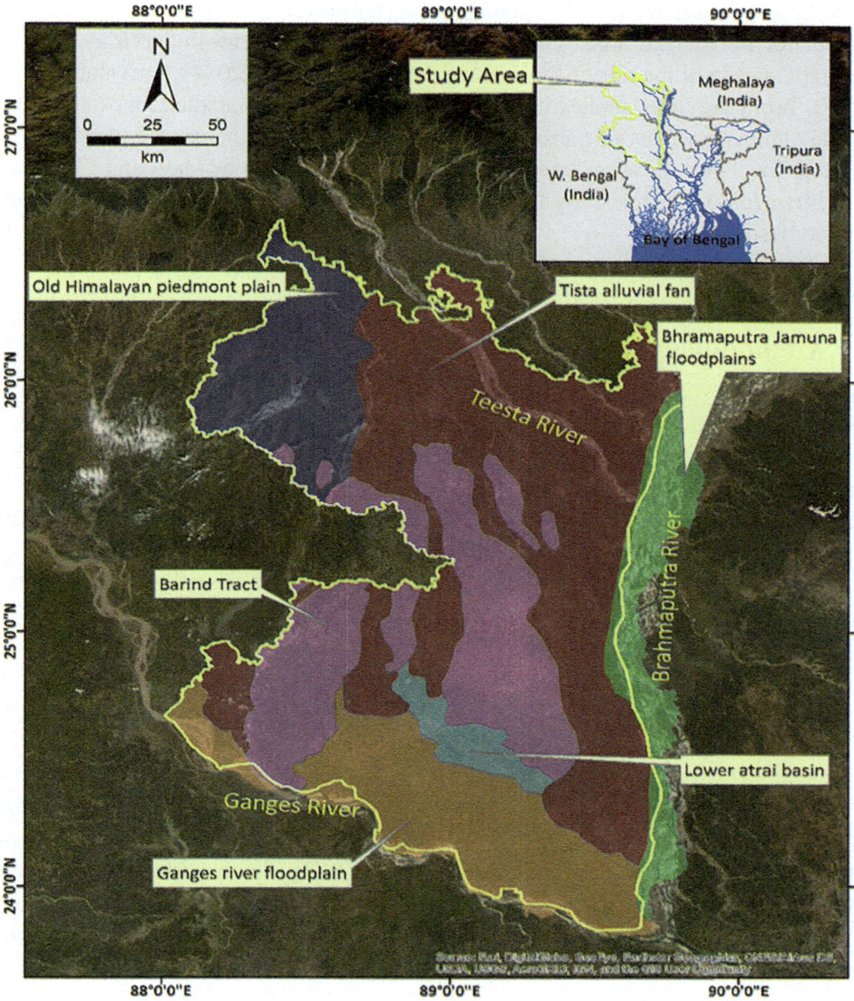

Fig. 1 Physiographic units of the study area. (SRDI 2007)

subjected to flash flood flowing down the Karatoya and Nagar River after heavy rainfall. Tista alluvial fan region has three major alluvial rivers called Tista, Dharala and Dudkumar shallowly flooding their adjacent areas during monsoon. Rest of the area is mainly flooded by rainfall water. Generally, water flows including high river floods come with alluvial soil and depositions over the landscape. NW region contains only a few parts of the Bhramaputra Jamuna floodplain landform and it is an active flood plain. This area is seasonally flooded by the river with silts and thick new silt depositions. The NW part of the Ganges river floodplain is flooded mainly by ponded rainwater and local run-off or raised GWL. The Barind tract area covers the central part of NW region with a poor drainage system. A large part of the area lies wet in the rainy season. Flooding occurs in the deposition sites throughout the area, up to 60 cm in the rainy season.

Single to multiple aquifer systems are present in the NW region of Bangladesh except for the Barind area. The Barind area contains s semi-impervious clay-silt aquitard system of the recent-Pleistocene period (thickness 3.0–47.5 m) (Jahan et al. 2007). From previous studies it is found that the aquifer characteristics of the area reveal (1) lower values of transmissivity (<500 m^2/day) in the central part which is suitable for domestic water supply, (2) medium (500–1000 m^2/day) and (3) higher (>1000 m^2/day) transmissivity values in the other parts that are suitable for irrigation and domestic needs.

3 Methodology

Currently, annual natural replenishment of Bangladesh aquifers occur mainly during the monsoon through the infiltration-percolation process and indirectly through high stage river discharging, provided there is storage space during the period of June to September. Therefore, a conservative data approach has been taken to investigate water table fluctuation hydrographs in shallow (depth < 150 m) unconfined aquifers with annual rainfall records. The details of the methods are described in the following sections:

3.1 Data Collection

Weekly (GWL) data were collected from the Bangladesh Water Development Board (BWDB). BWDB measures the (GWL) of each station from a common horizontal datum that is known as public works datum (mPWD) (Shamsudduha et al. 2009). Quality was assured and longer records of 312 monitoring wells were picked out of a total of 434 monitoring wells available at the NW region. Particularly, station records that showed less than 10% missing values were finally selected. Daily rainfall data was collected both from BWDB and Bangladesh Metrological Department (BMD). BMD has 6 rainfall measuring stations in the NW region, and BWDB has 74 measuring stations. All rainfall data collected from BWDB has been checked by the corresponding BMD rainfall station data. Median value was calculated for every year from these rainfall data. Also, information on irrigation census is collected from the Bangladesh Agricultural Development Corporation (BADC) for the years 1993, 2005 and 2017.

3.2 Trend Analysis

The nonparametric Sen's slope estimator test was employed for trend analysis. A time series of equally spaced data was required for calculating Sen's slopes. The estimation of the slope of N pairs of data involved the following process (Rahman et al. 2016).

$$Q_i = \frac{X_i - X_k}{j - k} \quad (1)$$

Where $j > k$ and $x_i$and x_k are data valued at times j and k, respectively. Where N was odd, the Sen's slope estimator was given by the median slope as:

$$Q_{med} = Q_{[(N+1)/2]} \quad (2)$$

Moreover, where N was odd. Moreover, the median slope was also estimated as follows:

$$Q_{med} = \frac{1}{2}(Q_{(N/2} + Q_{[(N+2)/2]} \quad (3)$$

where N is even. 'R' statistical language platform (R Core Team 2013; Rai et al. 2017) was used for statistical analysis using "Kendal" and "ZTP" packages. Here, positive values of the slope indicate an increasing trend and vice versa.

3.3 Correlation Observation

The Data period used for long-term daily rainfall data of 74 rainfall stations was between the years 1993–2017. Corresponding 74 weekly groundwater monitoring stations were chosen depending on (a) distance not exceeding 11 km from rainfall stations, (b) missing data not exceeding 15%, (c) number of observation points not less than 10, (d) correlation, $r > 0.5$. Groundwater and surface water interactions were observed for the corresponding available river water level monitoring points located within 5 km distance from the observation wells that show high correlation with rainfall stations.

4 Results and Discussions

4.1 Spatio-Temporal Distribution and Changes of Annual Rainfall

Highest annual median rainfall during the last 27 years is found to be 2700 mm, whereas the lowest annual median rainfall is recorded as 1006 mm. As illustrated in

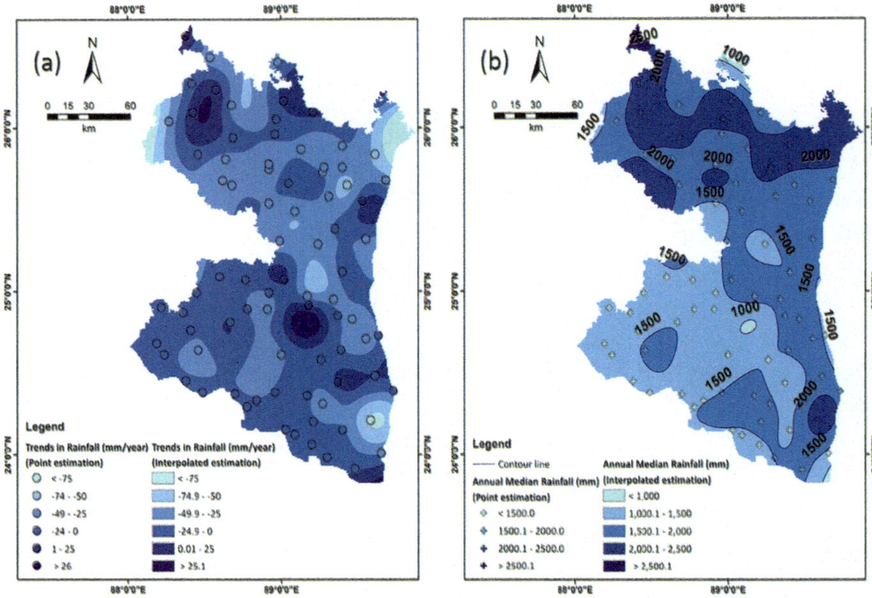

Fig. 2 Spatial distribution maps of (**a**) Trend (mm/year) (**b**) Contour of annual median rainfall during 1993–2017

Fig. 2b, the highest and the lowest annual median rainfall is found in the northern part and the southern part of the study area, respectively. Spatial distribution of long term (1993–2017) trends in rainfall is shown in Fig. 2a. Except for a few stations, most of the monitoring stations show declining trends over the study area. The rate of change varies from −77 to 26.91 mm/year. The lower part of the Barind Tract has the lowest annual rainfall with the declining trend up to −25 mm/year.

4.2 *Spatio-Temporal Distribution of GWL*

The water level of the shallow GW during the dry season in 1993 ranges from 1.89 to 16.8 m below ground surface (bgl) with a median GWL of 6.0 m bgl. The relatively lower level of GW is observed in the northwestern part of the area, to be specific, at Kurigram, Lalmonirhat, Nilphamari, Panchagarh, Rangpur and Thakurgaon districts (administrative unit). These districts belong mainly to the old Himalayan piedmont plains and having a few areas within the Tista alluvial fan landform (Fig. 1). In contrast, the deepest GWL in 1993 is found around Naogaon, Nawabganj, and Rajshahi districts. The landform of these districts fall under the Barind Tract which is developed over the Pleistocene Clay. Notably, Barind Clay is formed primarily with a type of gray heavy clay. GWL ranging from 8.8 to 16.8 m within the Barind Tract area also indicates that abstraction of GW is very high here, and replenishment is very low compared to other areas. On the other hand, the GWL of the shallow aquifer during the wet season in 1993 ranges from 0 to 9.2 m with a median value of 1.5 m bgl shown in Fig. 3d. The median of the water table fluctuation in 1993 is 4.56 m shown in Fig. 3g. Figure 3 shows the spatial distribution of GWL and the water table fluctuation for shallow aquifers for the years 1993, 2005, and 2017. The GWL during the dry season in 2005 ranges from 1.7 to 23.2 m bgl and the median GWL is found to be 6.7 m which is deeper than 1993. Overall, GWL is also found to be deepened shown in Fig. 3b. The GWL of the wet season in 2005 ranges from 0 to 17.5 m bgl with a median GWL of 1.8 m bgl, which also indicates an overall lower recovery rate than that of 1993 shown in Fig. 3e. Increase in the median (= 4.7 m bgl) GW fluctuation in 2005 indicates higher GW pumping than 1993 shown in Fig. 3h. Likewise, in 2017, (a) the dry season GWL ranging between 1.7 and 33.7 m bgl with a median of 7.5 m bgl; (b) the wet season GWL varying between 0 and 27.66 m bgl showing a median of 3.27 m bgl; and (c) the median GW fluctuation of 4.5 m (in Fig. 3i) imply more and more expansion in irrigation water withdrawal.

4.3 Trends in Dry Season GWL

For shallow aquifers across the study area long-term GWL time series starting from 1993 up to 2017, in general, shows declining trends except for a few areas marked with blue color in Fig. 4a. The Sen's slope trend rates vary from −0.12 to 1.19 m/year, where the calculated median trend is found to be 0.06 m/year. Highest (0.31–1.3 m/year declining) rate of change is observed in western Barind area. Otherwise, GWL in other parts of the NW area shows slow decreasing trends (0.01–0.3 m/year) during the dry season. In contrast, the rising trends (i.e., the subclass of −0.13–0 m/year in Fig. 4) are detected in lands near the rivers, for example, few pockets near the Brahmaputra-Jamuna and the Ganges. Change of land use pattern (Rai et al. 2017) or cropping pattern (BBS 2018), for instance, shifting from water consuming rice production to Rabi crops or alternate source of irrigation might be the reason(s) behind such rising of dry season GWL trends.

4.4 Trends in Wet Season GWL

Spatial distribution of long-term (1993–2017) trends in wet season GWL is shown in Fig. 4b. In general, likewise, the dry season trends, declining (0.01–1.3 m/year) is observed for wet season recovery rates with a few exceptions in some areas. The median rate of change is 0.17 m/year with higher scores (0.6–1.1 m/year) found in the lower Barind area (Fig. 4). Notably, the low infiltration rate is the main characteristic of this high Barind area having 45% clay loam type soil texture combined with 75% deep gray Terrace Soils of 120–140 m depth. The high positive trend is

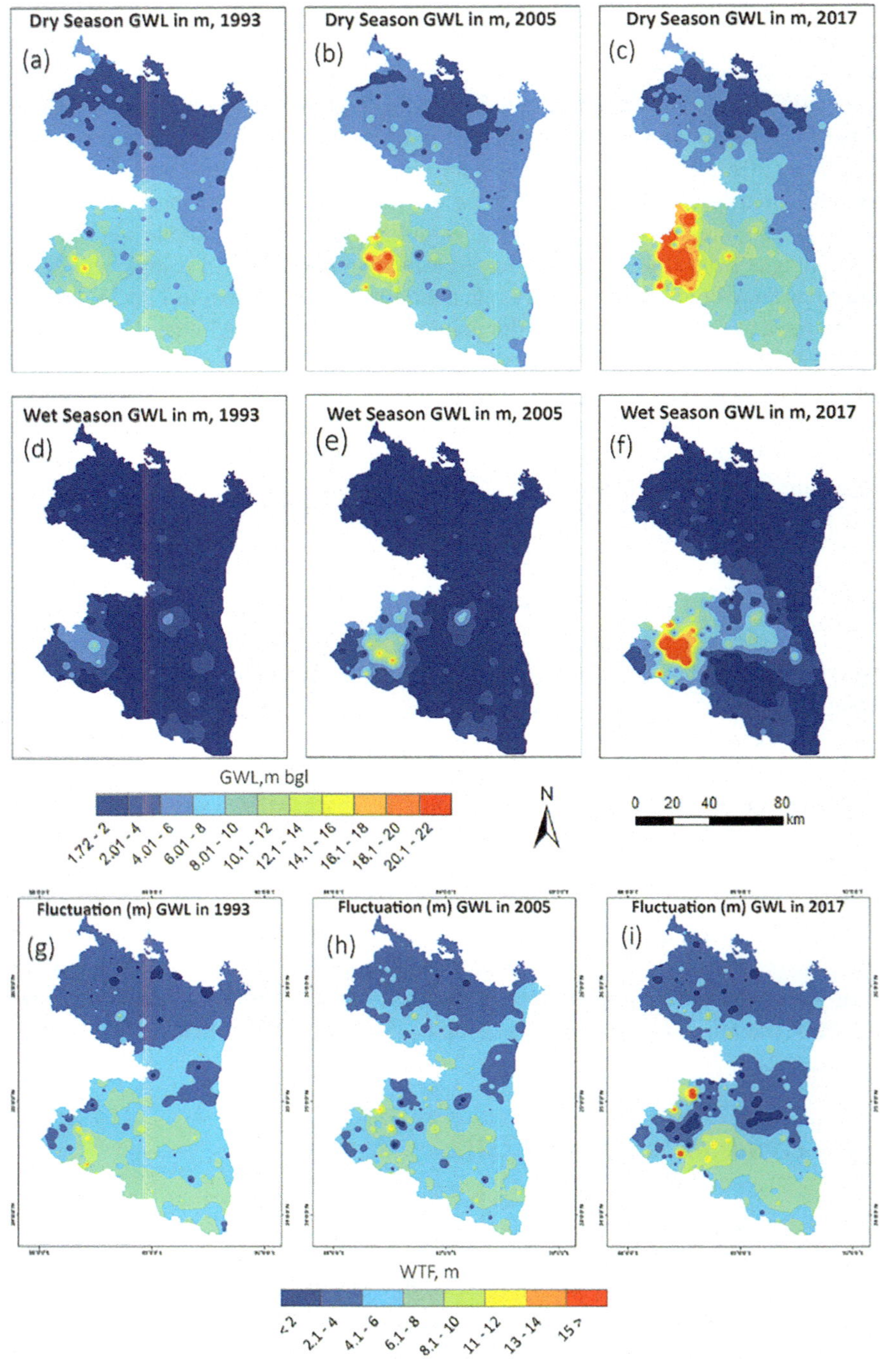

Fig. 3 Spatio-temporal distribution of GWL and Water Table Fluctuation. GWL position during the dry period in (**a**) 1993 (**b**) 2005 and (**c**) 2017; during the wet period in (**d**) 1993, (**e**) 2005, and (**f**) 2017; and water table fluctuations in (**g**) 1993, (**h**) 2005 and (**i**) 2017

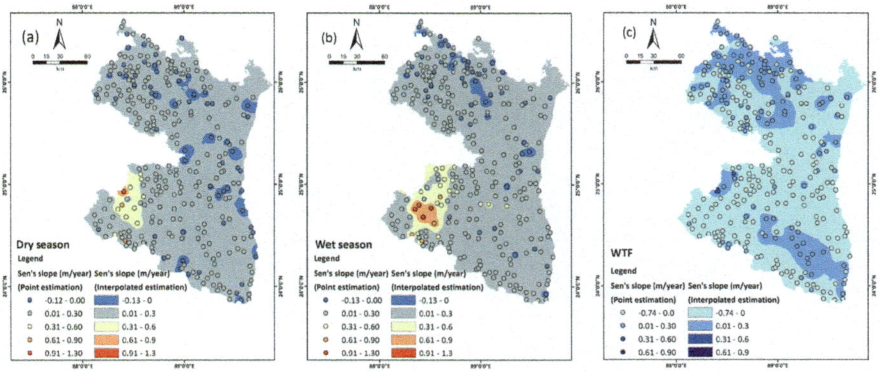

Fig. 4 Spatial distribution map of trends (1993–2017) in (**a**) GWL during the dry season, (**b**) GWL during the wet season and (**c**) water table fluctuation

also observed in Naogaon, Nawabganj, and Rajshahi Districts, indicating low recovery rates of the water table.

4.5 Trends in Water Table Fluctuation

Magnitudes of changes of WTF trends range from −0.74 to 0.6 m/year with the median change of −0.032 m/year in Fig. 4c. Figure 4c shows the spatial distribution of trends of WTF in this study area for the years 1993, 2005, 2017. Similar suppressing WTF is also observed by Nowreen (2017). Such suppressed seasonality is often the case where the Pleistocene aquifer is observed. Apart from low rainfall percolation under the Barind Tract, mining of GWL often causes shifting of the water table from semi-confined to unconfined resulting in suppressed seasonality as explained by Nowreen (2017). Negative trends in seasonality along with high positive trends in both the dry and wet seasons (as explained in Sects. 3.2 and 3.3) reveal that GWL under Barind Tract's Pleistocene aquifer is going deep to deeper and is not replenished properly. On the other hand, low positive trends (0–0.3 m/year) in WTF is found mostly under Old Himalayan piedmont plains and Ganges river floodplains with few irregularly scattered monitoring wells under Tista alluvial fan landform in Fig. 4c. This phenomenon here indicates more annual increase in irrigation pumping. Records shown in Fig. 5 also support intensified use in GW withdrawal over surface water.

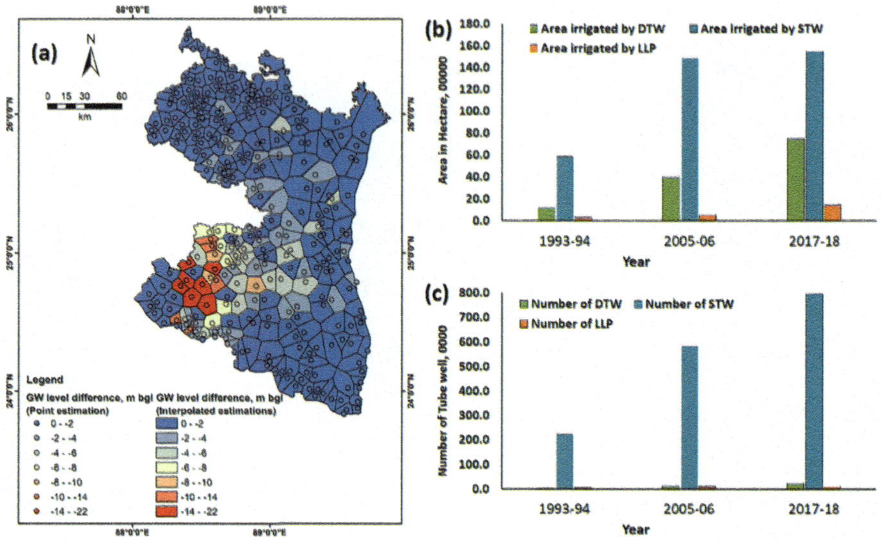

Fig. 5 (**a**) GWL difference between 1993 and 2017 (**b**) irrigated areas by deep tubewell (DTW), shallow tubewell (STW) and low lift pump (LLP)

4.6 Temporal Response of Rainfall and Groundwater Stations

A simplistic statistical approach was conducted between water table fluctuation and (nearest) rainfall data with the assumption that the annual rise in GW hydrograph is strongly related to annual total rainfall amount, particularly in case of a shallow unconfined aquifer. However, from the correlation analysis, it is found that high positive correlation (score more than 0.5) exists to only a set of 26 stations (Fig. 6). As presented in Table 1, these 26 monitoring stations are highly correlated (where $r > 0.5$) and statistically significant (where, $p < 0.05$). Therefore, it is expected that the surrounding areas of these sets of 26 stations encompass with permeable soil that is favorable for vertical infiltration and percolation. That implies, if the rainfall increases, the GW hydrograph rises because rain water easily infiltrates to the earth underneath and recharges to GWL, ultimately increasing the position of GWL. In these areas, the median WTF is at about 4.00 m, and annual rainfall varies from 750–3200 mm. In general, the groundwater replenishment is done by various processes like rainfall percolation, irrigation return flow, indirect recharge from river water, etc. (Ebrahimi et al. 2016). Various research studies in Bangladesh consider Rahman and Ravenscorft (2003) and BDP 2100 that infiltration-percolation acts prominent to replenish Bengal aquifers and rainfall provides a significant influence to the GW table (Kumar and Seethapathi 2002). Stations with high correlation also supports this hypothesis. Notably, sandy and loamy top-soil texture is observed near those 26 stations which may also exacerbate such higher statistical correlation

Table 1 Stations showing the correlations (having scores more than 0.5) and corresponding significant (p) values between groundwater fluctuation and rainfall

Station ID				Station ID			
Rainfall	Groundwater	R-value	P value	Rainfall	Groundwater	R-value	P value
CL6	BO033	0.65	0.003	CL167	RA-05	0.53	0.03
CL166	DI031	0.74	0.004	CL14	RJ033	0.53	0.032
CL16	RJ144	0.69	0.001	CL195	RJ109	0.53	0.014
CL163	RA-81	0.68	0.001	CL218	RA-35	0.52	0.019
CL178	RA-69	0.65	0.002	CL188	RA-50	0.51	0.024
CL34	PA013	0.63	0.025	CL157	DI083	0.49	0.059
CL3	RJ025	0.63	0.008	CL215	RJ079	0.49	0.034
CL210	DI044	0.60	0.005	CL168	DI025	0.48	0.029
CL182	RA-37	0.60	0.005	CL192	RJ142	0.48	0.039
CL186	RA-43	0.58	0.009	CL220	DI071	0.46	0.068
CL166	DI028	0.57	0.023	CL206	RA-28	0.46	0.038
CL172	RJ091	0.57	0.012	CL23	RJS08	0.45	0.173
CL201	DI067	0.53	0.038	CL38	PA008	0.42	0.086

permitting higher diffused recharge to aquifers compared to other top-soil types. According to lithology, formation aquifers of these areas are formed with old gravelly sand, young gravelly sand and alluvial sand. These types of deposition are very suitable to percolate and capture groundwater storage. Topographical slopes are also very low (less than 3%) for these areas, hence, allow more duration to percolate and capture rainfall water. Thus rainfall along with favorable soil surface geology may have great influence to the recovery of groundwater hydrographs (Adeleke et al. 2015). However, the correlation between GWL and rainfall also considers the impact of other factors like soil condition, top soil characteristics, or some local effects (such as pumping, stream leakage effect, etc.). Thereby, the areas with poor correlation might have other factors playing dominating roles over rainfall. Here, a certain portion of monsoon rainfall, even if not the whole annual rainfall, may cause rise in the ground water table. Therefore, the relationship between the selected variables used for computing correlation for annual years (1993–2017) is observed as poor (Fig. 6).

4.7 *Interaction Between Groundwater and Surface Water Stations*

In this part, nine surface water level (SW) stations were selected that are nearest to the GW stations that have been observed to have good correlations with rainfall stations, as presented in Table 1. Out of these nine rivers, four rivers, viz., Atrai, Buri Teesta, Kharkhuria and Dharala are losing streams contributing to groundwater all year around (Fig. 7). In contrast, the Dharala and the lower part of the Ghagot river

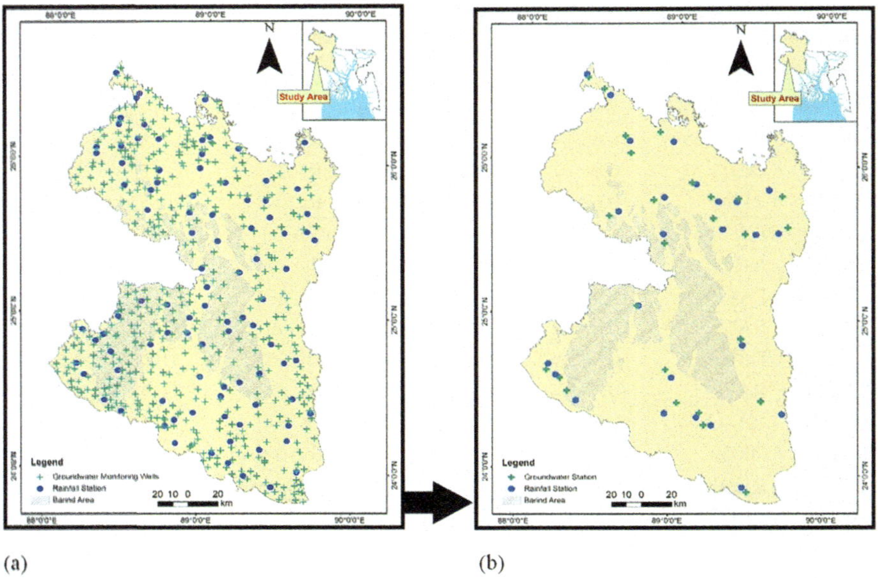

Fig. 6 Correlations between GWL and annual rainfall (a) Existing grondwater monitoring wells and rainfall stations of the North-west region (b) Stations that show significant correlations between groundwater level and annual rainfall

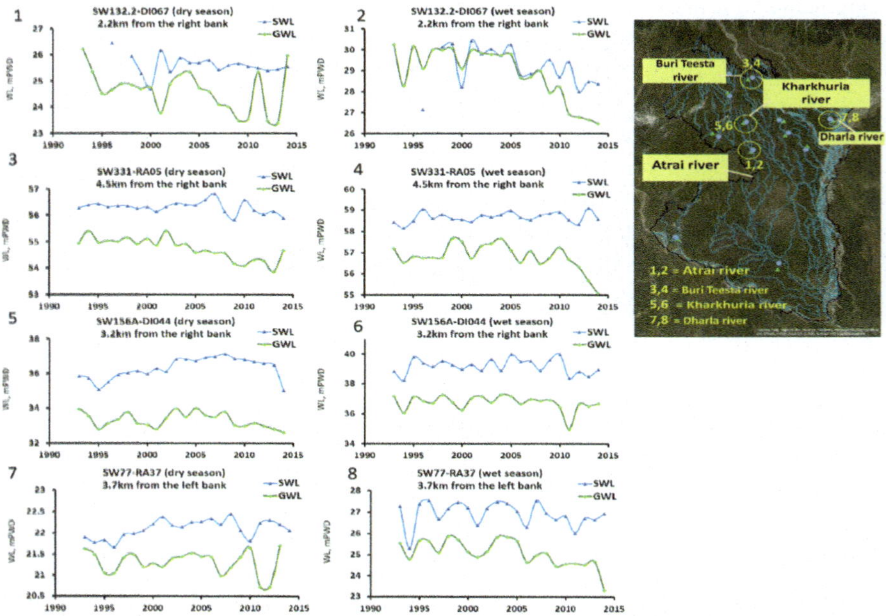

Fig. 7 Records that show loosing streams contributing to groundwater table whole the year round

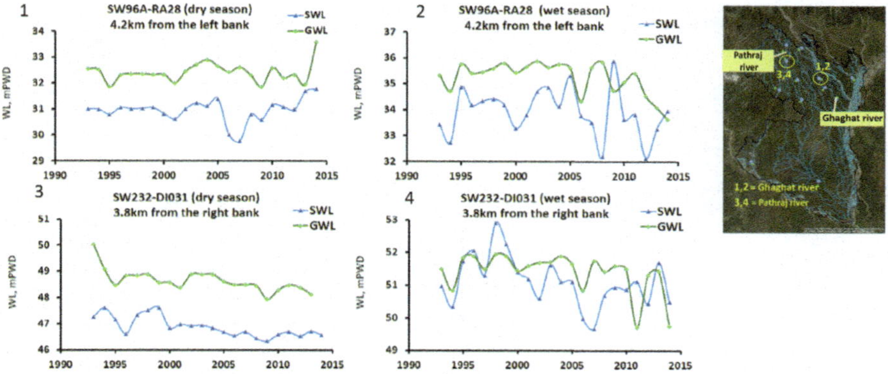

Fig. 8 Records that show gaining streams where groundwater table discharge most part of the years

are observed as gaining streams where groundwater discharges to the river in most years (Fig. 8). On the other hand, it is observed that SW and GW interaction is reverse to each other in different seasons for the Mahananda river as well as the upper and middle part of the Ghagot river. In general, for Bangladesh, this type of interacting rivers are common. Because during the dry season farmers in the NW mainly cultivates Boro rice mostly using groundwater. As a consequence, during this period, over-extraction for irrigation water results in rapid GWL drops. These anthropogenic phenomena cause declination of GWL more than the SW level. That is why during the dry season river water starts flowing towards the GW to replenish the over-extraction of subsurface water. On the other hand, the process gets reversed during the wet season. Here, 80% rainfall occurs in wet season along with a huge up-stream flow passing through the rivers. As indicated in the existing high correlation observation between annual rainfall and the water table rise, rainfall accelerates the GW replenishment process during the wet season. As a result, pressure over GW becomes almost none and rainfall contributes greatly, filling up the water demand in agriculture and other needs. Accordingly, the GWL reaches very close to the ground surface with a position higher than the SW level. Thereby, during this wet season, GW provides water to the river channel. Figure 9 here represents such common SW-GW interaction phenomena of the NW region.

5 Conclusion

Groundwater is used intensively in Bangladesh. National estimation of groundwater assumes that recharge in Bangladesh derives solely from the direct infiltration of rainfall and slightly from the indirect linkages of river channels. Based on the observational records and analysis thereof, the study attempts to revisit the existing correlation statistics between annual rainfall and GWL. This research also explored

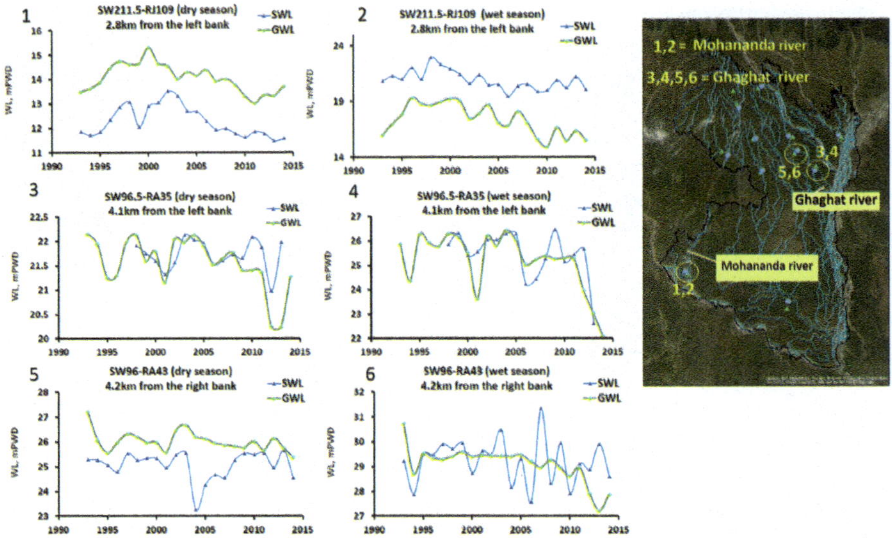

Fig. 9 Records that show interactions between Surface water level (SWL) and Groundwater level (GWL) are reverse to each other in different seasons of the years

the interaction between selective groundwater and surface water levels. The close proximity between groundwater and surface water exhibits interchangeable behaviors, e.g., losing streams during the wet season turns into a case of gaining streams during the dry season.

Trends in recovery rate during the wet season is found lower than that of dry season water level declination, indicating less replenishment rate to meet the high demand for extraction. On the other hand, lower Atrai Basin along with upper parts of Tista and Old Himalayan floodplain exhibit a greater water table fluctuation trend indicating plausibility of induced recharge. Lowering or no trend for WTF in the rest of the NW region can be attributed by the higher GW withdrawal causing aquifer properties to change from semi-confined to unconfined.

The spatial and temporal statistical analyses for the unconfined shallow aquifers in the NW region indicates that there is a good hydraulic connection between GWL and the annual rainfall. Rainfall is found to have reasonable correlations with the GWL, suggesting a significant contribution of direct recharge from the rainfall as a source. However, there exist some stations that do not have strong correlation proposing the existence of other controlling factors, e.g., localized pumping, over irrigational abstraction, etc. The study also notices that Pleistocene soil under the Barind Tract shows a high drop of GWL, and its requirement needs to be addressed in the future for preventing further declination of GW. Notably, poor statistics are found in cases where natural recharge condition might be absent due to the strong influence of irrigational abstraction of GW or the closer proximity of an influent river. However, due to a shortage of available data and monitoring stations, only an initial level of assessment (i.e., recharge or discharge contribution) could be done.

Further, the method does not consider the impact of other factors like soil condition, top soil characteristics, or some local effects (such as pumping, stream leakage effect, etc.). Albeit this backdrop, it is expected that this intensive study will open a window of opportunity for an initial level assessment of recharge for unconfined shallow aquifers for the North West region of Bangladesh. Such comprehensible knowledge on recharge mechanism will thus help in promoting better future prediction of the highly dynamic shallow aquifer system.

References

Adeleke, O. O., Makinde, V., Eruola, A. O., Dada OF, Ojo, A. O., & Aluko, T. J. (2015). Estimation of groundwater recharges in Odeda local government area, Ogun State, Nigeria using Empirical Formulae. *Challenges, 6*(2), 271–281. https://doi.org/10.3390/challe6020271.

Adhikary, S. K., Sharif, A. A., Das, S. K., & Saha, G. C. (2014). *Geostatistical analysis of groundwater level fluctuations in the shallow aquifer of Northwestern Bangladesh.* In Proceedings of the second International Conference on Civil Engineering for Sustainable Development (ICCESD-2014), 14–16 February, KUET, Khulna, Bangladesh. isbn: 978-984-33-6373-2.

Ahmadian, M., & And Chavoshian, M. (2012). Spatial variability zonation of groundwater table by use geo-statistical methods in Central Region of Hamadan Province. *Annals of Biological Research, 3*(11), 5304–5312.

Ali, M. H., Abustan, I., Rahman, M. A., & Haque, A. A. M. (2012). Sustainability of groundwater resources in the North-Eastern region of Bangladesh. *Water Resources Management, 26*(3), 623–641. https://doi.org/10.1007/s11269-011-9936-5.

Bartolino, J. R., & Cunningham, W. L. (2003). *Ground-water depletion across the nation.* U.S Geological Survey Fact Sheet, pp. 103–103.

BBS. (2018). *45 Years agriculture statistics of major crops (Aus, Amon, Boro, Jute, Potato & Wheat).* Bangladesh Bureau of Statistics: Statistics and Informatics Division, Ministry of Planning, Government of the People's Republic of Bangladesh.

BDP. 2100 Baseline Study of Water Resources Management, Volume 1(B). Bangladesh Planning Commission: General Economics Division, Ministry of Planning, Government of the People's Republic of Bangladesh.

Dey, N. C., Alam, M. S., Sajjan, A. K., Bhuiyan, M. A., Ghose, L., Ibaraki, Y., & Karim, F. (2012). Assessing environmental and health impact of drought in the Northwest Bangladesh. *Journal of Environmental Science and Natural Resources, 4*(2), 89–97.

Ebrahimi, H., Ghazavi, R., & Karimi, H. (2016). Estimation of groundwater recharge from the rainfall and irrigation in an arid environment using inverse modeling approach and RS. *Water Resources Management, 30*(6), 1939–1951. https://doi.org/10.1007/s11269-016-1261-6.

Gleeson, T., Vandersteen, J., Sophocleous, M. A., Taniguchi, M., Alley, W. M., Allen, D. M., & Zhou, Y. (2010). Groundwater sustainability strategies. *Nature Geoscience, 3*(6), 378–379. https://doi.org/10.1038/ngeo881.

Hossain, D., Islam, M., Sultana, N., & Tusher, T. (2015). Assessment of iron contamination in groundwater at Tangail Municipality, Bangladesh. *Journal of Environmental Science and Natural Resources, 6*(1), 117–121. https://doi.org/10.3329/jesnr.v6i1.22051.

Jahan, C. S., Islam, M. D. A., Mazumder, Q. H., Asaduzzaman, M. D., Islam, M. D., Islam, M. D. O., & Sultana, A. (2007). Evaluation of depositional environment and aquifer condition in Barind Area, Bangladesh using gamme ray well log data. *Journal of the Geological Society of India, 70*(6), 1070–1076.

Jahan, C. S., Mazumder, Q. H., Islam, A. T. M. M., & Adham, M. I. (2010). Impact of irrigation in Barind Area, NW Bangladesh – An evaluation based on the meteorological parameters and fluctuation trend in groundwater table. *Journal of the Geological Society of India, 76*(2), 134–142. https://doi.org/10.1007/s12594-010-0085-x.

Kumar, C., & Seethapathi, P. (2002). Assessment of natural groundwater recharge in Upper Ganga Canal Command Area. *Journal of Applied Hydrology, 247667*(15), 1–9.

Livingston, A. (2009). *Towards pragmatic action for addressing groundwater overexploitation in India*. World Bank.

Mustafa, S. M. T., Abdollahi, K., Verbeiren, B., & Huysmans, M. (2017). Identification des Facteurs Influençant la Sécheresse et le Rabattement des Eaux Souterraines au Nord-Ouest du Bangladesh. *Hydrogeology Journal, 25*(5), 1357–1375.

Nowreen, S. (2017). *Mechanism of groundwater response to recharge and its quantification for shallow aquifers in Bangladesh*. PhD thesis, Bangladesh University of Engineering and Technology.

R Core Team. (2013). *R: A language and environment for statistical computing*. R Foundation for Statistical Computing.

Rahman, A. A., & Ravenscorft, P. (2003). *Groundwater resources and development in Bangladesh background to the arsenic crisis, Agriculture potential and the environment*. The University Press Limited (UPL).

Rahman, A. T. M. S., Kamruzzaman, M., Jahan, C. S., Mazumder, Q. H., & Hossain, A. (2016). Evaluation of spatio-temporal dynamics of water table in NW Bangladesh: An integrated approach of GIS and statistics. *Sustainable Water Resources Management, 2*(3), 297–312. https://doi.org/10.1007/s40899-016-0057-4.

Rai, R., Zhang, Y., Paudel, B., Li, S., & Khanal, N. (2017). A synthesis of studies on land use and land cover dynamics during 1930–2015 in Bangladesh. *Sustainability, 9*(10), 1866.

Sajjan, A. K. (1998). *Characterization of 1994–95 drought in North Westernregion of Bangladesh and its impact on agriculture*. MSc thesis. Bangladesh University of Engineering and Technology.

Sarkar, A., & Ali, M. H. (2009). Water table dynamics of Dhaka City and its long-term trend analysis using the 'MAKESENS' model. *Water International, 34*(3), 373–382. https://doi.org/10.1080/02508060903115183.

Shahid, S. (2011). Impact of climate change on irrigation water demand of dry season Boro rice in Northwest Bangladesh. *Climatic Change, 105*(3–4), 433–453. https://doi.org/10.1007/s10584-010-9895-5.

Shahid, S., & Behrawan, H. (2008). Drought Risk Assessment in the Western Part of Bangladesh. *Natural Hazards, 46*(3), 391–413. https://doi.org/10.1007/s11069-007-9191-5.

Shahid, S., & Hazarika, M. K. (2010). Groundwater drought in the northwestern districts of Bangladesh-1989-2006. *Water Resources Management, 24*(10).

Shamsudduha, M., Chandler, R. E., Taylor, R. G., & Ahmed, K. M. (2009). Recent trends in groundwater levels in a highly seasonal hydrological system: The Ganges-Brahmaputra-Meghna Delta. *Hydrology and Earth System Sciences, 13*(12), 2373–2385. https://doi.org/10.5194/hess-13-2373-2009.

SRDI. (2007). *Soil physiography map of Bangladesh. Soil and Land Classification Survey Section, Soil Resource Development Institute (SRDI)*. Dhaka, Bangladesh.

Analysis on Flow and Water Balance Parameters of Teesta River Basin due to Climate Change and Upstream Intervention

Reazul Zannah, Khaled Mohammed, Afeefa Rahman, and Anika Yunus

1 Introduction

The fourth-largest trans-boundary river of Bangladesh, Teesta has become an important issue due to limited water availability for water sharing and diversion at upstream of catchment, political issues as well as climate change. Teesta River encompasses 12,159 km^2 of which 2004 km^2 area lies within Bangladesh (Prasai and Surie 2013). The construction of Gazoldoba barrage in the upstream reach of Teesta within India has reduced the water availability at the downstream reach (Khalid 2013). Through this barrage, the dry flow of Teesta is controlled mainly for the purpose of, slowing down the river flow within Bangladesh (Khalid 2013; Khan 2018). Moreover, it is a Himalayan River originating from glaciers in Sikkim up North and Sikkim has recently faced several sudden and devastating glacial lake outburst floods (Prasai and Surie 2013). Thus, climate change induced snowmelt and increased precipitation may lead to frequent river flooding during of the Teesta river basin during the monsoon season.

This study mainly focuses on water availability in the Bangladesh region as it relates to flow in this river which in turn is highly dependent on the amount of water released by the Gazoldoba barrage in India. A comparative analysis of flow and the water balance parameter has been made for future climate change scenarios, considering the impact it has for the people of five northern districts of Bangladesh. The study applies a physically-based hydrological model for the Teesta river basin to assess the potential impact of climate change on water availability and water balance

R. Zannah (✉) · A. Rahman · A. Yunus
Department of Water Resources Engineering (WRE), Bangladesh university of Engineering and Technology (BUET), Dhaka, Bangladesh

K. Mohammed
Department of Civil Engineering, University of Sherbrooke, Sherbrooke, Canada

A. Haque, A. I. A. Chowdhury (eds.), *Water, Flood Management and Water Security Under a Changing Climate*, https://doi.org/10.1007/978-3-030-47786-8_19

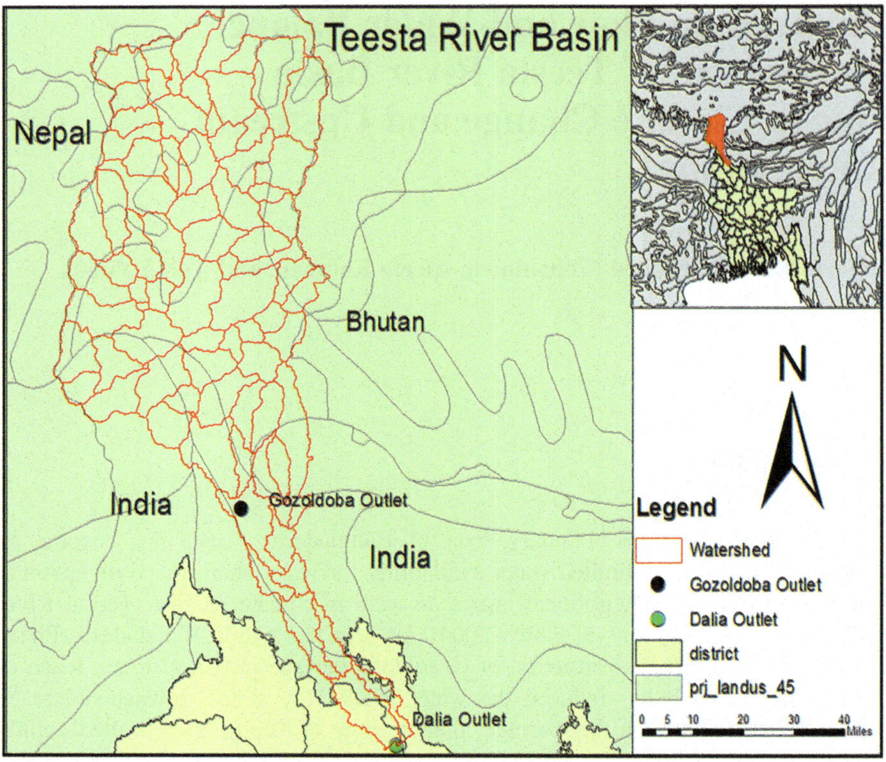

Fig. 1 Study area showing Teesta River basin

parameters in Teesta river basin using Soil Water Assessment Tool (SWAT). The developed model has been calibrated and validated for climate normal period of 2001–2016. Figure 1 shows the study area of the Teesta basin.

2 Methodology

2.1 Model Description

As Soil Water Assessment Tool (SWAT) is a physically-based hydrological model, it uses physiographical data such as elevation, soil use, land use, meteorological data, and river discharge (Arnold and Allen 1996). The basin of the concerned area is divided into a number of sub-basins and the sub-basins are further divided into several Hydrological Response Units (HRU) that are assumed spatially uniform based on land use, soil use and management practice. These subdivisions of the watershed are useful to reflect differences in evapotranspiration for various crops and soils of the model. In SWAT, the water balance equation is used to simulate the hydrologic cycle.

$$SW_t = SW_o + \sum_{(t=1)}^{i} \left(R_{day} - Q_{surf} - E_a - W_{seep} - Q_{gw}\right)$$

Where, SW_t is the final soil water content, SW_0 is the initial soil water content on day i, t is the time in days, R_{day} is the amount of precipitation on day i, Q_{surf} is the amount of surface runoff on day i, Ea is the amount of evapotranspiration on day i, W_{seep} is the amount of water entering the zone from the soil profile on day i and Q_{gw} is the amount of return flow on day i (All units except the time are presented as mm of water).

2.2 *Data Collection for the Setup of Hydrological Model of Teesta River Basin in SWAT*

To develop the hydrologic model of Teesta river basin, a Digital Elevation Model (DEM) of the study area, having 90 m grid resolution, downloaded from SRTM (Shuttle Rudder Topography Mission), was used in watershed delineation. Soil map (1 km grid resolution) of the selected area provides soil properties such as particle-size distribution, bulk density, organic carbon content, available water capacity, saturated hydraulic conductivity, etc. Land use map (1 km grid resolution) of the basin area, consisting of different classes of the land use type, was collected from the U.S. Geological Survey (USGS). To simulate the hydrological process, daily values of precipitation, maximum and minimum temperature (the minimum required daily data to run SWAT model) were collected from WFDEI (WATCH forcing datasets) for the period of 1972–2016. Discharge data at Dalia within Bangladesh was collected for the period of 2001–2016 from the Bangladesh Water Development Board (BWDB).

2.3 *Model Setup and Sensitivity Analysis*

The first step in the model setup was watershed delineation where the filling sink, defining flow direction and accumulation were done automatically through the user interface and the basin was divided into 83 sub-basins. To extract the land use and soil information of the Teesta river basin, soil and land use maps were loaded onto SWAT and these were overlaid to define the hydrologic response units (HRUs). Daily precipitation, maximum and minimum temperature data are loaded with a period of 1972–2016. To estimate surface runoff volume, the Soil Conservation Service (SCS) curve number procedure was applied. Using Hargreaves and Muskingum methods, the potential evapotranspiration (PET) estimate and channel routing was performed respectively. For rainfall distribution, the skewed normal distribution method has been used. The model has been simulated from the year 2001–2016 with a daily time step and the first 2 years of the simulation have been considered as warm-up period.

In this study, the automatic sensitivity analysis was carried out using SWAT-CUP to understand the degree of sensitivity of each parameter for a small change in the model input. As a watershed is divided into HRUs, the smallest unit of spatial disaggregation based on elevation, soil and land use, a distributed spatial parameter such as hydraulic conductivity, bulk density, or CN2 (Initial SCS surface runoff curve number for moisture condition II) can potentially be defined for each HRU. Because of unavailability, it is a difficult task to collect a large number of input parameters and so an alternative approach for the estimation of distributed parameters is to lump them based on soil type, land use type, location slope or a combination of these. In this study, All 27 parameters are divided into six types of parameterization scheme. These are general watershed attribute (.bsn), land and water management practice (.mgt), HRU input file (.hru) which refer to topographic characteristics, water flow, depression storage area etc., main channel input file (.rte) which summarizes the physical characteristics of the channel which affect water flow and transport of sediment, nutrients and pesticides, groundwater input file (.gw) which initialize the properties governing water movement into and out of the aquifers, and the soil input files (.sol) which define the physical properties for all layer in the soil. Out of 27 selected parameters SCS runoff curve number (management parameter) and base flow alpha factor for bank storage (.rte parameter) affected the flow highly.

2.4 Calibration and Validation

Using the Sequential Uncertainty Fitting (SUFI-2) implemented in SWAT-CUP, the model was calibrated from 2001 to 2007 and validated from 2008 to 2016, with monthly observed stream flow data at Dalia station to improve the fit between the observed and simulated flow. The Sequential Uncertainty Fitting (SUFI-2) implemented in SWAT-CUP (Abbaspour et al. 2004) was used to perform the auto-calibration. Figure 2 shows the graphical representation of monthly observed and simulated flow for both the calibration and validation period depicting good match with the observed and simulated flow during both the monsoon and dry season.

Nash–Sutcliffe efficiency value (NSE) and the coefficient of determination (R^2) help to evaluate the model performance statistically. NSE can interpret the model performance in replicating individually observed values while the R^2 is only useful for measuring the deviation from the best fit line. The systematic difference between the model and observations are indicated by percentage difference (PBIAS) and the ratio of the root mean square error between simulated and observed values to the standard deviation of the observations (RSR) has been used (Moriasi et al. 2007). The statistical model performance is given in Table 1. The NSE values are 0.8 and 0.7 for calibration and validation period respectively. These are in the range of very good and good (very good for $0.75 < NSE < 1.00$ and good for $0.65 < NSE < 0.75$). The co-efficient of determination (R2) values are 0.81 for calibration and 0.75 for

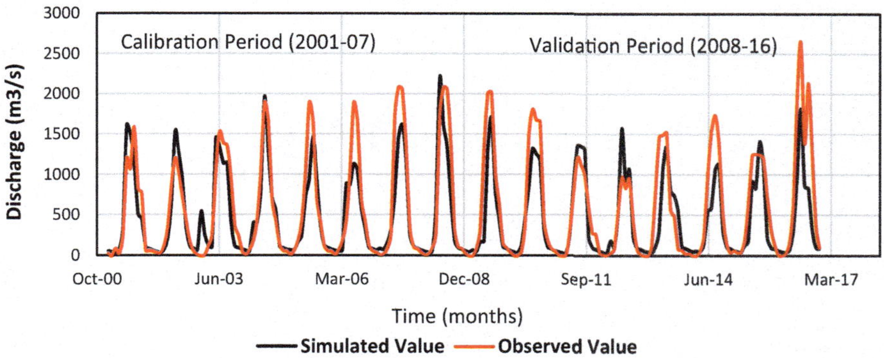

Fig. 2 Observed and simulated reservoir flows for the calibration and validation period

Table 1 Model performance statistics for calibration (2001–07) and validation (2008–16) of the Teesta river basin

Period	Observed mean (m^3/s)	Simulated mean(m^3/s)	Model performance			
			NSE	R^2	PBIAS	RSR
Calibration	494	545	0.8	0.81	11.53	0.31
Validation	570	502	0.7	0.75	24.79	0.43

validation period. These are also in satisfactory range (Satisfactory for R2 > 0.5) (Moriasi et al. 2007). The PBIAS and RSR values are also in the range of good to very good for both calibration and validation.

3 Result and Discussion

3.1 Assessment of the Impact of Climate Change on Flow Availability of Teesta River Basin

For GCMs, RCP scenarios were analyzed to assess temperature and precipitation changes (Alam 2015). For three separate periods i.e. 2010–2039 (2020s), 2040–2069 (2050s) and 2070–2099 (2080s), results are obtained for RCP 2.6 and RCP 8.5. Monthly precipitation and temperature data for each model were averaged and compared with the base period data (2001–2016). In this study, MIROC-ESM-CHEM and HADGEM2-ES model are used as climate change scenarios. Generally MIROC-ESM-CHEM and HADGEM2-ES models are suitable for Indian subcontinent, and popular to use as they give the results not so underestimating or overestimating of spatio-temporal variability of aerosol species (Sanap et al. 2014; Alam 2015).

Climate change scenarios of the models including MIROC-ESM-CHEM RCP 8.5, MIROC-ESM-CHEM RCP 2.6, HadGEM2-ES RCP 8.5 and HADGEM2-ES

RCP 2.6 for the 2020s, 2050s, and 2080s with respect to that of the climate normal were analyzed. It was found that the discharge at Dalia of Teesta River is increasing for all the selected models over the twenty-first century except HADGEM2-ES RCP 2.6 at the 2050s. The enhanced precipitation usually causes an increase in discharge while the enhanced temperature causes a decrease in discharge by increasing evaporative losses.

3.1.1 Mean Monthly Dry Period Flow Analysis for Future Regime

Mean dry season (Dec–April) flow was found to slightly increase for all the scenarios in the 2020s, 2050s and 2080s. In the 2020s, the maximum projected increase in discharge was found for MIROC-ESM-CHEM (RCP 8.5) with 56.54% change and the minimum projected change was −4.83% which was obtained for HADGEM2-ES (RCP 2.6). From the graph of Fig. 3a, it is observed that the line of MIROC-ESM-CHEM RCP 8.5 shows the abrupt change in discharge and gives the maximum value around above 1700 cumecs. In the 2050s, the maximum projected increase in discharge was found for MIROC-ESM-CHEM (RCP 8.5) with 82.54% change and the minimum projected decrease was 4.97% which was obtained for HADGEM2-ES (RCP 8.5). From the graph of Fig. 3b, it is observed that the line of MIROC-ESM-CHEM RCP 8.5 shows the abrupt change in discharge and gives the maximum value around above 1900 cumecs. In the 2080s, the maximum projected increase in discharge was found for MIROC-ESM-CHEM (RCP 8.5) with 187%

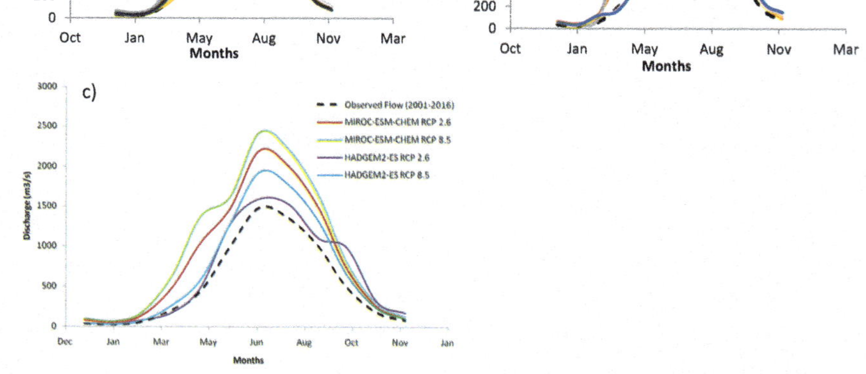

Fig. 3 Prediction of flow due to different climate model for future regime for scenarios RCP2.6 and RCP8.5: (**a**) 2020s (**b**) 2050s (**c**) 2080s

change and the minimum projected decrease was −22.05% which was obtained for HADGEM2-ES (RCP 8.5). From the graph of Fig. 3b, it is observed that the line of MIROC-ESM-CHEM RCP 8.5 shows the abrupt change in discharge and gives the maximum value around above 2600 cumecs. Figure 3a, b, c show the details of prediction of flow due to different climate model for scenarios RCP2.6 and RCP8.5 for future regime.

3.1.2 Mean Monthly Wet Period Flow Analysis for Future Regime

In the 2020s, the mean wet season (May-Nov) flow was found to increase for almost all the scenarios.

Severe flow alteration (around 40–200 m^3/s) occurs in the month of December to February in the Teesta River which is the main source of water in northern drought-prone region of Bangladesh. But in the months, May and November the flow is around more than 300 m^3/s. So these 2 months May and November are also included in wet season especially for Teesta River basin (Mullick et al. 2010).

The highest increase in projected flow was found for HADGEM2-ES (RCP 8.5) with 8.03% change. The highest decrease in discharge was found for the driest scenario MIROC-ESM-CHEM (RCP 8.5) with −4.62% change. In the 2050s, mean wet season (May-Nov) flow was found to increase for almost all the scenarios except HADGEM2-ES (RCP 2.6). In this scenario, flow decreased around 10.87% during the wet season.

The highest increase in projected flow was found for HADGEM2-ES (RCP 8.5) with 32.76% change. In the 2080s, mean wet season (May–Nov) flow was found to increase for almost all the scenarios. The highest increase in projected flow was found for HADGEM2-ES (RCP 8.5) with 47.6% change. Figure 3a, b, c show the prediction of flow availability due to different climate models for scenarios RCP 2.6 and RCP8.5 in the 2020s, 2050s, and 2080s.

3.2 Mean Annual Stream Flow Analysis

The percentage changes in the mean annual stream flow from the climate normal (2001–2016) with respect to climate change scenarios have been simulated by the model. Figure 4 shows boxplots of differences in the mean annual average stream flow simulated by the model. A gradual increase in annual average flow is found from the 2020s to 2080s (Exception: HADGEM2-ES RCP 2.6, 2050s). The range of 75th and 25th percentile has increased from the 2020s to 2080s, which indicates that uncertainty associated with the projected stream flow of Teesta River at Dalia catchment is likely to increase in the future.

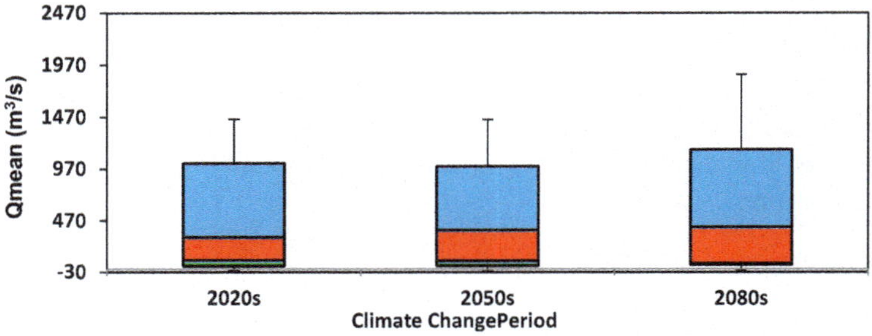

Fig. 4 Comparison among future flow regime for HADGEM2-ES RCP 2.6. (Blue boxplots show 75th percentile, Red boxpots show 50th percentile and Green boxplots show 25th percentile values)

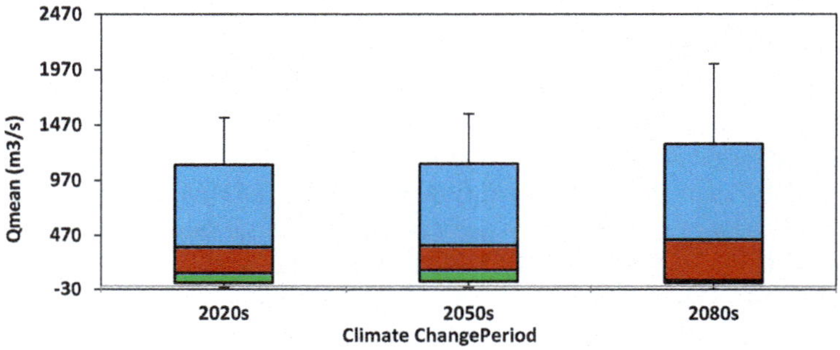

Fig. 5 Comparison among future flow regime for HADGEM2-ES RCP 8.5. (Blue boxplots show 75th percentile, Red boxepots show 50th percentile and Green boxplots show 25th percentile values)

3.2.1 Scenario for HADGEM2-ES RCP 2.6

The scenario of HADGEM2-ES for RCP 2.6 is unique. Figure 4 shows the boxplots of comparison in the mean annual average stream flow simulated for the different future regimes by the model for HADGEM2-ES RCP 2.6. In the 2050s, the average value for the flow of wet and dry season is higher than in the 2020s, but the maximum flow is lower than 2020s, so the maximum value in whisker plot shows smaller value. The range of variability is higher for the 2080s compared to the early twenty-first century. The range for 25th and 75th percentile flow is very high for the 2080s compared to others. The comparatively median value range is higher during the 2050s, whereas it is very low in the 2080s. The range between 75th and 25th percentile is relatively high for the 2080s.

3.2.2 Scenario for HADGEM2-ES RCP 8.5

The mean dry season stream flow increase is found to be gradual from the 2020s to 2080s. Figure 5 shows the boxplots of comparison in the mean annual average stream flow simulated for the different future regimes by the model for HADGEM2-ES RCP 8.5. Similar to the mean annual stream flow, the range of variability is higher for the 2080s compared to the early twenty-first century. The range for 25th percentile flow is relatively very low for the 2080s compared to others, where the 75th percentile is high range. The comparatively median value range is found higher during the 2050s whereas it is very low in case of 2020s. The range between 75th and 25th percentile is relatively high for the 2080s, with significant variation between the highest and the lowest value. The increasing rate from 2050s to 2080s is relatively low. The figure shows the comparison among future flow regimes.

3.2.3 MIROC-ESM-CHEM RCP 2.6

In case, MIROC-ESM-CHEM RCP 2.6, the mean dry season stream flow increase is found to be gradual from the 2020s to 2080s while the rate of increase enhanced from 2050s to 2080s. Comparable to the mean annual stream flow, the range of variability is higher for the 2080s compared to the early twenty-first century. Figure 6 shows the boxplots of comparison in the mean annual average stream flow simulated for the different future regimes by the model for MIROC-ESM-CHEM RCP 2.6. The range for 25th percentile flow is relatively very low for the 2080s compared to others. The comparatively median value range is found higher during the 2080s whereas it is very low in case of 2020s. Mean wet season flow showed a slight increasing pattern in the 2020s to 2050s which then decreased from 2050s to 2080s. The range between 75th and 25th percentile is relatively high for the 2080s with significant variation between the highest and the lowest value.

3.2.4 MIROC-ESM-CHEM RCP 8.5

In case of RCP 8.5, the mean dry season stream flow increase is found to be gradual from the 2020s to 2080s, while the rate of increase again enhanced from 2050s to

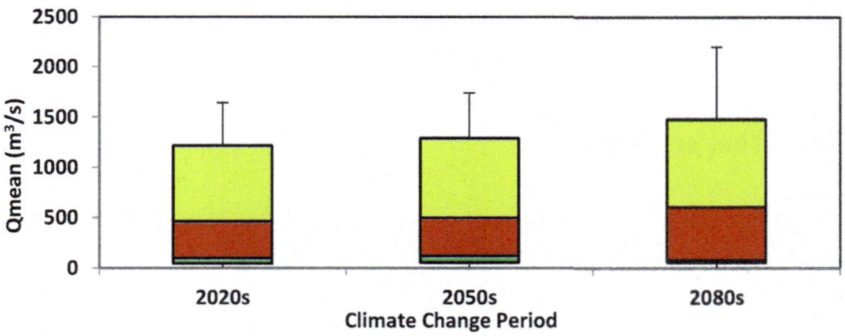

Fig. 6 Comparison among future flow regime for MIROC-ESM-CHEM RCP 2.6. (Blue boxplots show 75th percentile, Red boxplots show 50th percentile and Green boxplots show 25th percentile values)

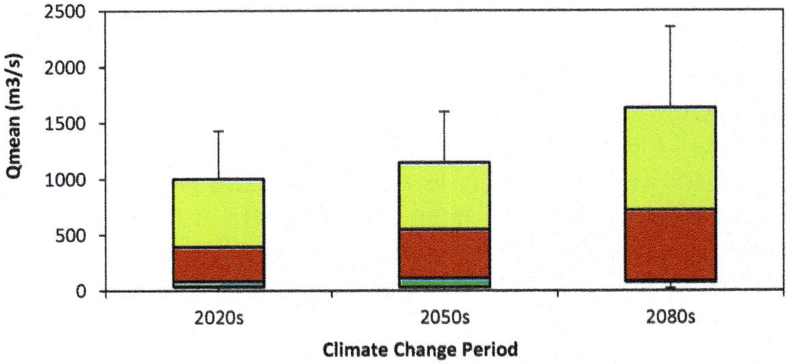

Fig. 7 Comparison among future flow regime for MIROC-ESM-CHEM RCP 8.5. (Blue boxplots show 75th percentile, Red boxplots show 50th percentile and Green boxplots show 25th percentile values)

2080s. Figure 7 shows the boxplots of comparison in the mean annual average stream flow simulated for the different future regimes by the model for MIROC-ESM-CHEM RCP 8.5.

Similar to the mean annual stream flow, the range of variability is higher for the 2080s compared to the early twenty-first century. Mean wet season flow showed a slight increasing pattern between the 2020s and 2050s, which then highly increased from 2050s to 2080s. The range between 75th and 25th percentile is relatively small for the 2080s, despite of significant variation between the highest and the lowest value. The figure shows the comparison among future flow regime.

3.3 *Flow Diversion Scenario*

As mentioned above, there are a number of constructed structures which affect the flow of the Teesta river basin. The nearest structure, Gazoldoba Barrage at Gazoldoba in India, (upstream of the Dalia, Bangladesh) has been considered for simulation (Khalid 2013; Prasai and Surie 2013; Khan 2018).

3.3.1 Diversion at Gazoldoba

The model has been simulated for six different flow diversion scenarios, each considering a portion of upstream flow diverted through Gazoldoba side canal to simulate the changes in water availability for the Bangladesh portion of Teesta river basin due to upstream developments. Within Table 2, the changes in mean monthly flow at Dalia point for different flow diversion scenarios have been presented. It can be seen from Table 2 that with the increasing rate of diversion, the mean monthly flow at Dalia point decreases. Since the flow diversion activates in only the lean

Table 2 Mean monthly flow of the Teesta River at Dalia due to different flow diversion

Mean monthly flow at Dalia (% flow diversion through Gozozldoba)								
Month	Base(2001–2016)	15%	25%	35%	45%	55%	65%	Remarks
Jan	38.46	32.87	29.33	25.79	22.36	19.51	16.88	Diversion
Feb	27.23	22.75	20.36	18.25	16.33	15.04	13.16	
Mar	65.09	55.07	50.22	46.31	42.93	40.83	37.03	
Apr	220.20	185.26	165.15	149.06	134.82	120.47	110.98	
May	478.66	478.56	478.60	478.42	478.27	478.22	478.56	No diversion
Jun	1101.17	1101.1	1101.16	1101.16	1101.16	1101.16	1101.16	
Jul	1642.91	1642.9	1642.91	1642.91	1642.91	1642.91	1642.91	
Aug	1501.01	1501.0	1501.01	1501.01	1501.01	1501.01	1501.01	
Sep	1115.16	1115.1	1115.15	1115.15	1115.15	1115.15	1115.15	
Oct	547.37	465.45	415.46	363.91	319.18	284.41	246.22	Diversion
Nov	200.11	171.03	152.85	132.80	116.53	97.15	79.17	
Dec	89.69	76.69	68.40	60.36	53.61	43.44	36.45	

Table 3 Percent reduction of Flow at Dalia due to different flow diversion scenarios at Gozoldoba

% reduction of flow at Dalia compared to base							
month	15	25	35	45	55	65	Remarks
Jan	14.51	23.73	32.92	41.84	49.25	56.1	Diversion
Feb	16.43	25.21	32.95	40.01	44.74	51.65	
Mar	15.38	22.83	28.85	34.03	37.26	43.1	
Apr	15.86	25	32.30	38.77	45.29	49.6	
May	0.02	0.013	0.05	0.081	0.092	0.021	No diversion
Jun	0	0	0	0	0	0	
Jul	0	0	0	0	0	0	
Aug	0	0	0	0	0	0	
Sep	0	0	0	0	0	0	
Oct	14.96	24.09	33.51	41.68	48.04	55.01	Diversion
Nov	14.53	23.61	33.63	41.76	51.45	60.43	
Dec	14.48	23.72	32.7	40.22	51.56	59.35	

Table 4 Maximum percent reduction of flow at Dalia due to different flow diversions

% Flow through diversion Gozoldoba	Max reduction in flow at Dalia %
15	−16.43
25	−25.21
35	−33.63
45	−41.84
55	−51.56
65	−60.43

season (October–April), the monsoon flow for all the diversion scenarios remain the same. In the month of February, the mean flow decreases at its lowest and with 65% flow diversion the mean flow reduces to 13.16 cumecs for 65% diversion. The situation also slowed down in January where for 65% flow diversion, the mean monthly discharge decreased to 16.83 cumecs only. This amount of water diversion has a substantial impact on the Teesta river basin within Bangladesh. During the month of January, February, and March, water availability appears to be the lowest. The reduction of discharge at Dalia point due to different flow diversion scenarios is presented in percentage reduction compared to the base condition in Table 3. From Table 3, it can be seen that only during 15% flow diversion, the flow at Dalia reduces by 15%. Whereas, in the other scenarios, the flow decreases lesser than the amount diverted through Gazoldoba. In the lean period, the highest amount of reduction compared to base flow occurs during the month of January when for 65% flow diversion, the flow reduction amounts to 56.12%. The maximum reduction of flow at Dalia point due to different flow diversion percentages is shown in Table 4.

From Table 4, simple linear Regression analyses was carried out to find the reduction of flow at Dalia point. The following equation has been derived from the regression analysis-

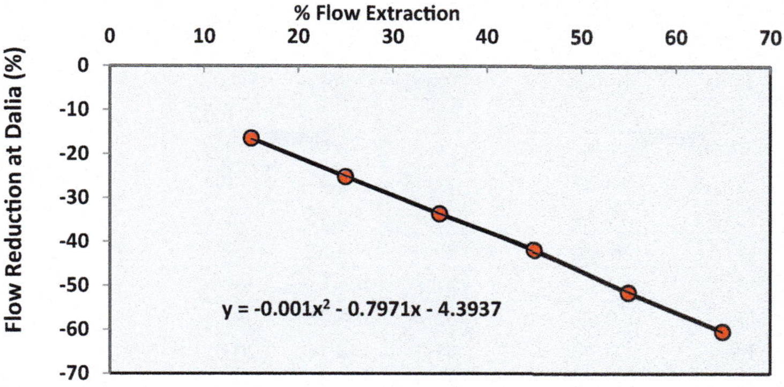

Fig. 8 The flow reduction at Dalia due to different flow extraction scenarios at Gazoldoba

$$\mathbf{Q_d = -(0.001 * Q_g{}^2 + 0.797 * Q_g + 4.393)}$$

Here, Q_d is the maximum percent flow reduction at Dalia point and Qg is the percentage flow diversion at Gazoldoba. For example, if all the water at Gazoldoba is to be diverted, then Qg = 100, and from the above equation Qd = −94.639 flow reduction from the base condition may occur at the Dalia point due to 100% flow diversion through Gazoldoba. The flow reduction at Dalia due to different flow extraction scenarios at Gazoldoba has been presented graphically in Fig. 8.

3.4 Yearly Comparison on Water Balance Parameters

To analyze and compare the water balance, the main four segments are taken - Evapotranspiration (ET), Percolation (PERC), Surface Runoff (SURQ), Surface runoff contribution to streamflow during time step (mm H_2O) and Precipitation (PRECIP) such as Rainfall (Khan 2018). Table 5 shows the present monthly average value of availability of these water balance parameters (WYLD = Water Yield (mm H_2O), LAT_Q = Lateral flow) Pursuant to the SWAT analysis for GCMs for future regimes, all these parameters will increase deliberately. But precipitation will decrease in the dry season only in case of HADGEM2-ES. Precipitation will increase in the wet and dry season both for MIROC. MIROC-ESM-CHEM shows the severe result that means high precipitation (which may cause severe flooding), high surface runoff, evapotranspiration and percolation compared to others. In the case of both GCM, RCP 2.6 and RCP 8.5 are similar but RCP 8.5 shows very severe results. Such as in the regime for the 2080s for MIROC-ESM-CHEM RCP 8.5, comparing the lean season flow, we can see that the total precipitation in the lean season is increased to 261.01 mm (in base condition 220 mm). Notwithstanding, annual precipitation increased by 58.9%. This means that the precipitation in the Teesta basin is predicted to increase in the wet and dry seasons both. All these variations are

Table 5 Average monthly water availability of Teesta River Basin

Month	PRECIP (mm)	SNOWMELT (mm)	Et (mm)	PERC (mm)	SURQ (mm)	WYLD (mm)	LAT_Q (mm)
Jan	13.56	3.5	4	0.34	0.62	31.03	0.98
Feb	19.01	6.9	9.1	0.91	1.9	21.24	1.98
Mar	51.03	13.23	12.23	2.9	9.4	12.11	7.3
Apr	128.81	19.33	39.9	10.55	29.7	25.08	26.1
May	147.18	9.68	39.69	83.52	57	91.76	41.31
Jun	359.98	1.1	51.1	101.62	180	453.79	79.1
Jul	471.31	0.57	52.9	198.76	281	391.96	112.14
Aug	396.1	0.7	41.7	148.44	201.11	301.17	82.50
Sep	251.57	1.44	32.9	126.14	99.1	264.62	71.16
Oct	76.02	7.03	19.1	33.67	51.97	161.27	20.09
Nov	11.04	6.27	9.8	3.40	1.34	83.64	3.02
Dec	8.31	1.26	4.2	0.23	0.37	43.75	0.6

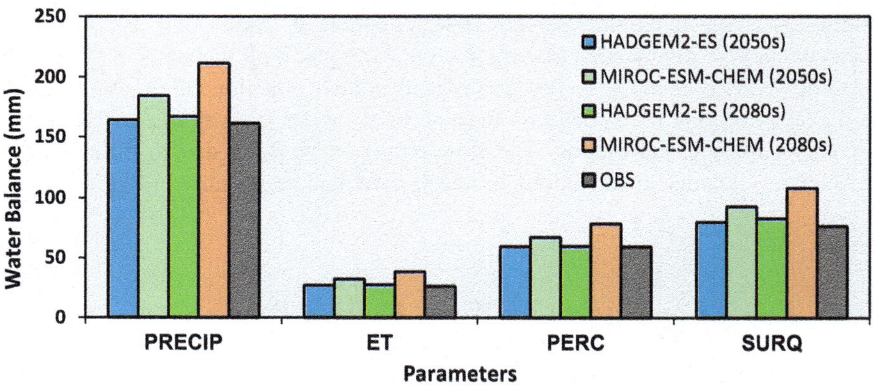

Fig. 9 Annual average comparison between MIROC-ESM-CHEM RCP 2.6 and HADGEM2-ES RCP 2.6 (2080s and 2050s)

shown in Figs. 9 and 10 for RCP 2.6 and RCP 8.5 respectively (both for MIROC-ESM-CHEM and HADGEM2-ES).

4 Conclusion

In this study, the potential impact of climate change on the future stream flow of the Teesta river basin has been assessed on the basis of a calibrated and validated physically based hydrological model which shows significant compliance with observed data. This flow and water balance parameter analysis helped identify the future condition of the Teesta river basin at Dalia catchment within Bangladesh and

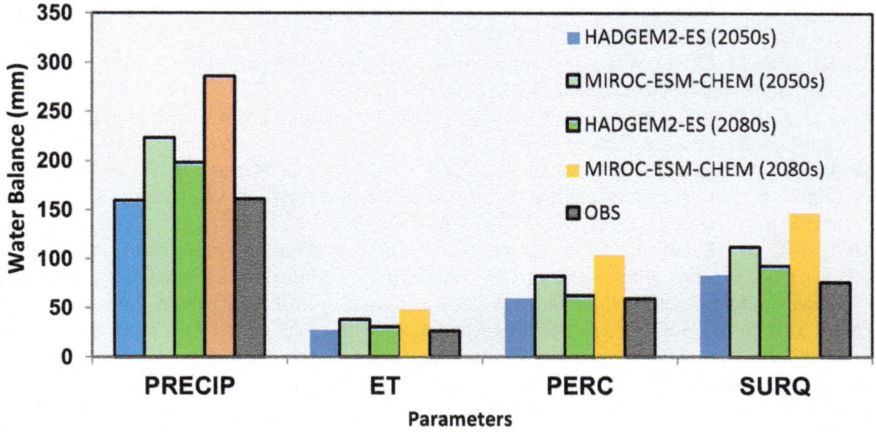

Fig. 10 Annual average comparison between MIROC-ESM-CHEM RCP 8.5 and HADGEM2-ES RCP 8.5 (2080s and 2050s)

this model can also be used for further research. According to mean monthly dry flow analysis, in the 2080s, the maximum projected increase in discharge was found for MIROC-ESM-CHEM (RCP 8.5) with 187% change and in case analysis of mean monthly wet flow, the highest increase in projected flow was found for HADGEM2-ES (RCP 8.5) with 47.6% change, which indicates potential for substantial impact from climate change. Besides this, based on these simulation results of mean annual stream flow of Teesta, under the impact of potential climate change, a high probability of flood and drought in the future can be expected. Flow diversion at Gazoldoba further exacerbate the risk of increasing floods and droughts. It is expected that result of this study will help identify and undertake take possible steps to minimize the impact of climate change in and around major river basins in Bangladesh.

Acknowledgement Authors would like to express their thanks to one another for their timely cooperation, guidance and productive criticism throughout the research. They intend to express gratitude to Bangladesh Water Development Board (BWDB), Institute of Water Modeling (IWM) for providing with necessary hydrological data, Shuttle RADAR Topography Mission and Institute of Water and Flood Management (IWFM) for giving necessary primary and secondary data sets.

References

Abbaspour, K. C., Johnson, C. A., & Van Genuchten, M. T. (2004). Estimating uncertain flow and transport parameters using a sequential uncertainty fitting procedure. *Vadose Zone Journal, 3*(4), 1340–1352.

Alam, S. (2015). *Impact of climate change on future flow of Brahmaputra river basin using swat model.*

Arnold, J. G., & Allen, P. M. (1996). Estimating hydrologic budgets for three Illinois watersheds. *Journal of Hydrology, 176*(1–4), 57–77.

Khalid, A. R. M. (2013). Water negotiations in the Ganges-Brahmaputra-Meghna Basin-"multilateral" is the way forward. *Hindu, 309*, 327.

Khan, I. (2018). *Effect in the water balance of the Teesta river basin due to different climate changes and upstream development.*

Moriasi, D. N., Arnold, J. G., Van Liew, M. W., Bingner, R. L., Harmel, R. D., & Veith, T. L. (2007). Model evaluation guidelines for systematic quantification of accuracy in watershed simulations. *Transactions of the ASABE, 50*(3), 885–900.

Mullick, R. A., Babel, M. S., & Perret, S. R. (2010, September). *Flow characteristics and environmental flow requirements for the Teesta River, Bangladesh.* In Proceedings of International Conference on Environmental Aspects of Bangladesh (ICEAB10), pp. 159–162.

Prasai, S., & Surie, M. D. (2013). *Political economy analysis of the Teesta River basin* (p. 37). New Delhi: The Asia Foundation.

Sanap, S. D., Ayantika, D. C., Pandithurai, G., & Niranjan, K. (2014). Assessment of the aerosol distribution over Indian subcontinent in CMIP5 models. *Atmospheric Environment, 87*, 123–137.

Selection of Appropriate Shifting of Tidal River Management

Rocky Talchabhadel, Hajime Nakagawa, and Kenji Kawaike

1 Introduction

The Bengal Delta has been home to a dynamic interaction of water, sediment and land for millennia. The SW coastal part of Bangladesh shown in Fig. 1 is characterized by an active tidal river system. Before the construction of earthen embankments enclosing *beel*s and keeping the main tidal rivers outside (polderization hereinafter), tidal rivers used to inundate vast regions of lowlands twice a day. And large volumes of sediment would deposit on the lowlands. Historically, people used indigenous practices to save their crops from tidal intrusion by making temporary low earthen embankments during the eight dry months of the year (*ostomashi bundh*s). After the harvest, those *ostomashi bundhs* were either dismantled or swept away by natural river floods during monsoon (Mutahara et al. 2017).

To meet the increasing agricultural demand under amplified population pressure, and to protect land and livelihoods from floods and salinity intrusion, the coastal embankment project (CEP) constructed 123 polders in the SW coastal region of Bangladesh in the 1960s. Daily tidal intrusions became a thing of the past, and all-round the year cultivation was then possible. With two to three harvests in a year, CEP produced positive results for a couple of decades. The flood and sediment dynamics in the enclosed polder region need to be understood for an effective management of water and land.

As embankments delinked the huge natural floodplains and restricted the gradual process of natural deposition inside the polders, they accelerated silt deposition on the river outside the polders and ultimately closed the exits of the sluiced gates. The natural flow was seriously altered and rivers lost their navigability (several tidal channels died within a few years to few decades), as a result of which they could not drain the nearby lands and polders anymore. During monsoon season, pluvial

R. Talchabhadel (✉) · H. Nakagawa · K. Kawaike
Disaster Prevention Research Institute, Kyoto University, Kyoto, Japan

A. Haque, A. I. A. Chowdhury (eds.), *Water, Flood Management and Water Security Under a Changing Climate*, https://doi.org/10.1007/978-3-030-47786-8_20

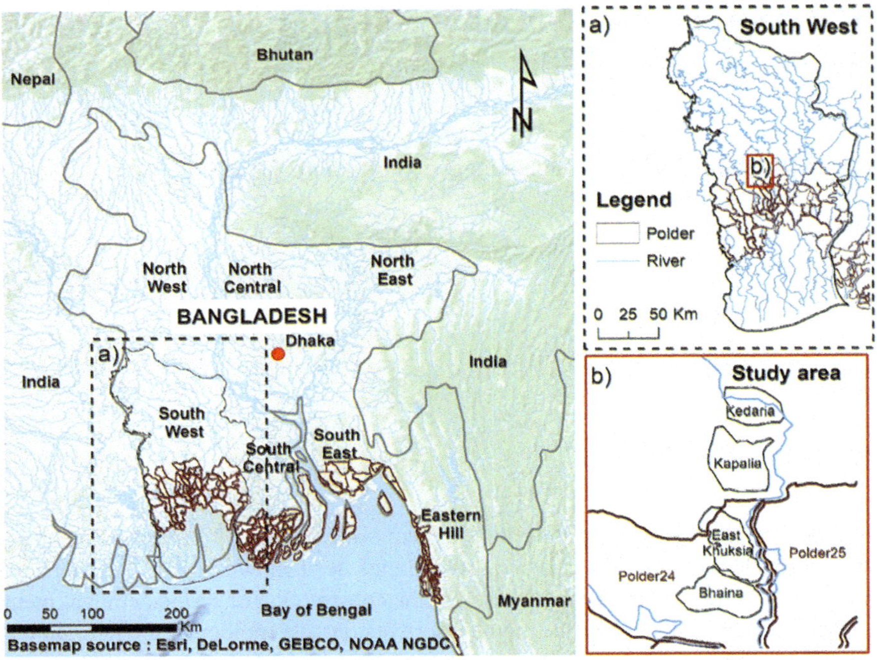

Fig. 1 Map showing coastal polders in SW region of Bangladesh: *Beels* Kedaria, Kapalia, east Khuksia and Bhaina are four selected *beels* in this study

inundation and then prolonged waterlogging has now become a serious issue in the SW region of Bangladesh. Almost 90% of rainfall occurs during six months from May to October (Talchabhadel et al. 2018a) which is normally trapped due to the aggravated river bed and poor drainage. Dredging and excavation of sediment from tidal rivers are not cost-effective and environment-friendly.

Temporary de-poldering and then controlled flooding allows tidal movement in a selected tidal basin, popularly known as tidal river management (TRM), is one of the most effective measures to address this issue. TRM is not a new way of flood/ sediment management but it involves taking full advantage of the natural movement of tidal streams up and down. Detailed information on poldering and de-poldering could be found in Talchabhadel et al. (2018b). TRM allows entering muddy water during every tide, depositing a major portion of suspended sediments and finally, eroding the river bed by relatively clearer water flowing back to the sea during every low tide. Many studies have conducted qualitative/quantitative analyses of TRM and its relationship to disasters, ecosystem, environment, agriculture, etc. and suggested the appropriate design of TRM. Few researchers (Ibne Amir et al. 2013; Rahman et al. 2015; Shampa and Pramanik 2012) have analysed the effectiveness with different interventions like crossing dam, dredging, and compartmentalization. Some (Stive and Rakhorst 2008; Talchabhadel et al. 2017) have suggested optimum opening sizes of link canals connecting the tidal basin and tidal river.

Importantly, the closure of the embankment after operation of TRM will again create a similar issue. To solve the problem in the long run, such processes are to be shifted/rotated effectively in different *beels*. By shifting TRMs, farmers of one tidal basin do not have to suffer for a long time. The plan of shifting of TRMs has been included in long-term policies for the SW region. For instance, Hari-Mukteswari river system: Kedaria, east Khuksia, Kapalia (supposed to operate), Baruna, Payra, west Khuksia and Singha by 2047 (IWM 2010a); Kobadak river system: Pakhimara (currently operational), Hariharnagar, Raruli, Rajapur, Harinkhola, Delua and Jalalpur by 2045 (IWM 2010b).

Beel Bhaina was operated during 1997–2001, then *beel* Kedaria at the top in Fig. 1b was operated during 2002–2005 and finally east *beel* Khuksia near to *beel* Bhaina was operated during 2006–2012. In fact, *beel* Kapalia was planned to operate as a tidal basin after east *beel* Khuksia, but due to some socio-economic-political and technical causes, it did not happen. People from various groups like farmers, fishermen, landowners, etc. have varied interests. Some of the technical hindrances in this regard are as follows: – (1) The people of *beel* Kedaria were very much in favour of TRM to raise their land, but it did not happen as per their expectations. The landowners lost their interest to operate TRM by submerging their land without any yield and compensation for three years; (2) Even compensation was started in east *beel* Khuksia and was operated for almost seven years, the initial plan was to only open it for 3–4 years. This created a lack of trust in government; (3) Non-uniform sediment deposition occurred acorss the tidal basin. The *beels* which have been selected to operate for TRM or in future the *beels* which are going to be selected for TRM are very sensitive to technical, social, economic and many other factors. This paper attempts to discuss the technical aspects.

Very few studies are realized based on numerical simulation of such shifting plans of the TRMs. This study attempts to examine the shifting of TRMs considering two virtual cases: (1) from d/s to u/s and (2) from u/s to d/s in a small stretch of Hari river with four selected *beels* shown in Fig. 1b. Out of four selected *beels*, only *beel* Kapalia has not experienced the operation of TRM. Each *beel* was simulated for four years of operation of TRM and subsequently, another *beel* was operated after the termination of one *beel*. This study focuses on the effectiveness of TRMs according to their shifting plans. An indicator taken in this study for evaluation of effectiveness is the total sediment deposition inside the *beel*.

2 Methodology

2.1 Data

The river channel is comprised of mud or fine sediment, originating from the Bay of Bengal. The mean size diameter of sediment ($d = 0.025$ mm) was used. Due to a lack of long-term time series of sediment concentration and flow discharge, we conducted a repetitive condition of tidal discharge (shown in Fig. 2) (Ibne Amir et al. 2013; Van Minnen 2013) at Ranai site located downstream (d/s) of *beel* Bhaina as a boundary

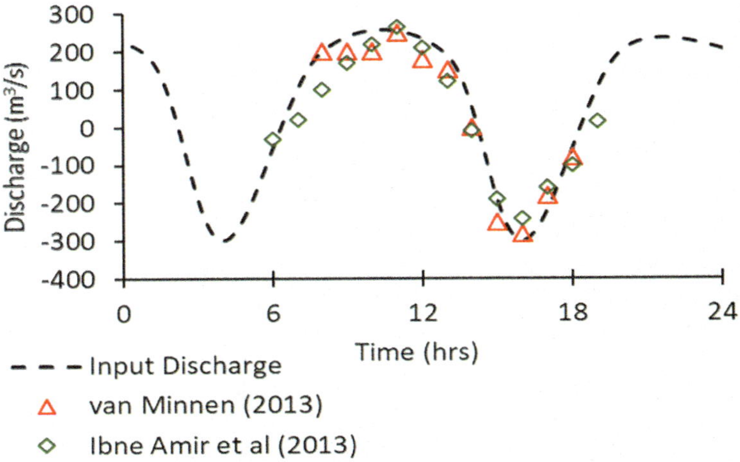

Fig. 2 Tidal discharge as a boundary condition for the simulation

Fig. 3 Unstructured triangular meshes in the study area. Smaller meshes were specified in the river channel

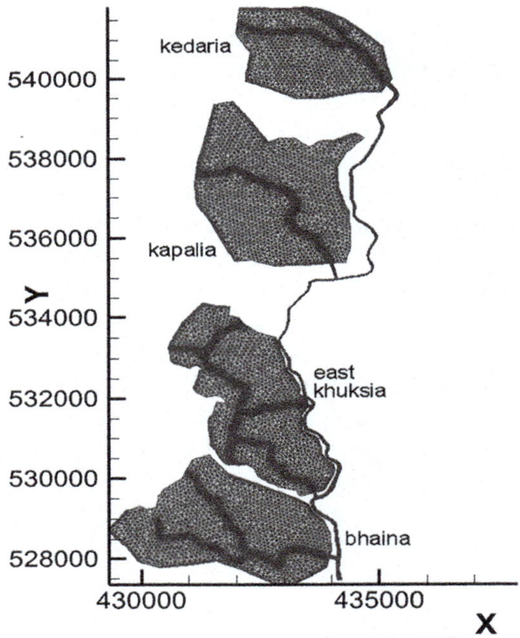

condition at d/s. During every high tide sediment supply of 1 kg/m^3 (an average value of measured suspended concentration at Ranai site at different time) was given. Ibne Amir et al. (2013) used hourly measured suspended sediment concentration for one tidal cycle (13 h.: 8:00 AM to 9:00 PM). The inflow of water and sediment from upstream (u/s) was not provided during the simulation assuming that either the u/s flow has completely been diverted or the available crossing dam did not allow u/s flow to pass through.

The ground was divided by a non-structured mesh shown in Fig. 3. The total mesh is 20,852 comprised of 11,483 nodes. The roughness coefficient for meshes at river channel was set to be 0.025 whereas it was set to be 0.04 for the remaining meshes. This study used two space-borne digital elevation models (DEMs) .biz. (1) ASTER GDEM (Advanced Space-borne Thermal Emission and Reflection Radiometer-Global DEM), and (2) SRTM (Shuttle Radar Topography Mission) for topographical data. The SRTM DEM (90 m resolution) was generated by C-band radar interferometry on board the space shuttle during February 2000. ASTER GDEM (30 m resolution data taken 2000–2008) used stereoscopic correlation techniques. Even ASTER GDEM has a limitation of the presence of artefacts but has more realistic values over water bodies compared to SRTM. The elevations of river channels were manually decreased to let the tides easily flow. The elevation data of Hari river based on a survey of March 2007 conducted by the Institute of Water Modelling (IWM) was used to estimate the profile and gradient of the rivers.

The absolute magnitudes of elevation in ASTER GDEM are quite high whereas the SRTM provides reasonable values in the study area. Figure 4a and b show the elevation based on two space-borne DEMS across the study area. Even the magnitude of SRTM revealed reasonable values suggesting it performed poor over water bodies/inundated area. The elevation values of most of the areas of *beel* Bhaina and east *beel* Khuksia are constant throughout, which is not realistic. This study used the

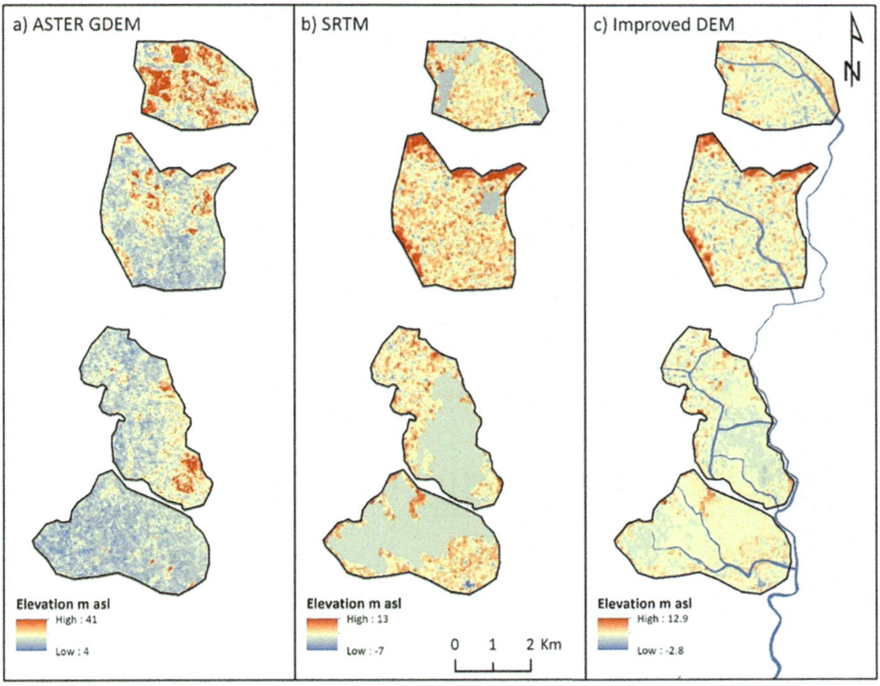

Fig. 4 Elevation in m asl (above sea level) based on (**a**) ASTER GDEM, (**b**) SRTM, and (**c**) improved DEM

variability of ASTER GDEM over those areas to fill the data gaps of SRTM. The improved DEM, shown in Fig. 4c, was then used to generate mesh elevation data.

2.2 Numerical Model

A two-dimensional (2D) flood simulation model based on a shallow water equation and a suspended sediment transport was used to simulate the mechanism of sediment transport and deposition. The underlying governing equations are as follow:

$$\frac{\partial h}{\partial t}+\frac{\partial M}{\partial x}+\frac{\partial N}{\partial y}=0 \tag{1}$$

$$\frac{\partial M}{\partial t}+\frac{\partial (uM)}{\partial x}+\frac{\partial (uM)}{\partial y}=-gh\frac{\partial H}{\partial x}-\frac{gn^2u\sqrt{u^2+v^2}}{h^{1/3}} \tag{2}$$

$$\frac{\partial N}{\partial t}+\frac{\partial (uN)}{\partial x}+\frac{\partial (vN)}{\partial y}=-gh\frac{\partial H}{\partial y}-\frac{gn^2v\sqrt{u^2+v^2}}{h^{1/3}} \tag{3}$$

where, h is the water depth, M and N are fluxes in the x and y directions respectively, u and v are velocities in the x and y directions respectively, H is the water level and g is the acceleration of gravity. To solve this equation, leap-frog difference scheme was employed. Suspended sediment transport was simulated using the equations:

$$\frac{\partial (Ch)}{\partial t}+\frac{\partial (CM)}{\partial x}+\frac{\partial (CN)}{\partial y}=D\left(\frac{\partial^2 (Ch)}{\partial x^2}+\frac{\partial^2 (Ch)}{\partial y^2}\right)+E-Cw \tag{4}$$

$$w=\sqrt{\frac{2}{3}\left(\frac{\sigma}{\rho}-1\right)gd+\frac{36\nu^2}{d^2}}-\frac{6\nu}{d} \tag{5}$$

$$E=wC^* \tag{6}$$

$$C^*=0.015\frac{dT^{1.5}}{aD_*^{0.3}} \tag{7}$$

$$D_*=d\left[\frac{\left(\frac{\sigma}{\rho}-1\right)g}{\nu^2}\right]^{1/3} \tag{8}$$

$$T=\frac{\tau-\tau_c}{\tau_c} \tag{9}$$

where C is the concentration of sediment, D is a coefficient of diffusion, E is the parameter of flowing up, w is the settling velocity, σ is the density of sediment particles, ρ is the water density, d is the diameter of sediment particles and ν is the coefficient of kinematic viscosity of water, C^* is the equilibrium concentration of sediment using Van Rijn (1984) [12], a is the reference level taken $0.05H$, D_* is the particle size parameter, T is an excess bed shear stress parameter, τ is the bottom shear stress and τ_c is the critical bottom shear stress according to Shields. D was set to 0.1 m^2/s. The bed level changes were computed from the information suspended load transport rates using the mass-balance equation.

$$(1-\lambda)\frac{\partial z_b}{\partial t} = Cw - E \tag{10}$$

where λ is the sediment porosity.

Please see Talchabhadel et al. (2017), Talchabhadel et al. (2018c) for the detailed explanation about boundary condition, sediment properties, and numerical simulation. The numerical model has also been checked and validated with experimental results (Talchabhadel et al. 2017). At first, the model was applied to the east *beel* Khuksia which was operational for 2006–2012 and compared with observed data. Then it was applied to two virtual cases – (1) From d/s to u/s and (2) From u/s to d/s shown in Fig. 5b and **c** respectively. Each *beel* was simulated for four years of operation of TRM and subsequently, another *beel* was operated after the termination of one *beel*. It is assumed that during the operation of TRM, crossing dam u/s of the embankment cut was available, the selected tidal basin was enclosed with peripheral embankments, and there was no inflow from u/s. The rate of sedimentation was carried out during a non-monsoon period (Oct–May i.e. eight months in a year).

3 Results

3.1 Topographical Data Correction

Average elevations of 500 m × 500 m grid were generated. Figure 6 shows an inter-comparison of average elevations based on ASTER GDEM and SRTM. Even though the absolute magnitudes are dissimilar, they captured the spatial variability of topography. Since the absolute values of ASTER GDEM are unrealistically high, SRTM was used as base DEM. As mentioned above, there was some data limitation for SRTM especially in water bodies/inundated areas (can be found in Fig. 4b). These data gaps were modified by using the variabilities of ASTER GDEM and the magnitude of surrounding SRTM DEM. The black lines in Fig. 6 show the elevation based on the improved DEM.

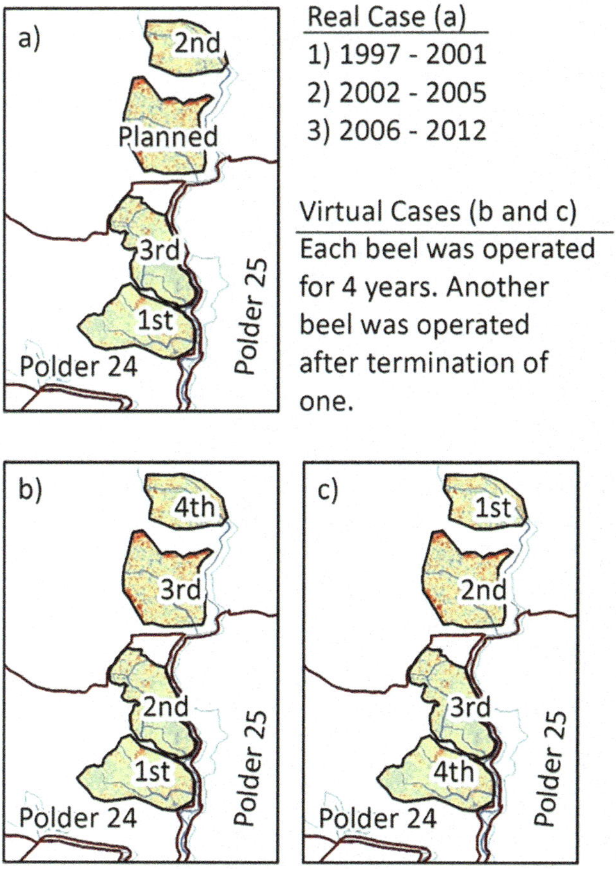

Fig. 5 Different cases: (**a**) a real case of TRM shifting in the study area, (**b**) a virtual case used for simulation from d/s to u/s, and (**c**) another virtual case for simulation from u/s to d/s. The improved DEM is superimposed as a background for which legend scale is the same as that of Fig. 4c

3.2 *Patterns of Sediment Deposition*

Figure 7 shows the spatial distribution of water depth in high tide during the operation of TRM in different *beel*s. The flow velocity and spatial distribution of water are highly dependent on the topographical setting (Hashimoto et al. 2018). Hashimoto et al. (2018) highlighted that the spreading of substance and its distribution are highly affected by the extent as well as the patterns of flooding. The flow velocity and direction are mainly governed by the topographical condition and boundary condition. In general, spatial distributions of water are higher nearer to the river channel. A slightly reduced water depth (nearly 1 m) in the main channel is observed during the operation of TRM in *beel* Kedaria compared to other cases. The same inflow boundary condition at d/s was given during the operation of TRM in

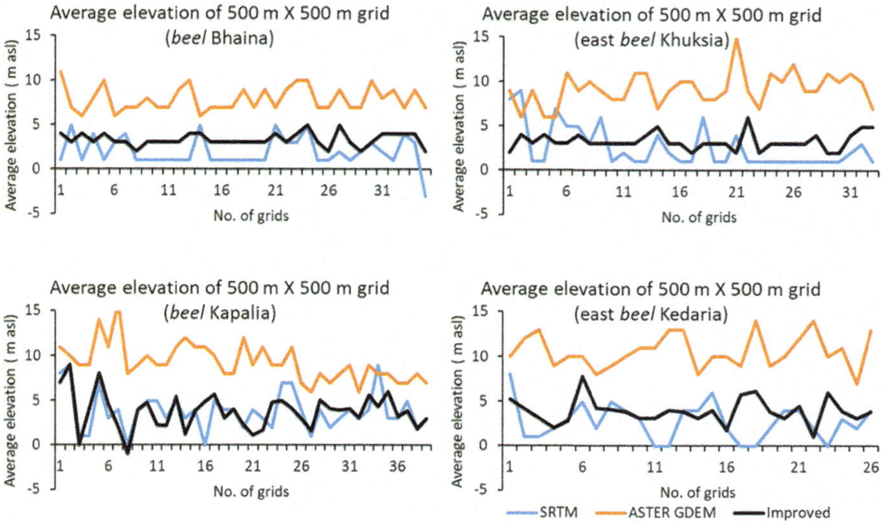

Fig. 6 Average elevation of 500 m × 500 m grid based on ASTER GDEM, SRTM and improved DEM

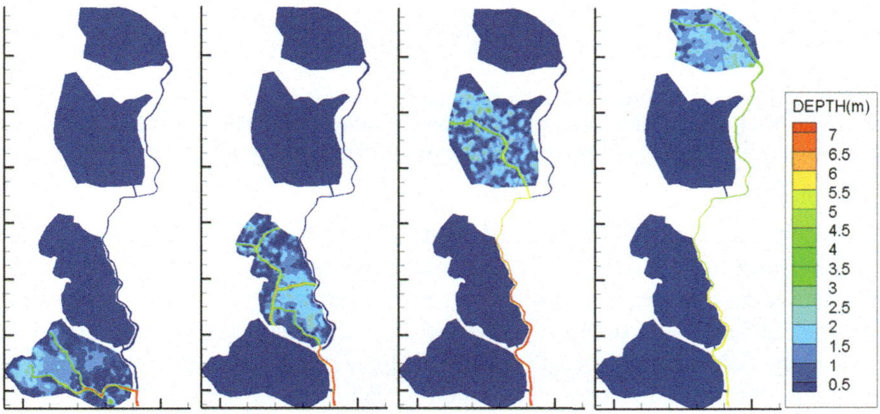

Fig. 7 Spatial distribution of simulated water in high tide during the operation of TRM in different *beel*s. (Bhaina, east Khuksia, Kapalia, Kedaria from left)

different *beel*s. Since the proximity of Kedaria *beel* is noticeably higher compared to other three cases, the increase in surface area reduces the water depth (indicating that under the constant water volume, the water depth is reduced for increased surface area condition). When the shear stress is smaller than the resisting force generated by settling velocity of sediment particles, a portion of transported suspended sediment gets deposited. The depressions inside the *beel*s are prone to more sediment deposition. During low tide, relatively clearer water goes back to the sea. In addition, it takes away deposited sediment from the river bed due to increased shear stress

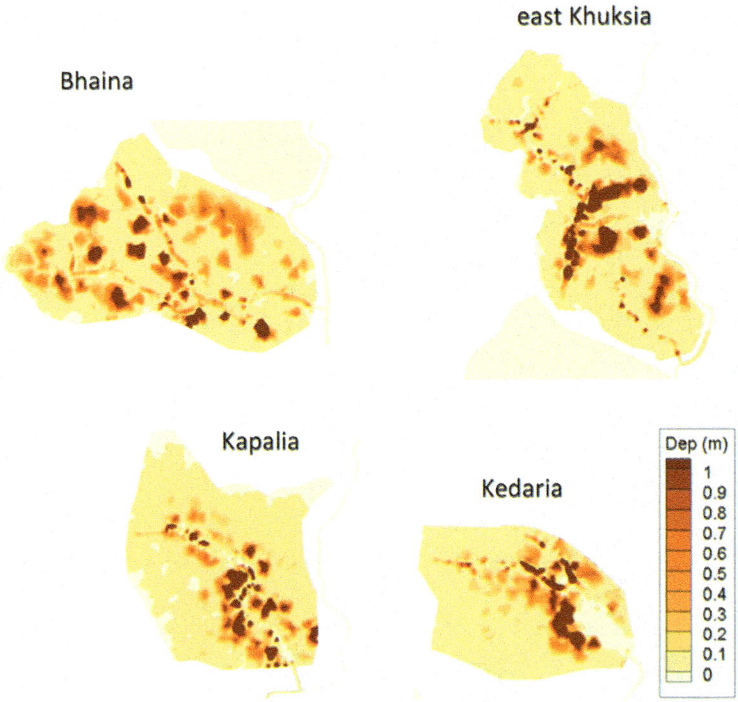

Fig. 8 Representative spatial distribution of simulated sediment deposition after four years of operation of TRM in different *beel*s (Scales are in proportion). Operation of TRM is sequential so the results of four different simulations are placed here in a single domain

especially near the outlet of the tidal basin. A repetitive process as such enhances land heightening inside the tidal basin and deepening of the river bed or/and width. Currently, the numerical simulation does not consider lateral widening of the river channel. In a real case, some protective measures like concrete blocks are employed around the link canal to prevent further lateral widening (Talchabhadel et al. 2017).

Figure 8 shows the spatial distribution of deposited sediment in different *beel*s after four years of operation of TRM. D/s *beel*s (Bhaina and east Khuksia) have relatively larger sediment deposition compared to u/s *beel*s (Kapalia and Kedaria). In general, the spreading of sediment and their deposition follow the topographical settings. The sediment deposition analysis showed different spatial extents for two individual DEMs (not shown) which suggested checking the accuracy of space-borne DEMs with local surveyed data. Importantly, the regions which are closer to the river channel have larger deposition compared to remote areas. To ensure uniform distribution, channelization needs to be carried out. This study focuses on the effectiveness of TRMs according to their shifting plan.

Figure 9 shows the anticipated flooding scenario and sediment deposition in *beel* Kapalia, proposed for the operation of TRM in the near future. As expected, simulated results show that higher deposition is expected near the link canal and

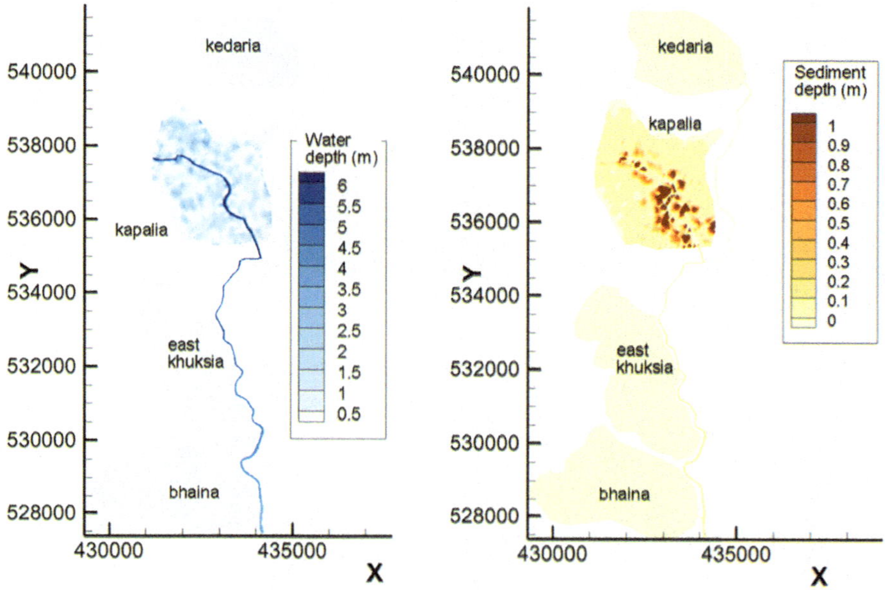

Fig. 9 Spatial distribution of simulated water during high tide [left] and final simulated sediment deposition after four years of operation of TRM [right] in *beel* Kapalia

river channel. For effective distribution of sediment deposition, further channels should be constructed on both sides of the main river channel. Compartmentalisation inside the tidal basin could also be done to prioritize deposition from one compartment to, another (Ibne Amir et al. 2013). Currently, the simulation considered repetitively same tidal prism throughout which is not the case in reality. Depending on the tidal prism, multiple embankment cuts/openings could be made. Thus, the preliminary results of this study can be helpful to assess the size of the tidal basin, requirements of channelisation and to provide a picture of anticipated regions of heterogeneity of sediment deposition.

3.3 *Total Sediment Deposition*

Figure 10 shows the temporal distribution of total sediment inside different *beel*s on an annual scale for two cases: (1) shifting from d/s to u/s and (2) shifting from u/s to d/s. In general, *beel* Kedaria has almost half the sediment deposition of two d/s *beel*s (Bhaina and east Khuksia). Due to higher proximity from the tidal source, tidal influence is significantly reduced. As a result, the conveyance capacity of sediment also decreases. The overall tendencies of annual variation of sediment deposition in both cases are analogous, but there exists small fluctuation.

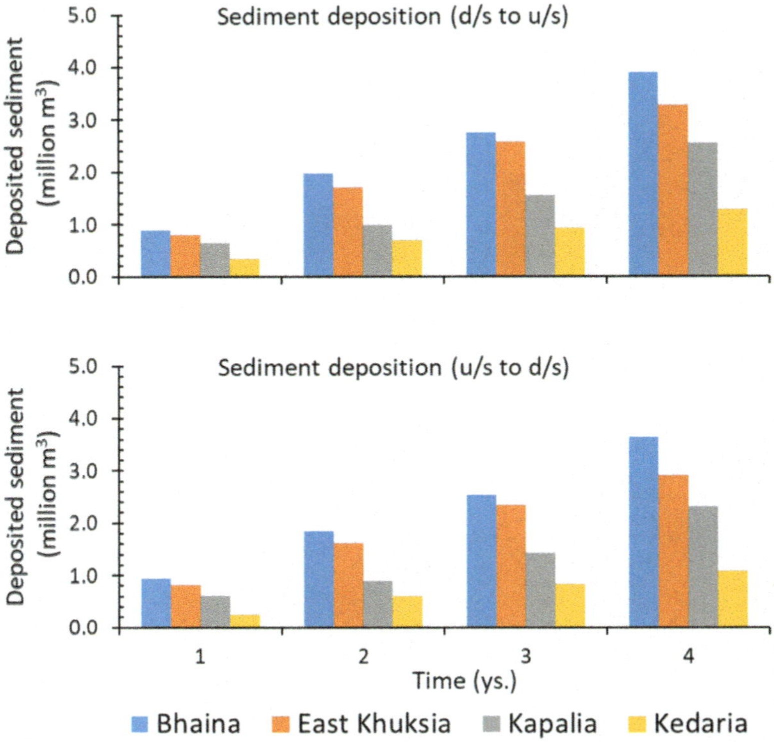

Fig. 10 Simulated sediment deposition in four years for different *beel*s for two cases (**a**) from d/s to u/s [top], and (**b**) from u/s to d/s [bottom]

The effects of shifting of TRMs are considered in the numerical simulation by riverbed evolution due to deepening effects in every low tide during the operation of TRM. In reality, due to changes in variations in cross-sections of the river (both width and depth), tidal prisms significantly change with time. Our future work would be considering the effect of change in tidal asymmetry and tidal prism.

Figure 11 shows the deviation between total sediment deposition inside the *beel*s for the two above mentioned cases. A negative value represents the case of shifting of TRMs from u/s to d/s has a higher deposition whereas positive value represents the case of shifting of TRMs from d/s to u/s has the higher deposition. There is a clear tendency of positive deviations which means shifting of TRMs from d/s to u/s has a higher deposition. During the first two years, it is a slight negative for *beel* Bhaina. It means for the d/s *beel*, higher deposition occurs if TRM shifts from u/s. By doing so, the deepening of river bed around the d/s *beel* starts already from the effect of u/s operation of TRM. At the same time, getting significant effects in most u/s TRM takes longer time. The greater positive deviation as visible in Fig. 11 is observed in *beel* Kedaria. For u/s *beel*s, the case of shifting of TRM from d/s to u/s showed greater sediment deposition. This is due to the fact that the deepening effect of river channel develops gradually as the operation of TRM shifts u/s. Overall

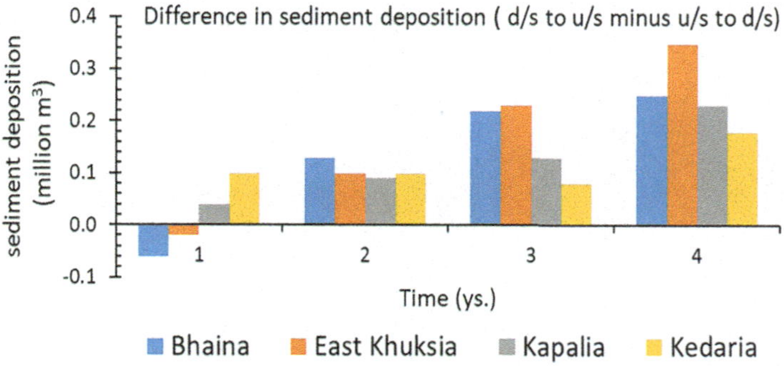

Fig. 11 Difference between sediment depositions for two cases. The values are derived by subtracting the final deposition from the case of d/s to u/s to the case of u/s to d/s. A negative value represents the case of u/s to d/s has a higher deposition whereas positive value represents the case of d/s to u/s has the higher deposition

results for the study areas show that shifting the TRMs from d/s to u/s produced a slightly better result in view of the indicator taken as total sediment deposition.

4 Discussion

A single intervention from the place of TRM implementation can increase drainage capacity to a long distance of up to 20 km of the lower stream at d/s (Al et al. 2018). It takes full advantage of controlled beneficial flooding with little human intervention. It is essential to understand the sediment dynamics upon the closure of embankment. The closure of embankment after designated land heightening is a critical factor that needs to be considered during the planning of shifting of TRMs. If another tidal basin is not operated, then it is likely to again create the similar issue of river bed aggradation and waterlogging. Therefore, to solve this problem in the entire basin, various *beel*s are to be rotated one after another. The tidal basin concept keeps the river alive. In parallel, this concept strongly needs strong public participation by the stakeholders with a great deal of sacrifice for a designated period (about five years depending on tidal prism and area of the *beel*) during which their lands would remain flooded. This concept is the example of building with nature contributing land heightening, flood resistance, and food security and is a resilient measure for addressing waterlogging, drainage congestion, and river siltation (Rezaie and Naveram 2013).

This study discusses the effectiveness of shifting of TRMs taking a simple indicator of total sediment deposition inside *beel*s. Several other criteria should also be considered. During our simulation, there were a number of assumptions i.e. a repetitive tidal prism of the same magnitude, the operation of TRM for four years for all *beel*s irrespective of sizes and many others. There is again a clear

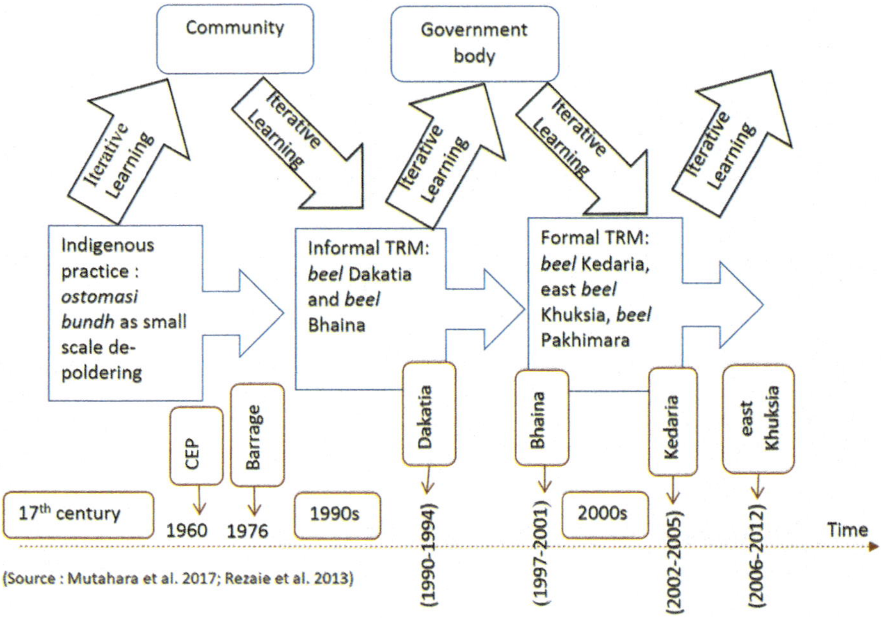

Fig. 12 Evolution of tidal basin operation, social learning and participation of stakeholders

need for an approach to TRM implementation that considers multi-dimensional (social, economic and environmental) consequences along with technical consequences.

Bangladesh Water Development Board (BWDB) is trying to revive the scheduled operation of TRM in *beel* Kapalia by addressing the socio-economic problems in the area. In the meantime, BWDB has replicated the TRM process in other receptive areas such as Kobadak river system where Pakhimara is currently operational from July 2015. Figure 12 shows how TRM has evolved into an environmentally accepted water/sediment management practice based on indigenous knowledge (Al et al. 2018). At the same time, the evolution of stakeholder's participation clearly shows a clear shift from only local communities, non-governmental organizations (NGOs), and some social/political activists to multidisciplinary actors such as government bodies, research organizations, and donors including communities (Mutahara et al. 2017). The level of technical planning, infrastructure development is significantly improved.

Even though the mechanism of compensation has started, conflicts among local stakeholders (shrimp-farm owners, local farmers, landless people, daily wage laborers, NGOs, civil society organizations) as well as between local stakeholders and government institutions (local government, *Union Parishad*, *Upazila Parishad*, district administration, BWDB) seem to be prevailing. The conflicts are generally related to compensation, procedure of compensation, land use practices, employment, rotation of TRMs and coordination. A process of iterative learning (evaluation of success and failure of operated TRMs) and growing awareness is continuously guiding the effective operation in next *beel.*

The Bangladesh Delta Plan (up to 2100), which is a long term strategy document for Bangladesh, has highly acknowledged the tidal basin concept for solving waterlogging problems and reviving tidal rivers. Restoration of tidal plains increases the tidal prism and provides sufficient room for the rivers. Temporary de-poldering raises the low lying lands (i.e. sediment-starved flood plain) and manages the sediment in a sustainable way. Some of the important things for a sustainable way forward are (1) intensive consultation and a clear agreement from concerned communities and affected landowners as a precondition for selection of *beel* for next rotation, (2) open, transparent and inclusive planning, implementation, operation, maintenance, (3) regular supervision by government institutions, (4) continuous information and motivational campaign, and so on.

At present, BWDB and many organizations are playing significant roles in terms of scientific research, TRM planning. Besides these, local water management bodies are continuing their efforts to illustrate the authentic conditions and demands of the stakeholders. Although an embankment along the coast is required, sediment management in parallel is also required for sustainability. TRM, evolved from traditional wisdom, blended with scientific knowledge seems to be a sustainable way forward. Also, there needs to be regular continuous updating of approaches based on monitoring results.

5 Conclusion

This study examines the shifting of TRMs considering two virtual cases in a small stretch of *Hari* river: (1) from d/s to u/s and (2) from u/s to d/s. We used two space borne DEMs for topographical data. At first, an inter-comparison of elevations based on two selected DEMs was made. Then we proposed a methodology to best use the available space-borne DEMS. The error of one DEM could be reduced by using another DEM or locally surveyed information. A 2D unstructured mesh flood simulation model based on a shallow water equation and a suspended sediment transport was used to simulate the mechanism of sediment transport and deposition during the operation of TRM. The effectiveness of shifting of TRMs is analysed in the view of an indicator of total sediment deposition inside the *beel*s. It is recommended from our preliminary study to shift the *beel*s from d/s to u/s. The numerical model still needs a lot of improvement like consideration of salinity, flocculation effect, cohesiveness etc. Better topographical data is also needed for better representation of sediment transport and deposition. A long term measurement of the tidal prism, suspended sediment at multiple locations is likely to make the numerical model more robust.

Tidal basin concept strongly contributes to flood resistance, food security and land heightening of sediment-starved floodplains. This can serve as a resilient measure for addressing issues of waterlogging, drainage congestion, and river siltation. Since it involves the requirement of strong participation of local people, there is a clear need for an approach that considers multi-dimensional (social, economic, political and environmental) consequences along with technical consequences. This study might serve as a step forward for the appropriate shifting of TRM.

Acknowledgement This study was supported by JST/JICA SATREPS Program on Disaster Prevention/Mitigation Measures against Floods and Storm Surges in Bangladesh (PI: Prof. Dr. Hajime Nakagawa).

References

Al, M., Naher, N., Azadi, H., & Van Passel, S. (2018). Sustainability impacts of Tidal River management: Towards a conceptual framework. *Ecological Indicators, 85*, 451–467.

Hashimoto, M., Yoneyama, N., Kawaike, K., Deguchi, T., Hossain, M. A., & Nakagawa, H. (2018). Flood and substance transportation analysis using satellite elevation data: A case study in Dhaka City, Bangladesh. *Journal of Disaster Research, 13*(5), 967–977. https://doi.org/10.20965/jdr.2018.p0967.

Ibne Amir, M. S. I., Khan, M. S. A., Kamal Khan, M. M., Golam Rasul, M., & Akram, F. (2013). Tidal River sediment management – A case study in southwestern Bangladesh. *International Journal of Civil Architectural Structural and Construction Engineering, 7*, 183–193.

IWM. (2010a). *Feasibility study and detailed engineering design for long term solution of drainage problems in the Bhabodah Area*.

IWM. (2010b). *Feasibility study for sustainable drainage and flood management of Kobadak River Basin under Jessore and Satkhira Districts*.

Mutahara, M., Warner, J. F., Wals, A. E. J., Khan, M. S. A., & Wester, P. (2017). Social learning for adaptive Delta management: Tidal River management in the Bangladesh Delta social learning for adaptive delta management: Tidal River management in the Bangladesh Delta. *International Journal of Water Resources Development*, 1–21. https://doi.org/10.1080/07900627.2017.1326880.

Rahman, M. Z., Islam, M. S., & Khan, Z. H. (2015). *Tidal River Management (TRM)-An innovative scientific approach for sustainable sediment management*. In: International conference on recent innovation in Civil engineering for Sustainable Development (IICSD-2015), pp. 954–959.

Rezaie, A. M., & Naveram, U. K. (2013). Tidal River management: An innovative approach for terminating drainage congestion and raising land through sedimentation in the Bhabodaho area, Bangladesh. *Advances in River Sediment Research*, 1363–1375.

Shampa, & Pramanik, M. I. M. (2012). Tidal River management (TRM) for selected coastal area of Bangladesh to mitigate drainage congestion. *International Journal of Scientific & Technology Research, 1*, 1–6.

Stive, M. J. F., & Rakhorst, R. D. (2008). Review of empirical relationships between inlet cross-section and tidal prism. *Journal of Water Resources and Environmental Engineering, 23*, 89–95.

Talchabhadel, R., Nakagawa, H., Kawaike, K., Hashimoto, M., & Sahboun, N. (2017). Experimental investigation on opening size of Tidal Basin management: A case study in southwestern Bangladesh. *Journal of Japanese Society of Civil Engineers Ser B1 Hydraulic Engineering, 73*, 781–786.

Talchabhadel, R., Nakagawa, H., & Kawaike, K. (2018a). Spatial and temporal variability of precipitation in southwestern Bangladesh. *Journal of Japanese Society of Civil Engineers Ser B1 Hydraulic Engineering, 74*, 289–294.

Talchabhadel, R., Nakagawa, H., Kawaike, K., & Sadik, M. S. (2018b). Polder to De-polder: An innovative sediment management in Tidal Basin in the southwestern Bangladesh. *DPRI Annuals, 61B*, 623–630.

Talchabhadel, R., Nakagawa, H., & Kawaike, K. (2018c). *Sediment management in Tidal River: A case study of East Beel Khuksia, Bangladesh.* In: E3S web of conferences riverflow 2018, 40, 1–10. https://doi.org/10.1051/e3sconf/20184002050.

Van Minnen, J. N. (2013). *Sediment transport in Tidal Basin, South West Delta Bangladesh a historical and physical perspective*. MSc thesis, Wageningen University.

Van Rijn, L. C. (1984). Sediment transport, part II: Suspended-load transport. *Journal of Hydraulic Engineering, ASCE, 110*(11), 1613–1641.

Vulnerability Assessment of Bangladesh Coastline Using Gornitz Method

S M Samiul Islam, Ahad Hasan Tanim, and Md. Reaz Akter Mullick

1 Introduction

Bangladesh coast is widely recognized as vulnerable under accelerated sea level rise (SLR), coastal flood, shoreline erosion, tropical cyclone activity etc. Since 1484 coastal disasters devastated more than 2 million human lives in this region (Alam and Dominey-Howes 2015). Whilst many structural measures are adopted in Bangladesh to offset damages from the coastal hazard, there is still a lack of a comprehensive coastal vulnerability analysis combining both the physical and the forcing factors of coastal hazards. Therefore, in order to get a comprehensive coastal management plan, the coastline is made into a frontier of a coastal zone that is subjected to complex oceanic, atmospheric and anthropogenic environments. The coastline is exposed to climate change induced effects from the beginning of the twenty-first century and this emerges as a fundamental step to minimize the potential impact of coastal hazards. In the last century, the global mean SLR was 1.7 ± 0.2 mm/year (IPCC 2014), although the Bangladesh coast has experienced a higher SLR rate of 5–15 mm/year (Mullick et al. 2019). Under such circumstances it is urgent to explore the coastline vulnerability of Bangladesh under threatening forces such as tropical cyclone activity, shoreline change, SLR etc.

S. M. S. Islam (✉) · A. H. Tanim
Center for River, Harbor and Landslide Research, Chittagong University of Engineering and Technology, Chattogram, Bangladesh
e-mail: ahtanim@cuet.ac.bd

M. R. A. Mullick
Department of Civil Engineering, Chittagong University of Engineering and Technology, Chattogram, Bangladesh
e-mail: reazmullick@cuet.ac.bd

A. Haque, A. I. A. Chowdhury (eds.), *Water, Flood Management and Water Security Under a Changing Climate*, https://doi.org/10.1007/978-3-030-47786-8_21

The resilience of a coastal community can be measured with social, economic, environmental and physical conditions and with corresponding vulnerability (Wu et al. 2002). Amongst them, the degree of vulnerability of a coastline can be directly measured with physical vulnerability using the widely applied matrices, namely the coastal vulnerability index (CVI). Identifying responsible factors of physical vulnerability is the first step to obtain CVI. There are two types of dominating factors and those are: coastal characteristics and coastal forcing factors (Sahoo and Bhaskaran 2018). The coastal characteristic factors are the coastal slope, geomorphology, land cover, bathymetry, etc. (Hegde and Reju 2007; Kunte et al. 2014). On the contrary, oceanographic factors like tide, sea level rise, wave etc. are coastal forcing factors (Rueda et al. 2017). Tidal range can also create adverse effect during a storm surge particularly in the head of the Bay of Bengal due to dominated non-linear tide-surge interaction in the surge generation process (Antony and Unnikrishnan 2013). Again, coastline slope is another important parameter for coastal vulnerability assessment. When the coastline slope is gentle, it has a greater chance to propagate the storm surge and coastal flood more easily, in comparison to a steep slope. Another coastal forcing factor is the cyclone activity which can be measured with the cyclone track density of a particular area. From the compilation of those studies the responsible physical vulnerability factors can be identified as SLR, SWH, cyclone track density, coastal slope, mean tidal range, and shoreline change. Using those factors the CVI can be quantified applying a suitable coastal vulnerability assessment approach.

The degree of vulnerability of any coast can be calculated adopting two different methods as proposed by Gornitz (1991); Mclaughlin and Cooper (2010) and Gornitz (1991) described the index as the product of factors' square root whereas the latter approach computed the sum of the factors' weight. The product-based index calculation provides higher index value with a distinctive classification range that can identify the vulnerability condition of any shoreline segments easier than the sum approach i.e., McLaughlin and Cooper method. The Gornitz method is an established method that has already been used in different coastal vulnerability studies e.g. Arun Kumar and Kunte (2012); Islam et al. (2016). With the advent of geospatial techniques, responsible coastal vulnerability factors can be quantified and spatially represented in the ArcGIS environment.

Very few studies took into account the entire Bangladesh coastline as a whole when assessing the physical vulnerability. For coastline vulnerability analysis, the SLR, tropical cyclone activity and significant wave height is overlooked in previous studies, although those factors put a separate emphasis in the context of Bangladesh coasts. Previous studies such as Islam et al. (2016) considered the GDC (Ganges Deltaic Coast) of Bangladesh for physical vulnerability analysis. Considering the knowledge gap, this study aims to identify the coastline vulnerability, combining responsible coastal characteristics and forcing factors. To achieve this, the responsible factors were identified and quantified using the remotely sensed; the model derived and digitized data assimilation technique in ArcGIS environment. Finally, the CVI (Coastal Vulnerability Index) throughout the coastline was calculated by the Gornitz method.

2 Study Area

The Bangladesh coast is situated from 89° to 92°20′ East longitudes and 20°30′ to 20°50′ North latitudes. The coastline of Bangladesh stretches about 600 km and has principle components of geologic formation with silt, sand, and unconsolidated clay. The coastline encompasses 12 administrative districts. The approximate population densities of those coastal districts are 800–850 per square km, which is one of the highest numbers in the world. Approximately 275 tropical cyclones hit the coast since the eighteenth century (Alam and Dominey-Howes 2015). The entire Bangladesh coast can be classified under three major categories: (a) cliff coast, (b) the estuarine coast and (c) Ganges Deltaic Coast (GDC) (Fig. 1). The Chittagong and Cox's Bazar coast are considered as cliff coast whereas the Meghna and Feni estuary fall under the estuarine coast. The GDC is located at the outlet of GBM (Ganges-Brahmaputra-Meghna) basin that discharges sediment flux into the Bay of Bengal. The GBM basin discharges a sediment load of about 1–2.4 billion tons per year which mainly shapes the geologic formation of GDC (Rahman et al. 2018). To commensurate with the data, the analysis was conducted at a 2 km strip line along the coast. The 2018 LANDSAT satellite images were used to determine the reference line of the coast.

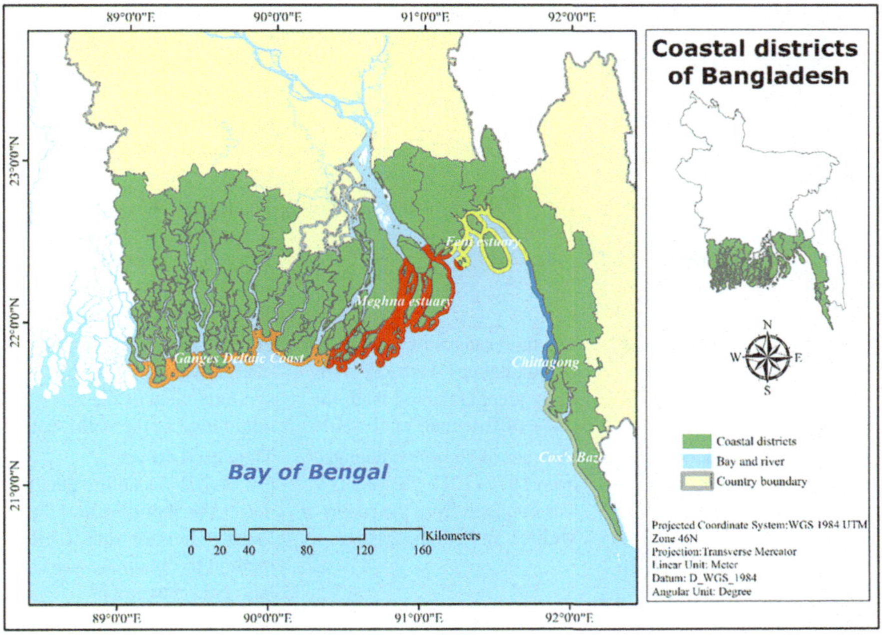

Fig. 1 Study area

3 Data and Method

This study compiled data from secondary sources to quantify the factors related to CVI. Table 1 represents the data source and the brief analytical procedure for each of the factors considered. The ArcGIS has been applied for digitizing, processing and analyzing the spatial information of those factors.

$$Cyclone\ track\ density = \frac{\sum_{i=1}^{i=n} X_i \times Y_i}{R} \quad (1)$$

Where, R denotes the radius of the inscribed circle (30 km), X_i is for the length of cyclone track inside the circle, and Y_i is for wind speed.

Table 1 Selected vulnerability factors and their data sources with brief analytical procedure

Type	Factors	Data sources	Analytical procedure followed
Coastal forcing	SLR (Sea Level Rise)	Dataset from Bangladesh Inland Water Transport Authority (BIWTA) during 1985–2017 CCC 2016)	The information was digitized in ArcGIS environment to create a raster layer of sea level rise.
	Cyclone track (TC) density	For the period of 1976–2017 the cyclone track was retrieved from JTWC best track was achieved.	Six hourly cyclone positions were digitized, and the extent of the coastline was kept only the landfall locations. Then, the cyclone track density was determined using Silverman (1986) equation[a].
	SWH (Significant Wave Height)	The SWH was obtained from the report of Mullick et al. (2019).	The mean SWH was obtained from the wave climatology study of Patra and Bhaskaran, (2016) and compiled by Mullick et al. (2019).
	Mean tidal range	The tidal ranges of the total 29 tide gauges were set in the study area. Obtained from the study of Bricheno et al. (2016)	Further, the IDW interpolation method was used to obtain a synoptic view of the mean tidal range of the entire coast.
	Shore line change	The rate of shoreline change from 1977–2017 (40 years) was obtained from the report of Mullick and Islam (2018).	They used 40 years (1977–2017) satellite data to detect the shoreline and the linear regression rate (LRR) method to obtain seaward or landward movement.
Coastal characteristics	Coastal slope	The bathymetry dataset of the Bangladesh NAVY was digitized in the ArcGIS environment and create a coastal slope layer.	The bathymetry dataset was digitized along the entire coastline. Finally, coastal slope information was processed through ArcGIS spatial analysis at a 10 m resolution.

Silverman (1986) equation

After estimating the vulnerability of the six vulnerability factors, the range of dataset was categorized using an equal interval method in five distinct classes i.e.; very low, low, moderate, high and very high. The main advantage of this classification method is that there is no chance of biasness in this term of categorization and each dataset was considered to be equally important as the others. Several studies recommend equal interval classification of the factors as more convenient for data classification of coastal vulnerability analysis (İkizler 2002; Mullick et al. 2019). The break values of SWH, mean tidal range, and SLR was obtained from the study of Mullick et al. (2019). The rate of shoreline change was chosen based on the study of Mullick and Islam (2018).

The detail of classification and rating is shown in Table 2. Finally, using the feature scaling by (Eq. 2) each of the vulnerability factors were reclassified and normalized in 0–1 scale (İkizler 2002).

$$\text{Normalized value} = \frac{\text{Value} - \text{Value}_{\min}}{\text{Value}_{\max} - \text{Value}_{\min}} \tag{2}$$

In this study, Gornitz's (1991) coastal vulnerability index was estimated at 10 m cell resolution to integrate spatial variability of vulnerability information. The Gornitz CVI method was computed using Eq. (3). Finally, the CVI obtained from Gornitz method was rescaled and normalized using Eq. 2 in 0–1 scale to make it consistent with the index value.

$$\text{CVI} = \sqrt{\frac{F_1 * F_2 * F_3 * F_4 * F_5 * F_6}{6}} \tag{3}$$

Here,

F_1 = Sea Level Rise.
F_2 = Cyclone track density.
F_3 = Significant Wave Height.
F_4 = Coastal slope.
F_5 = Mean tide range.
F_6 = Shore line change.

4 Result and Discussion

4.1 Sea Level Rise

SLR is recognized as a major climate induced threat for the coastal ecosystem and the environment. From the satellite altimetry, 17–25 mm/year SLR was observed at Hiron point and Salfapur (Fig. 2). Patharghata of the GDC segment experienced the highest SLR i.e., 35 mm/year. The majority of the Feni estuary is subjected to

Table 2 The range with index value and assigned rating of different responsible factors

Factor	Range	Index value	Rating	Rationale
Sea Level Rise (mm/year)	<8	0–0.2	Very low	SLR increases the rate of inundation probability.
	9–12	0.2–0.4	Low	
	13–16	0.4–0.6	Moderate	
	17–25	0.6–0.8	High	
	26–35	0.8–1	Very high	
Cyclone track density (cyclone intensity per year per 30 km radius)	<0.0022	0–0.2	Very low	The cyclone track density indicates the cyclone activity. It has a positive relation with coastal hazard.
	0.0022–.0044	0.2–0.4	Low	
	0.0045–0.0067	0.4–0.6	Moderate	
	0.0068–0.0090	0.6–0.8	High	
	0.0091–0.0011	0.8–1	Very high	
SWH (m)	<0.25	0–0.2	Very low	SWH is correlated with wave runup height.
	0.26–0.5	0.2–0.4	Low	
	0.51–0.75	0.4–0.6	Moderate	
	0.76–1	0.6–0.8	High	
	1–1.24	0.8–1	Very high	
MTR (m)	0–1	0–0.2	Very low	High tidal range can increase the surge height and thereby the extent of coastal inundation can be increased.
	1.1–2	0.2–0.4	Low	
	2.1–3	0.4–0.6	Moderate	
	3.1–4	0.6–0.8	High	
	>4	0.8–1	Very high	
Coastal slope (%)	0–0.45	0–0.2	Very high	The degree of coastal vulnerability decreases with the increase of the coastal slope.
	0.46–1.3	0.2–0.4	High	
	1.31–2.7	0.4–0.6	Moderate	
	2.71–5.4	0.6–0.8	Low	
	>5.41	0.8–1	Very low	
Shore line change (m/year)	>17.98 (accretion)	0–0.2	Very low	Accretion denoted as positive shoreline change and vice versa. The erosion in a coastline section indicates its vulnerability.
	4.79 to 17.98 (accretion)	0.2–0.4	Low	
	−4.78 to 4.79 (stable)	0.4–0.6	Moderate	
	−12.61 to −4.78 (Erosion)	0.6–0.8	High	
	< −12.61 (erosion)	0.8–1	Very high	

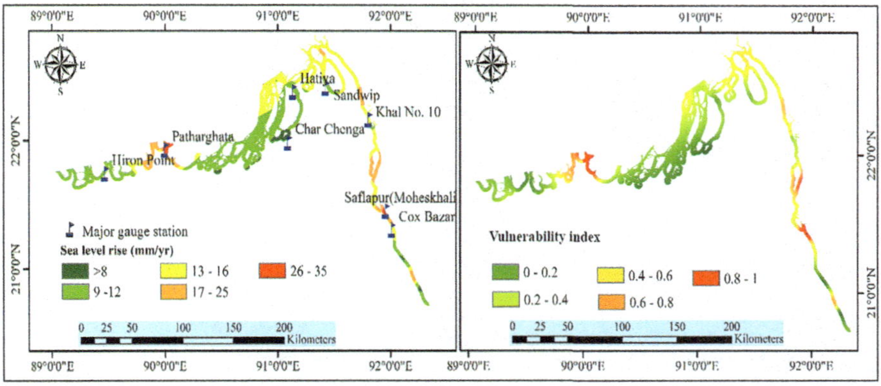

Fig. 2 The rate of SLR with its vulnerability index

moderate vulnerability due to SLR. Again, the majority of the Meghna estuary was found to be within low to moderate vulnerability class. However, the rate of SLR from the global perspective is relatively high, that exceeds global mean SLR 1.7 ± 0.2 mm/year (IPCC 2014). Figure 2 (right panel) shows the vulnerability due to SLR.

4.2 *Cyclone Track Density*

Historical track of the cyclone along the Bangladesh coast was categorized according to Suffir-Simpson scale, which depends on central pressure and wind speed (Schott et al. 2012). During 1960–2018, the coast of Bangladesh experienced a total of 37 cyclones in a cyclone scale varying from category 1–5, various tropical storms, and depression. Among them, the cyclone in 1991 and Sidr, the two category-5 cyclones are the most notable. With a total of 11 tropical cyclones, the coast of Chittagong has experienced the highest number of tropical cyclones. Based on the cyclone track density four coastal districts i.e., Patuakhali, Bhola, Barguna, and Chittagong are observed as the most vulnerable regions, in comparison to the rest of the coast. Figure 3 represents the TCs and vulnerability due to cyclone track.

4.3 *Mean Significant Wave Height*

When the mean SWH is more, the wave action in a section is more prone to high wave energy. As a result, the wave-surge interaction and the nearshore current can strongly impact the coastline section to generate high surge height or erosion. The mean SWH in Bangladesh coast was observed to be 0–1.24 m, with the highest range near the coast of Cox's Bazar (Fig. 4). The estuarine coast has comparatively lower

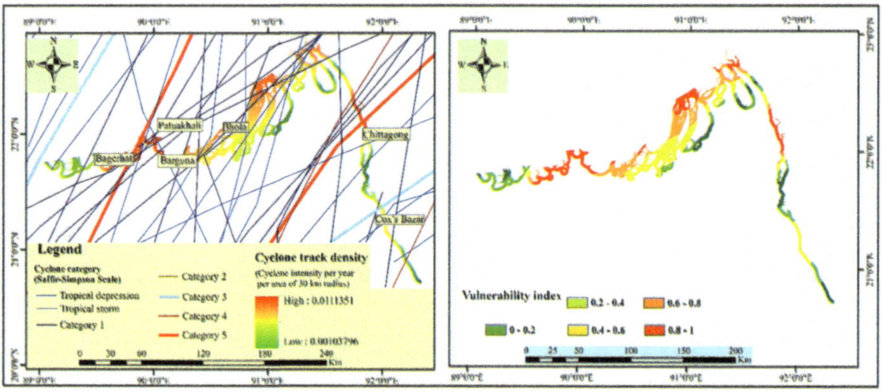

Fig. 3 The cyclone track density with its vulnerability index

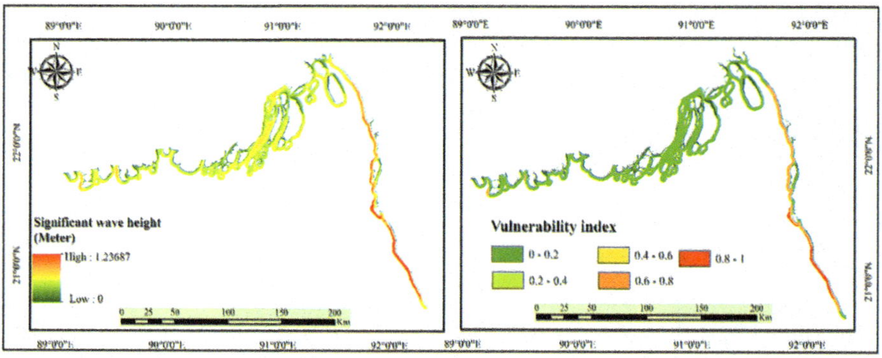

Fig. 4 SWH with its vulnerability index

SWH than the shallow bathymetry. At Cox's Bazar, the vulnerability index had a highest range than the rest of the coast (Fig. 4). At Chittagong, the mean SWH range was found to be 0.49–0.71, which is moderately vulnerable. Rest of the coastline was found low vulnerable as the mean SWH range is negligible i.e., 0.23 m (Fig. 4).

4.4 Mean Tidal Range

The mean tidal range over the Bangladesh coast was found to be 1–4 m throughout the coast. The tide in Bangladesh coast is semi-diurnal in nature with M2 dominant amplitude. The tide range gradually amplifies at the northeast part of the study area (Fig. 5). According to the tide range, the three significant classes are notable: microtidal (>2 m), macrotidal (>4 m), and mesotidal (2–4 m). The macrotidal tide range was observed in the Chittagong coast. The major portion of Cox's Bazar and

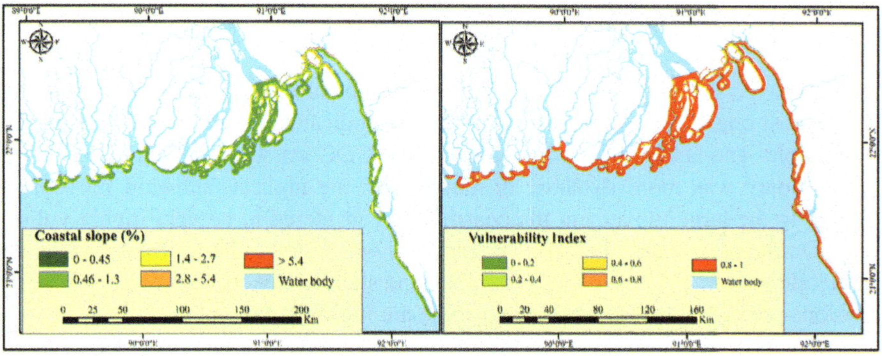

Fig. 5 Mean tidal range with its vulnerability index

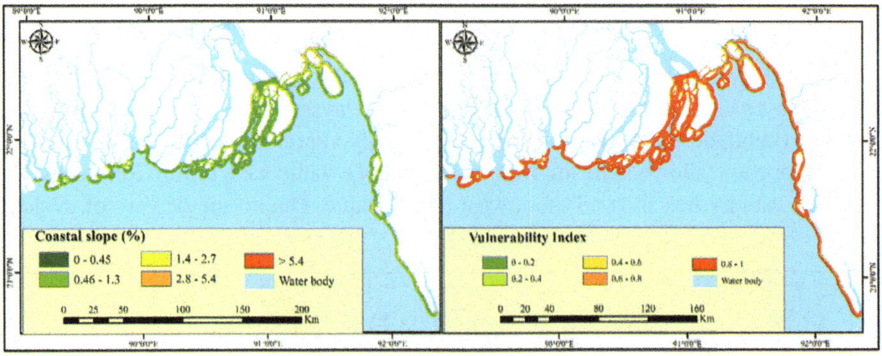

Fig. 6 Coastal slope with its vulnerability index

Feni estuary showed mesotidal tide range. The rest of the coast was classified as microtidal regime; accordingly, it was classified as low vulnerable. The highest vulnerability was observed at the Chittagong coast due to the highest tidal range (>4 m).

4.5 *Coastal Slope*

Coastal slope is considered as one of the important factors for coastal vulnerability as it is directly related to coastal flooding. Coastal flooding can become devastating when the gradient of coastal slope is very low or gentle. On the contrary, the steeper slope is less vulnerable as it obstructs the surge propagation in the adjacent coast. Overall Bangladesh coast has 0–0.23% coastal slope that can be classified as gentle slope (Fig. 6). The largest portion of the coast, except the portion of Feni estuary, has a mild slope which is accountable for the coastal vulnerability (Fig. 6). The estuary of Meghna was observed as very highly vulnerable whereas the Feni estuary has a low degree of vulnerability (Fig. 6).

4.6 *Rate of Shoreline Change*

The landward shoreline movement is another vulnerability indicator. The GDC had the highest rate of the landward movement that indicates that the shoreline is highly vulnerable. The net landward movement at the GDC was 0.27 m/year (Fig. 7). The Feni estuary was more dynamic in nature with the highest landward movement. Owing to the large movement the coastline is considered to be very highly vulnerable. On the other hand, the Cox's Bazar coast was found to be relatively stable and moderate vulnerable. The Chittagong coastline showed mixed vulnerability as some portions are moderate vulnerable and some are highly vulnerable.

5 Coastal Vulnerability

The coastal vulnerability was obtained using six coastal factors as stated earlier. The following coastal vulnerability map (Fig. 8) portrays that the major portion of the coast of Bangladesh is found as moderate to highly vulnerable, where the Chittagong coastal portion showed the highest average CVI value i.e., 0.56. Contrarily, the Meghna estuary has showed the lowest (0.17) value. Due to the density of cyclone

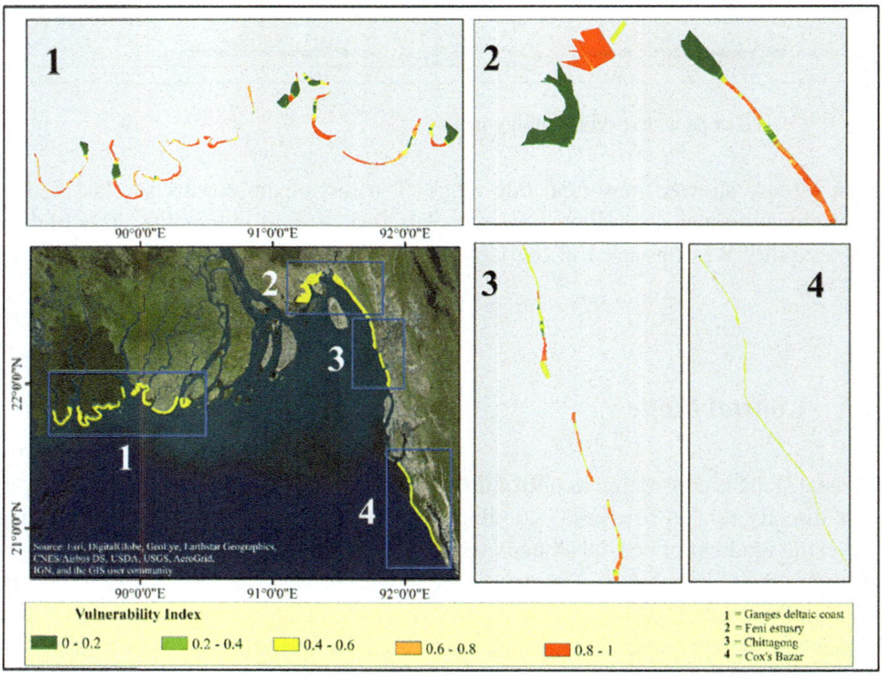

Fig. 7 The rate of shoreline changes with its vulnerability index

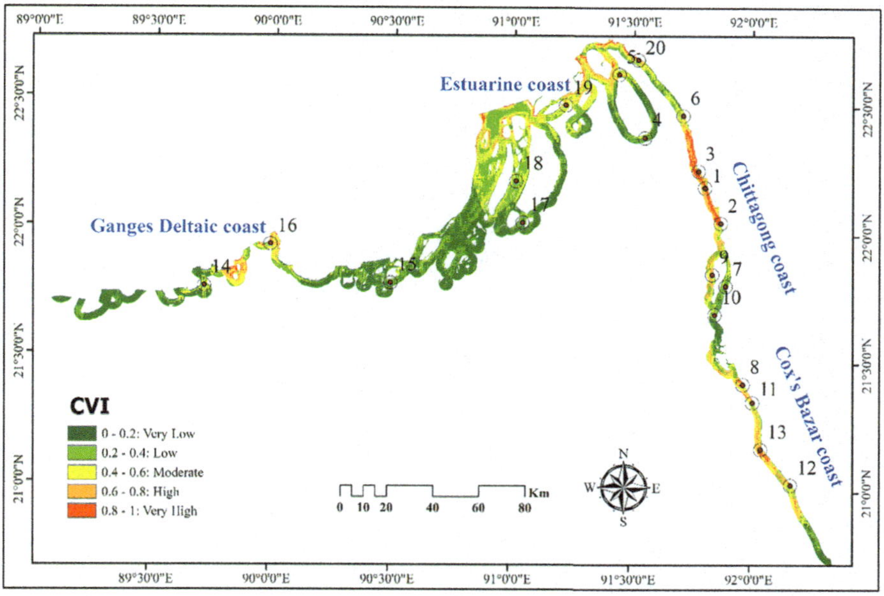

Fig. 8 The coastal vulnerability map of Bangladesh

track, Mean Tidal Range, and shoreline change has made Chittagong to be the most vulnerable coast. Following Chittagong, Cox's Bazar coast was found to be the second most vulnerable coast with an average CVI value of 0.46. Cyclone frequency and tidal range contribute to the vulnerability of the Chittagong coast whereas high range of SWH is dominated at Cox's Bazar. The study also exhibits that about 7% of the total coast of Bangladesh are at highly vulnerable conditions. Figure 8 represents the Bangladesh coastline with five vulnerability classes. The vulnerability at 20 key locations throughout the study area is summarized in the Table 3. Except for Mongla and Patharghata areas in the GDC (Table 3), the rest of the coasts demonstrated low and very low vulnerability. Presence of mangrove forests in the coast has notable contribution to reduce the coastline vulnerability. The forest acts as a potential barrier to decrease the CVI.

6 Conclusion

This study was aimed to develop a CVI map combining coastal characteristics and forcing factors. The Gornitz method was applied in GIS environment to obtain the CVI. Considering regional prospects, the density of cyclone track and the MTR are the most controlling factors of CVI in Bangladesh coast. The coastal slope is gentle throughout the coast, and except the Cox's Bazar coast the SWH has minimal contribution in CVI of the coast. The final output of CVI represents Chittagong as

Table 3 CVI at the key locations of Bangladesh

Location	Name	CVI value	Vulnerability class
1	Anwara	0.89	Very high
2	Banskhali	0.72	High
3	Chittagong port	0.55	Moderate
4	Sandwip 1	0.29	Low
5	Sandwip 2	0.47	Moderate
6	Sitakunda	0.61	High
7	Chakaria	0.26	Low
8	Cox's Bazar	0.74	High
9	Kutubdia	0.69	High
10	Maheskhal	0.55	Moderate
11	Ramu	0.38	Low
12	Taknaf	0.59	Moderate
13	Ukhiya	0.94	Very high
14	Mongla	0.47	Moderate
15	Galachipa	0.2	Very low
16	Patharghata	0.80	High
17	Hatiya	0.18	Very low
18	Manpura	0.34	Low
19	Noakhali Sadar	0.54	Moderate
20	Mirsharai	0.77	High

the most vulnerable coast. The cyclone activity and high tide range (>4 m) is responsible for the coastline vulnerability. From the study it is also apparent that the mangrove forest, Sundarbans has a significant role in reducing the vulnerability of Ganges deltaic coast. The study only considered the factors related to coastal forcing and coastal characteristics, not taking into account the socio-economic issues. The study is entirely based on secondary data obtained from several public and private sources. However, if some data can be checked from primary sources, it may enhance the strength of the work. Despite some lacking, this study can be a good tool for policy development and implementations in coastal areas. According to this study, it is highly recommended to emphasize the preservation of mangrove forests in the GDC. This study identified the vulnerable locations of Bangladesh coastline and their relative degree of vulnerability. From those outcomes, it is expected that the output of this vulnerability map may contribute to develop the coastal management policy as well as contributing to coastal protection work along the coastline.

Acknowledgement Authors acknowledge the Directorate of Research and Extension of Chittagong University of Engineering and Technology, Chattagram-4349, Bangladesh for providing research grant through a project, Grant ID CUET/DRE/2017-18/CRHLSR/001.

References

Alam, E., & Dominey-Howes, D. (2015). A new catalogue of tropical cyclones of the northern Bay of Bengal and the distribution and effects of selected landfalling events in Bangladesh. *International Journal of Climatology, 35*(6), 801–835.

Antony, C., & Unnikrishnan, A. S. (2013). Observed characteristics of tide-surge interaction along the east coast of India and the head of Bay of Bengal. *Estuarine, Coastal and Shelf Science, 131*, 6–11.

Arun Kumar, A., & Kunte, P. D. (Oct. 2012). Coastal vulnerability assessment for Chennai, east coast of India using geospatial techniques. *Natural Hazards, 64*(1), 853–872.

Bricheno, L. M., Wolf, J., & Islam, S. (2016). Tidal intrusion within a mega delta: An unstructured grid modelling approach. *Estuarine, Coastal and Shelf Science, 182*, 12–26.

CCC. (2016). *Assessment of sea level rise on Bangladesh coast through trend analysis*. Climate Change Cell (CCC), Department of Environment, Ministry of Environment and Forests, Bangladesh.

Gornitz, V. (1991). Global coastal hazards from future sea level rise. *Palaeogeography Palaeoclimatology Palaeoecology, 89*(4), 379–398.

Hegde, A. V., & Reju, V. R. (2007). Development of coastal vulnerability index for Mangalore Coast, India. *Journal of Coastal Research, 23*(5), 1106–1111.

İkizler, N. (2002). *Benefit maximizing classification using feature intervals* (Technical Report BU-CE-0208). Bilkent University.

IPCC. (2014). *Climate change 2014: Synthesis report. Contribution of Working Groups I, II and III to the Fifth Assessment Report of the Intergovernmental Panel on Climate Change* (Core Writing Team, R. K. Pachauri, & L. A. Meyer, Eds.). Geneva: IPCC, 151 pp.

Islam, M., Mitra, D., Dewan, A., & Akhter, S. (2016). Coastal multi-hazard vulnerability assessment along the Ganges deltaic coast of Bangladesh – A geospatial approach. *Ocean and Coastal Management, 127*, 1–15.

Kunte, P. D., Jauhari, N., Mehrotra, U., Kotha, M., Hursthouse, A. S., & Gagnon, A. S. (2014). Multi-hazards coastal vulnerability assessment of Goa, India, using geospatial techniques. *Ocean and Coastal Management, 95*, 264–281.

Mclaughlin, S., & Cooper, J. A. G. (2010). A multi-scale coastal vulnerability index: A tool for coastal managers. *Environmental Hazards, 9*(3), 233–248.

Mullick, M. R. A., & Islam, K. M. A. (2018). *Detection of Bangladesh coastline change due to climate change and sea level rise*. DRE Report.

Mullick, M. R. A., Tanim, A. H., & Islam, S. M. S. (2019). Coastal vulnerability analysis of Bangladesh coast using fuzzy logic based geospatial techniques. *Ocean and Coastal Management, 174*, 154–169. https://doi.org/10.1016/j.ocecoaman.2019.03.010.

Patra, A., & Bhaskaran, P. K. (2016). Trends in wind-wave climate over the head Bay of Bengal region. *International Journal of Climatology, 36*(13), 4222–4240.

Rahman, M., et al. (2018). Recent sediment flux to the Ganges-Brahmaputra-Meghna delta system. *The Science of the Total Environment, 643*, 1054–1064.

Rueda, A., et al. (2017). A global classification of coastal flood hazard climates associated with large-scale oceanographic forcing. *Scientific Reports, 7*(1), 5038.

Sahoo, B., & Bhaskaran, P. K. (2018). Multi-hazard risk assessment of coastal vulnerability from tropical cyclones – A GIS based approach for the Odisha coast. *Journal of Environmental Management, 206*(November), 1166–1178.

Schott, T., Landsea, C., Hafele, G., Lorens, J., Taylor, A., Thurm, H., Ward, B., Willis, M., & Zaleski, W. (2012). The Saffir-Simpson hurricane wind scale. *Natl. Hurric. Cent.* 1–4

Silverman, B. W. (1986). Density estimation for statistics and data analysis. *Biometrical Journal, 30*(7), 876–877.

Wu, S. Y., Yarnal, B., & Fisher, A. (2002). Vulnerability of coastal communities to sea-level rise: A case study of Cape May County, New Jersey, USA. *Climate Research, 22*(3), 255–270.

Government and NGO Provided Water Interventions and Its Effectiveness on Urban Poor: A Study in Gazipur Sadar Area

Saadmaan Jubayer Khan and Shamima Prodhan

Notes

Interventions: The action or process of intervening to improve water situation in the community (Hoque et al. 1996)

Effectiveness: Efficacy of establishing a viable source to fulfill the long-term basic water needs of the urban water insecure community

Secure Water: The capacity of a population to safeguard sustainable access to adequate quantities of and acceptable quality water for sustaining livelihoods, human well-being, and socio-economic development, for ensuring protection against water-borne pollution and water-related disasters, and for preserving ecosystems in a climate of peace and political stability (UN-Water 2013).

Urban Poor: A group of ten or more adjacent households whose housing structures are of visibly poor quality, and/or whose homes have been laid out in a non-conventional fashion without adherence to a ground plan (Mckinney 2016).

Safe Water: Safe water means water that will not harm you if you come in contact with it. The most common use of this term applies to drinking water, but it could also apply to water for swimming or other uses (USGS 2008).

S. J. Khan (✉)
Department of Disaster and Human Security Management, Bangladesh University of Professionals (BUP), Dhaka, Bangladesh

S. Prodhan
Institute of Disaster Management and Vulnerability Studies (IDMVS), University of Dhaka, Dhaka, Bangladesh

A. Haque, A. I. A. Chowdhury (eds.), *Water, Flood Management and Water Security Under a Changing Climate*, https://doi.org/10.1007/978-3-030-47786-8_22

1 Introduction

Southern Asia is one of the more swiftly urbanizing regions of the world. More than 2 billion people live in urban areas; with regional urban population's 30% living in slums. In the Asian economy, estimated $10–12 billion (US) dollar is contributed by the irrigation of groundwater. Conferring to United Nations World Water Development Report (2015) nearly half of world's use of groundwater is done by China, Pakistan, Nepal, India and Bangladesh. Around 4 million people lack access to safe water and 85 million people lack the chance for improved sanitation facilities worldwide (Water Org 2018). Bangladesh's main lifeline is considered to be about 700 rivers and its tributaries, which comprise approximately 15,000 miles of total waterway (Banglapedia 2012). Millions of people in Bangladesh are directly reliant on rivers for their regular livelihoods as it provides various sources of revenue options such as agricultural, fishing and fisheries as well as transportation practices; which are unswervingly reliant on the existence of rivers. Buriganga, Turag, Dhaleshwari, Shitalakhya and Balu are the major rivers that are surrounding the Dhaka metropolitan. Rivers near the capital have been polluted due to unplanned and rapid industrialization and urbanization. Habited by huge amount of people with substandard housing Turag has long been considered as one of the highly polluted rivers in Bangladesh (Tania 2014). Industries have been up-surging over the last decade around Dhaka city (Ahmad et al. 2010). Though it uses small amounts of water compared to public usage, quality of water is affected immensely (Munnaf et al. 2015). Additionally, solid wastes from both industrial as well as municipal areas are being dumped on the open space of the Turag river bank contributing factors combine to create a huge threat for people to secure fresh water from the river with toxic odor (Halder and Islam 2015). The population of soon to be megacity Gazipur is increasing, thus the consumption of water will also immensely escalate. As a result, dearth of safe water access is going to affect most people; from city-dwellers to the poorest people. Around 88% slum dwellers in Dhaka have no access to drinking water. Subsequently, people in the Gazipur region are using contaminated water from Turag river; causing various diseases including those of the colon. There are differences in terms of characteristics, capacity and resources between rural poor and urban poor, so their needs will also not be the same. Thus, implementation of the water policy or strategy for the urban poor needs to be tailored to their specific needs. Since a poor person is less able to afford individual interventions to secure safe water, government and NGOs are more likely to intervene for them. The identified group, prior to outside interventions, had no permanent sources of water to meet their daily water need, sometimes they collect unsafe contaminated water for drinking and to ensure their daily water need from household consumption they need permanent solution.

Regarding the comparative advantages of multiple intervention types, methods and dynamics which could contribute to impact the community as a whole – very limited knowledge is available. There has been little information regarding what has been irrefutably proved concerning different intervention approaches (Campos 2008). There have been so many literatures discussing the improved situation after implementation of interventions in the community to deal with diarrhea (Clasen et al. 2004). Studies addressing the situational analysis of endemic water-borne diseases in the global level as well (Colford et al. 2006). Water security for urban poor is a relatively new concept however if the implementation level was as robust as the policy level, for any country, water securing for urban poor ought not to be a concern (Singh and Desouza 1980; Meissner 2016; Pahl-Wostl 2016; Yang et al. 2016). So far there are not many studies completed regarding the Government and NGO interventions that in working on water security securing water for the urban poor from the context of Bangladesh. With the aforementioned arguments, specific objective of the present study is to assess the efficiency of Government and NGO provided water interventions for establishing a viable source to fulfill the long-term basic water needs of the urban water insecure in Gazipur Sadar areas, specifically Baimail Nadir Parr, Koroitola and Machimpur. This study would make a transparent depiction of the existing mechanism and their efficiency for securing water for the poor by the proper authorities.

2 Methodology

Based on the nature and set of objectives of the research, primarily qualitative and some required quantitative techniques were applied to measure the efficiency of the government and NGO provided interventions for the urban water insecure. A variable-based technique has been employed in order for the study to be supported by descriptive data obtained from varied primary and secondary sources. Focus Group Discussion (FGD), Case Study, In-depth Interview (IDI) as well as Key Informant Interview (KII) were used for gathering information about the condition and facilities provided by the interventions in Gazipur. FGDs are used for understanding the societal view towards the interventions, whereas the KIIs were developed to understand the position the NGO and the Government has in terms of installing, maintaining and evaluating the interventions. Associated variables and factors are understood through quantitative methods; household survey method has been used for collecting data from the field through purposive sampling. Qualitative data on the other hand has been collected via face-to-face interview method from the beneficiaries of those installed interventions.

Also there had to be criteria for population selection such as; the respondent would need to lack one or more of the followings –

- Permanent and durable housing that protects people from extreme climate conditions;
- Not more than three people sharing the same room or sufficient living space;
- Easy access to water (at an affordable price in sufficient amounts);
- Adequate sanitation access
- Direct user (or beneficiary) of either or both Government and NGO provided water supply interventions

For KIIs, respondents would need to be from a NGO that has worked in those aforementioned communities as well as involved in Government body (City Corporation, WASA); or involved in any leadership capacity in the community. For IDIs and Case Studies, focal person; people who are living in that area for many years or sufferers of any water-borne diseases are selected. For FGDs, participants are not selected randomly, rather they are chosen based on their involvement in different working sectors, so that various points of views would be shared during the discussion.

2.1 Study Area

This study is to be conducted near the regions of the Turag river. For this purpose, Gazipur district has been chosen as most of the Turag river flow goes through Gazipur sadar area, other factors include industrial and municipal waste being

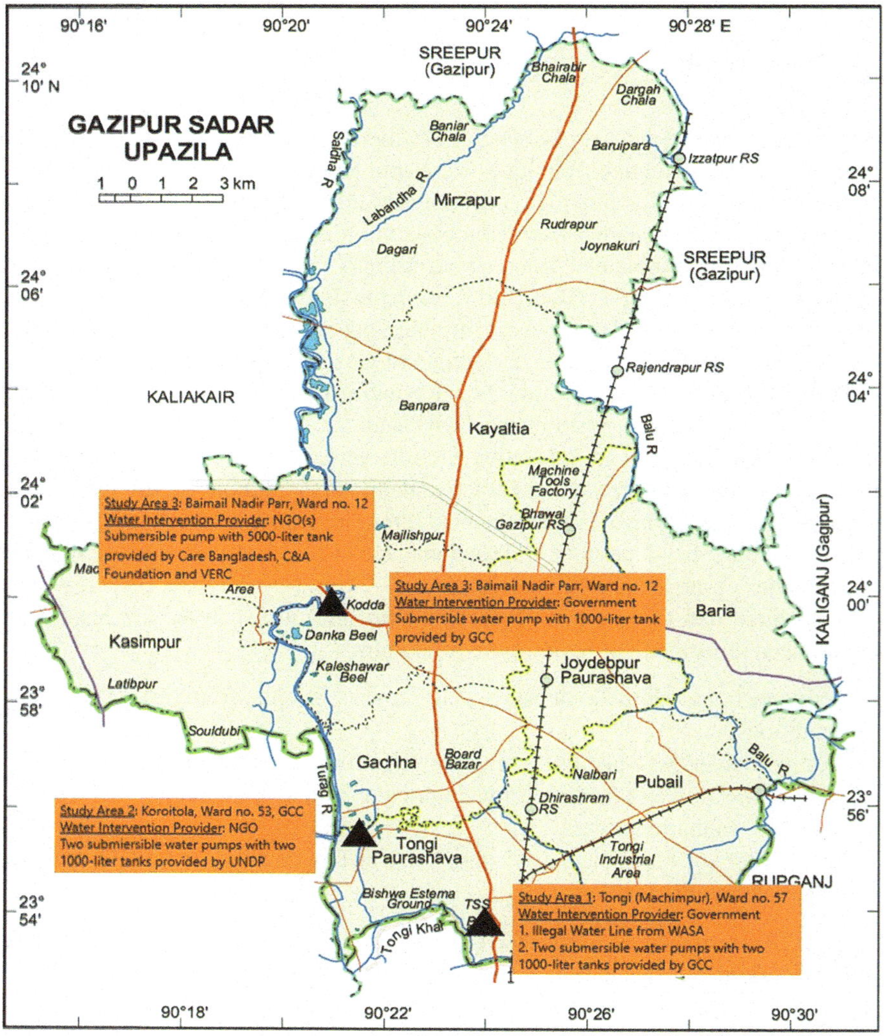

Fig. 1 Map showing study area (1, 2, and 3) in Gazipur

dumped into the river indicating that the urban poor are also dependent on the interventions, rather than on the river for safe water source (Fig. 1).

Data has been collected from three sites of Gazipur Sadar Upazila such as –

(i) Machimpur (Tongi): Ward number 57,
(ii) Koroitola: Ward number 53 and
(iii) Baimail Nadir Parr (Konabari): Ward number 12.

These wards have been selected as Turag river flows through along these areas. Hence water interventions provided by the Government and NGOs for the urban poor living near the Turag river region has been identified and studied.

2.2 Equation for Determining Sample Size

To determine sample size Gazipur City Corporation's ward number 12, 53 and 57 were selected based on the availability of the interventions as well as people's dependency on Turag river. Quantitative data collection was completed based on the systematic random sampling method.

To estimate the sample size Fisher's general formula is going to be used as the following:

$$n = \frac{z^2pq}{d^2}$$

Where,

n = Sample size
z^2 = Value of standard normal variable at 5% level of confidence interval (1.96)
p = Poverty rate in percentage (0.243) (Prodhan et al. 2017)
$q = (1 - p) = 1–0.243 = 0.757$
d = Degree of accuracy desired, set at 0.05 for this research

Based on Fisher's formula we get a sample size of approximately 283.

However, it was difficult to select 283 houses because the frequency of houses is different in Bangladesh as Bangladesh is a heterogeneous country. Most of the residents of the study area do not fall into the category of desired respondent of the research, and so 283 samplings were not applicable for this study. To avoid that, 210 surveys will be undertaken across the three study areas.

3 Results

3.1 Profile of the Respondents

A total of 210 respondents have been identified based on the quantitative random sampling method, where forty-five percent (45%) males and fifty-five percent (55%)

female respondents were interviewed. Highest representation of age group is between 31–40 years, which forty-three percent (43%) respondents belong to. All household surveys were conducted based on their day-to-day reliance on either or both Government and NGO established water interventions. All of the qualitative study participants were also stakeholders of previously mentioned water interventions and were chosen systematically. All of the respondents fulfill the set-out criteria of study population selection.

Average family size of surveys' respondents is 4.27 people, which is close to the national average of 4.4 persons per household according to Bangladesh Bureau of Statistics (Prodhan et al. 2017). Majority (67.1%) of those respondents live in a nuclear family household. Sixty-six percent (66%) respondents' occupancy status is Tenant. Sixteen percent (16%) respondents are found to have never attended school, whereas nineteen percent (19%) are found only to know how to sign their names. In primary education forty-six percent (46%) respondents are found to be educated, while around nineteen (19%) percent respondents are found to be schooled in secondary level. Illiteracy level of women (12.9%) is less than that of men (20.2%); however, dropout level after completion of primary education to secondary education, for women (22.3%) is much higher than men (15.5%). Most of the respondents, thirty-seven percent (37%) are directly involved in trade/business mostly in fish; followed by twenty-nine percent (29%) respondents working in garments factory as a worker/staff. In the study area around fifty-one percent (51%) respondents' income and household expenditure group has been identified between 16,000–20,000 taka. Plus, most of the monthly savings (52%) per household category falls in the category of 1000–2000 taka.

None of the respondents use Turag river's water as a main water source for drinking and cooking, only a minimal percentage (9.5%) people use it as an alternative water source for drinking and cooking, they only do so when it rains. There is a pattern of increase in Turag river use during the wet season, both as main and alternative water source for bathing and hygiene purposes. Bulk (87%) of the respondents believe that Turag river water is not safe, while the remaining thirteen percent (13%) respondents believe that during wet season river water is usable and safer than dry season.

3.2 Government and NGO-Established Water Interventions

Out of 210 respondents 112 (53%) respondents are beneficiaries of Government installed interventions; while 98 (47%) respondents are beneficiaries of NGO installed interventions in selected study areas.

3.2.1 Intervention Providers

A number of Government and non-government organizations have been working in the selected study sites to ensure water for the people. Various initiatives and activities undertaken by the intervention providers in the study site are discussed below:

I. **Gazipur City Corporation (GCC)**: Providers of multiple intervention facilities for the community, GCC has been undertaking initiatives to address safe water supply requirement of the people in Gazipur. Some of their facilities include:

   (a) Installing two 300 ft deep submersible pump, each having 1000-liter water tanks, that provide water for more than 100 families via public piped into dwelling method in Tongi (Machimpur), ward number 57 of GCC. First submersible was established back in 2011 while the second one was established in 2015. Families need to pay a monthly fee of 50 taka to use this facility (Fig. 2a).

   (b) Providing one 280 ft deep submersible pump with a 1000-liter water tank, that provides water for the people through public piped into yard method for 130 families in Baimail Nadir Parr, ward number 12 of GCC. Installed by Public Health Engineering Department under the project of "Creating supply source on district cities for providing water (submersible pump)", it was completed in 2014–15 fiscal year. Families need to pay a monthly fee of 30 taka to use this facility (Fig. 2b).

II. **Care Bangladesh, C&A Foundation and VERC**: Providing one 300 ft deep submersible pump with 5000-liter tank that supplies water to 85 fisherman households through public piped into dwelling method in Baimail Nadir Parr (Konabari).

    This intervention was funded by C&A Foundation, assisted by Care Bangladesh and completed by the Village. Education Research Center (VERC) under the "Building Resilience of the Urban Poor (BRUP)" project, it was installed in 2016. Households have to pay 80 taka per month to use this facility (Fig. 2c).

III. **UNDP**: United Nations Development Programme (UNDP) with permission from and partnership with GCC, installed a 300 ft submersible pump with two 1000-liter water tanks under the project name "Urban Partnerships for Poverty Reduction (UPPR)". This intervention was established in 2009 and it supplies water for around 200 families of Koroitola, ward number 53 of GCC. Each beneficiary household has to pay a monthly fee of 100 taka to use the facility (Fig. 2d).

3.2.2 Time Spent on Water Collection

Figure 3 pinpoints that NGO provided interventions take a lot less time for water collection when compared to government provided interventions. Forty-four percent (44%) respondents who are beneficiaries of NGO interventions answered that it takes them less than 5 min to collect water. While forty percent (40%) Government intervention beneficiaries responded that it takes them less than 5 min to collect water from the intervention source.

Fig. 2 Some phorograpg of Cazipur City Corporation area. (**a**) Water Intervention provided by GCC in Tongi (Machimpur). (**b**) GCC provided intervention in Baimail Nadir Parr. (**c**) NGO provided intervention in Baimail Nadir Parr. (**d**) NGO provided intervention in Koroitola

Sixteen percent (16%) respondents stated that it takes more than 60 min to collect water from government installed water sources. None of the NGO-intervention-beneficiary respondents have stated that it takes them more than 11–15 min to collect water from the source.

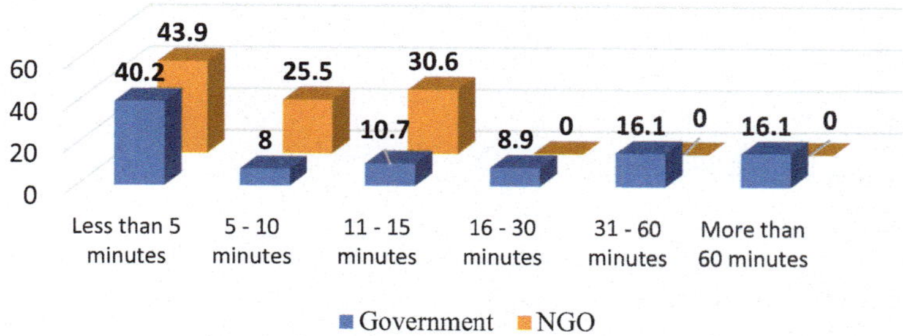

Fig. 3 Time taken to collect water from the interventions

Fig. 4 Percentage of respondents' satisfaction with the interventions

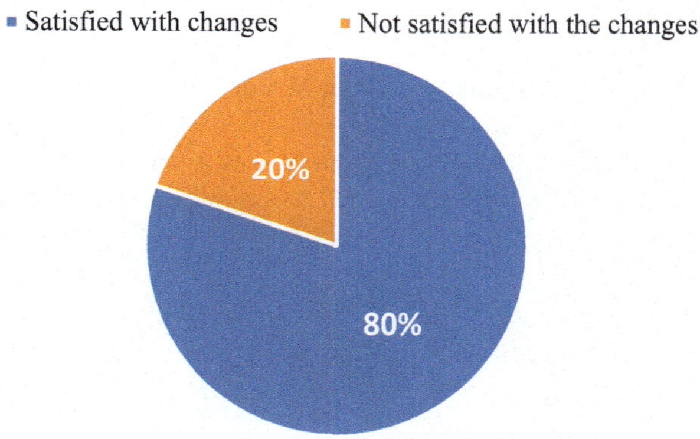

3.3 *Effectiveness of Water Interventions*

The core purpose of providing interventions in the community by the Government and NGOs is to supply safe water to the poor. Both Government and NGOs try to ensure maximum quality through utilization of their constrained resources.

3.3.1 Satisfaction with the Interventions

Any change in the community, whether good or bad, is bound to leave an impression on its inhabitants. While interventions from both Government and NGO ensued to address the immediate water needs of the people; some changes are bound to occur. Figure 4 shows that eighty percent (80%) respondents are happy with the changes brought on by these interventions.

3.3.2 Water Collection Time

As previously demonstrated in Fig. 3, NGO interventions take lot less time to collect water than that of Government provided interventions. Since the introduction of intervention in their community, maximum respondents (72%) think that it now takes less time to collect water. This is an emblem of success for interventions trying to achieve water security for the urban poor (Fig. 5).

3.3.3 Intervention Cost

With the arrival of outside intervention in the community, changes are bound to happen; cost for water supply could increase, decrease or remain the same. According to respondents' responses majority (41%) consider that cost has not increased, while thirty-eight percent (38%) respondents have confirmed that cost has increased after the inception of interventions in the community (Fig. 6).

3.3.4 Availability of Clean Water

In the survey of the present study, it is found that (Fig. 7) approximately seventy-two percent (72%) of respondents' households believe that water is now more accessible

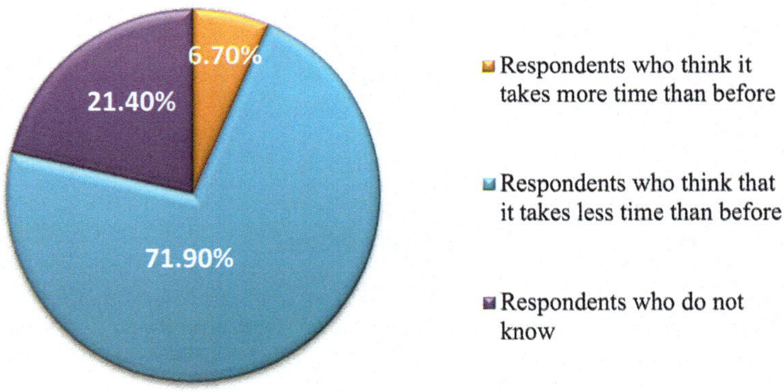

Fig. 5 Percentage of respondents' time consumption of interventions

Fig. 6 Percentage of respondents' identifying increase of cost

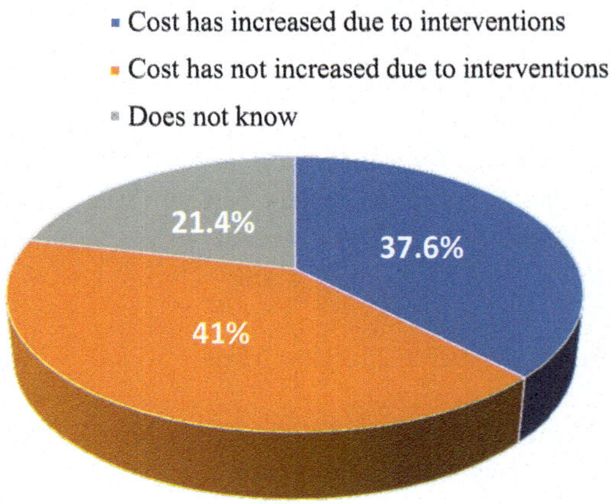

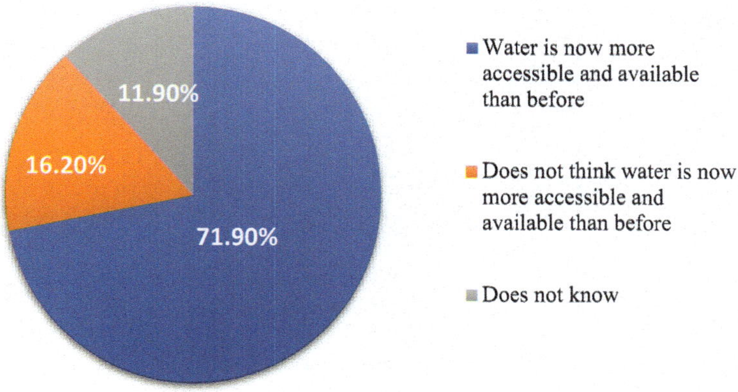

Fig. 7 Percentage of respondents' identifying water supply accessibility and availability

and available than before due to newly installed government and/or NGO interventions. While around sixteen percent (16%) respondents have reservations about that notion.

3.3.5 Water-Borne Diseases

Health risk is one of the reasons why people are usually concerned with having safe water source; better the quality of water, lesser the chances of suffering from water-borne diseases. Presently in three separate study areas: Diarrhea is the prevailing water-borne disease; closely followed by Chikunguniya. Majority (67%) of respondents' household has cited a new installed intervention as a reason for reducing water-borne diseases in the study sites.

3.3.6 Changes Brought by Interventions

Interventions are not always going to bring about positive changes, according to some respondents' household surveys, there are some negative changes brought on by these interventions as well. Majority (71%) of respondents' households believe that no negative changes have been brought on by the interventions installed by the Government and NGOs.

Bulks of the respondents agree that no negative changes have been brought by the interventions. By default, it means that majority of the households think that both the Government and NGO installed interventions have brought on positive changes in the society.

3.3.7 Government and NGO Interventions

The core purpose of providing intervention in the community by the Government and NGOs is to supply safe water to the water insecure. Both Government and NGOs

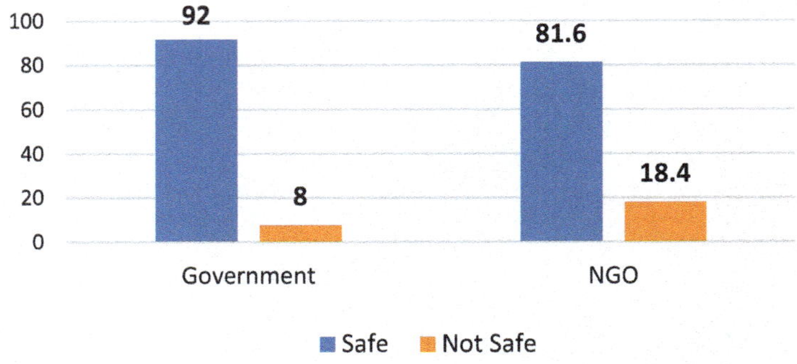

Fig. 8 Percentage of respondents' think intervention water is safe

try to ensure maximum quality through utilization of their restricted resources. However, trustworthiness among beneficiaries varies based on intervention providers. Figure 8 illustrates that most of the Government beneficiaries (92%) believe that supplied water from the Government provided intervention is safer, compared to NGO provided interventions, where around eighty-two percent (81.6%) NGO beneficiaries believe that their water is safe.

4 Discussion

4.1 Government and NGO Installed Water Interventions

Based on the aforementioned results from both qualitative and quantitative field level and secondary data, the following findings can be deliberated.

Majority respondents have identified 'less than 5 minutes' as the most common amount of time spent for water collection. In addition, NGO provided interventions take lot less time for water collection than government provided interventions. While none of the NGO-intervention-beneficiary respondents have answered that it takes them more than 11–15 min to collect water from the source, around sixteen percent (16%) beneficiaries have stated that sometimes it takes them more than 60 min to collect water from Government installed water interventions.

The cost of water supply for NGO interventions is higher than Government provided intervention support. Majority of the Government intervention beneficiaries, around sixty-three percent (63%) pay 50 taka per month, the remaining thirty-seven percent (37%) pay 30 taka. Whereas, most of the NGO intervention beneficiaries, seventy-two percent (72%) pay 100 taka and the remaining beneficiaries (28%) pay 80 taka per month for their services.

Ninety-two percent of government intervened users believe that supplied water which is provided by the government is safe, whereas around 82% NGO beneficiaries believe that that NGO provided water is safe. In the study area, people trust the Government more than NGOs Respondents who do not think intervention provided water is safe usually boil the water for safety (Table 1).

Table 1 Details about the intervention providers in three separate study sites

Name of the area	Provider of the intervention	Type of support provided	Method of water collection	Intervention for	Details about the intervention
Baimail Nadir Parr	Gazipur City Corporation (Government)	Submersible pump with 1000-liter tank	Public piped into yard	130 families	Creating water source on 37 district cities for providing water (submersible pump).
					Installed by: Public Health Engineering Department
					Completed in: 2014–15 fiscal year
	Care Bangladesh, C&A Foundation and Village Education Research Center – VERC (NGO)	Submersible pump with 5000-liter tank	Public piped into dwelling	85 families	**Project name**: Building Resilience of the Urban Poor (BRUP)
					Installed by: Care Bangladesh, C&A Foundation and VERC
					Installed in: 2016
Koroitola	United Nations Development Programme – UNDP (NGO)	Submersible pump with 2 (two) 1000-liter tank	Public piped into yard	200 families (approximately)	**Project name**: Urban Partnerships for Poverty Reduction (UPPR)
					Installed by: UNDP and Gazipur City Corporation
					Installed in: 2009
Tongi (Machimpur)	Gazipur City Corporation (Government)	Water line from WASA	Public piped into yard	20 gypsy families	With the permission of 57 number Ward Commissioner, taking a water line from WASA
					Installed by: Individual families
					Installed in: 2017
		2 submersible pumps with 2 (two) 1000-liter tank	Public piped into dwelling	100+ families	Provided by the Gazipur City Corporation
					Installed in: 2011 and 2015

Table 2 Assessing the indicators of water security with the collected data

Indicators		Yes (%)	No (%)	Comments
1.	Satisfied with change	80	20	Majority of respondents are satisfied with the changes brought on by the interventions
2.	Less time consuming water collection	71.90	6.70	Bulk of the respondents agree that water collection has become less time consuming than before
3.	Increase in cost	37.6	41	This data is inconclusive due to the fact that 21.4% respondents were not able to compare between the two
4.	Improved availability and accessibility of water	71.90	16.20	Most of the respondents settled that water is easily accessible and more available than before
5.	Reduction of water-borne diseases	67.1	32.9	Majority of the respondents came to an understanding that after the inclusion of interventions water-borne diseases have reduced.
6.	Negative changes brought on by the interventions	29	71	Most of the intervention users think that these interventions did not bring influx of negative impact on the community.

4.2 Indicators of Water Effectiveness

Based on the pre-set indicators of assessing Government and NGO-based water interventions in Gazipur Sadar areas to fulfill the long-term basic water needs of the urban water insecure, their efficacies are arbitrated below:

Aforementioned Table 2 illustrates that 5 out of 6 indicators have been met with positive feedback from the respondents' after the introduction of Government and NGO provided interventions. This means that the once susceptible water insecure community is now in a better shape than before. However, it is still far from achieving water security – but this movement is towards the right path nonetheless.

5 Conclusion

Majority of the people are becoming dependent on artificial water interventions (Government and NGO installed), as opposed to natural sources (Turag river), due to this they are now being contaminated in multiple self-destructive ways with no way for them to recover. Without Government and NGO water interventions access towards clean and accessible water for the urban poor of Gazipur sadar would have been problematic. However, it will not be possible to achieve water security through these interventions alone. The positive side of all these newly installed interventions is that people are responsive and satisfied with the visible social influences brought on. Forthcoming studies focusing on water security issues might be able to use this study as a baseline for the Government and NGO interventions scenario, perception of the people, and the importance of achieving an urban water secure community. Further researches can be done by using an established scientific method for assessing the effectiveness level and comparing them with the perception based data from this research.

Acknowledgement The lead author would like to express his heartiest gratitude, sincere appreciation and profound respect to the research supervisor and co-author, for her scholastic guidance, constant inspiration, valuable advice and constructive criticism throughout the research work.

He would also like to thank all participants and respondents in the study area, for sparing valuable time, sincere effort and honest answers to acquire the required data. Further, he would like to thank ward councilors for giving him the necessary permission to conduct the thesis in the selected areas, as well as NGO personnel for their valuable time and opinion in addition to assistance before and during the data collection process.

Finally, the lead author would like to express his utmost sincere gratitude to 7th International Conference on Water and Flood Management (ICWFM-2019) secretariat and his team, Institute of Water and Flood Management (IWFM), Bangladesh University of Engineering and Technology (BUET) for selecting their research paper for conference presentation as well as full paper submission for an opportunity in Springer publication.

References

Ahmad, M., Islam, S., Rahman, M., Haque, M., & Islam, M. (2010). Heavy metals in water, sediment and some fishes of Buriganga River, Bangladesh. *International Journal of Environmental Research, 4*(2), 321–332. Accessed on May 6, 2018. Retrieved from https://ijer.ut.ac.ir/article_24_bd9fc148ed573c706d417fdbb1b98e34.pdf.

Banglapedia: National Encyclopedia of Bangladesh. (2012). *River and drainage system*. Accessed on May 6, 2018. Retrieved from http://en.banglapedia.org/index.php?title=River_and_Drainage_System.

Campos, M. (2008). Making sustainable water and sanitation in the Peruvian Andes: An intervention model. *Journal of Water and Health, 6*(S1), 27–30.

Clasen, T., Roberts, I., Rabie, T., & Cairncross, S. (2004). Interventions to improve water quality for preventing diarrhea. *Cochrane Database of Systematic Reviews, 19*(3), CD004794.

Colford, J. M., Roy, S., Beach, M. J., Hightower, A., Shaw, S. E., & Wade, T. J. (2006). A review of household drinking water intervention trials and an approach to the estimation of endemic waterborne gastroenteritis in the United States. *Journal of Water and Health, 04*(Suppl 2), 71–88.

Halder, J., & Islam, N. (2015). Water pollution and its impact on the human health. *Journal of Environment and Human, 2*(1), 36–46. Accessed on May 6, 2018. Retrieved from https://pdfs.semanticscholar.org/6ffd/2e86026c3825adca39f453cd3a61782676ca.pdf.

Hoque, B. A., Juncker, T., Sack, R. B., Ali, M., & Aziz, K. M. (1996). Sustainability of a water, sanitation and hygiene education project in rural Bangladesh: A 5-year follow-up. *World Health Organization, 74*(4), 431–437.

Mckinney, P. (2016). *Phnom Penh: Multiple indicator assessment of the urban poor*. Phnom Penh, Cambodia: People in Need-Cambodia. Accessed on May 7, 2018. Retrieved from https://www.unicef.org/cambodia/PIN_URBAN_POOR_FA.PDF.

Meissner, R. (2016). Water security in Southern Africa: Discourses securitizing water and the implications for water governance and politics. *Handbook on Water Security*, 280–299.

Munnaf, A., Islam, M., Tusher, T., Kabir, M., & Molla, M. (2015). Investigation of water quality parameters discharged from textile dyeing industries. *Journal of Environmental Science and Natural Resources, 7*(1), 257–263. Accessed on May 6, 2018. Retrieved from https://www.banglajol.info/index.php/JESNR/article/view/22180/15258.

Pahl-Wostl, C. (2016). Water security, systemic risks and adaptive water governance and management. In *Handbook on water security* (pp. 91–104). Cheltenham: Edward Elgar Publishing.

Prodhan, S., Sikder, B. B., & Nasreen, M. (2017). Adaptation strategies undertaken by the community to reduce impacts of shrimp cultivation on agriculture: A study at Parulia Union, Satkhira (Bangladesh). *Asian Journal of Water, Environment and Pollution, 14*(4), 21–30. https://doi.org/10.3233/ajw-170033.

Singh, A. M., & Desouza, A. (1980). *The urban poor: Slum and pavement dwellers in the major cities of India*. New Delhi: Manohar.

Tania, F. (2014). Solid waste management of Dhaka City: A socio-economic analysis. *Banglavision Foundation, 13*(1), 91–100. ISSN:2079-567X. Accessed on May 6, 2018. Retrieved from http://bv-f.org/08.%20BV%20Final.-13.pdf.

United States Geological Survey (USGS). (2008). *Can you define safe water?* Accessed on May 7, 2018. Retrieved from https://www.usgs.gov/media/audio/can-you-define-safe-water.

UNESCO. (2015). The United Nations World Water Development Report 2015. pp. 74–77. Paris: United Nations Educational, Scientific and Cultural Organization. Accessed on May 24, 2020. Retrieved from https://unesdoc.unesco.org/in/documentViewer.xhtml?v=2.1.196&id=p::usmarcdef_0000231823&file=/in/rest/annotationSVC/DownloadWatermarkedAttachment/attach_import_c7942a28-e816-45af-817ca86722fd8f1f%3F_%3D231823eng.pdf&locale=en&multi=true&ark=/ark:/48223/pf0000231823/PDF/231823eng.pdf#%5B%7B%22num%22%3A60%2C%22gen%22%3A0%7D%2C%7B%22name%22%3A%22XYZ%22%7D%2Cnull%2Cnull%2C0%5D.

UN-Water. (2013, May 8). *Water security and the global water agenda*. Accessed on May 7, 2018. Retrieved from http://www.unwater.org/publications/water-security-global-water-agenda/

Water Org. (2018). *Bangladesh water crisis – Clean water in Bangladesh*. Accessed on May 6, 2018. Retrieved from https://water.org/our-impact/bangladesh/

Yang, H., Liu, J., & Xia, J. (2016). Water security – China perspective. In *Handbook on Water Security* (pp. 300–316). Cheltenham: Edward Elgar Publishing.

Virtual Water Information for Water and Food Security in Peri-Urban Khulna: Present Status and Future Requirement

Sadia Ashraf, Uthpal Kumar, Sharmistha Roy, and Dilip Kumar Datta

1 Introduction

Bangladesh is an agricultural country, where unpredictable weather and climate variability are threats for sustainable agricultural production and food security. Therefore, weather and climate information services are major requirements to manage agricultural activities and inform decision-making in order to ensure agricultural production and food security. Agricultural information plays a key role in agricultural development and production, and their effective communication will facilitate mutual understanding among farmers, agriculturist and extension workers (Alam and Haque 2014). Among many other agricultural information, weather and climate information are key for reducing risks and uncertainties in the production system. Weather and climate information includes short-term (up to week), medium-term (up to month) and long-term (seasonal or longer scale) forecast information. ICT has the potential to help farmers in the entire cycle of production, i.e., from production to sales by providing weather and climate information services, and ICT has an effect on both observable and unobservable transaction costs (Bhatnagar 2008). Most efforts to make ICT available to rural farmers have sought to improve the availability and quality of information either indirectly through producer associations, and extension workers, or directly through broadcast radio information, and mobile short messaging services (SMS) (Bertolini 2004). In Sri Lanka, farmers were

S. Ashraf (✉) · S. Roy · D. K. Datta
Environmental Science Discipline, Life Science School, Khulna University, Khulna, Bangladesh

U. Kumar
Water System and Global Change Group, Wageningen University and Research, Wageningen, The Netherlands
e-mail: uthpal.kumar@wur.nl

A. Haque, A. I. A. Chowdhury (eds.), *Water, Flood Management and Water Security Under a Changing Climate*, https://doi.org/10.1007/978-3-030-47786-8_23

able to improve their incomes through simple mobile phone applications that helped reducing waste through a feedback system (Silva and Ratnadiwakara 2008). ICT platforms could reduce farmers' risks and uncertainties through weather and climate information in advance. Before doing that, existing agricultural information services and needs of the ICT based weather and climate information should be analysed. This study was conducted to:

(a) analyse existing agricultural information services through ICT and traditional platforms and their usefulness in farming practices, and
(b) explore farmers' need for a virtual platform (using ICT) on weather and climate information for sustainable agriculture in the study area.

2 Methodology

2.1 Description of the Study Area

Khulna (22°12′–23°59 N; 89°14′–89°45′E) is the divisional city of southwest coastal Bangladesh. Khulna District covers a total area of about 4389.11 sq. km. which is about 2.97% of Bangladesh. It comprises of 9 upazilas and has a population of about 547,000. Batiaghata and Rupsa upazila (sub-districts) of Khulna are regarded as the two important peri-urban food producing hub of Khulna. Major crops grown in the upazilas are paddy, jute, sesame, betel nut and vegetable. Integrated farming practices are now increasing in Khulna due to high economic return from a single piece of land and the rise in adaptability to weather, water and salinity environment within Khulna. Figure 1 shows the study area and sample village location in the map.

2.2 Sample Selection and Data Collection

An initial desk research was conducted through reviewing and analysing available information services on agriculture and weather information through online sources. Following which, two upazilas (Batiaghata and Rupsa) of Khulna district were selected for the purpose of this research based on peri-urban land use characteristics. Peri-urban area is defined as an area having a mixture of land uses associated with a range of urban and rural livelihoods (Parkinson and Tayler 2003).

To conduct this study, a total of 200 farmers, selected via random sampling, were interviewed from six villages located in Rupsa and Batiaghata Upazila. Sample size was calculated by following the equation of Berenson and Levine (1992). Out of the six villages, three villages (Sreefaltala, Domra and Peyara) were selected from Rupsa Upazila and three from Batiaghata (Jharvanga, Rangemari and Sachibunia) of Khulna District. Preliminary field visits and formal discussions were held with extension officers before selection of the study sites. Figure 2 shows schematic diagram and population size at the study villages in Khulna, Bangladesh. To

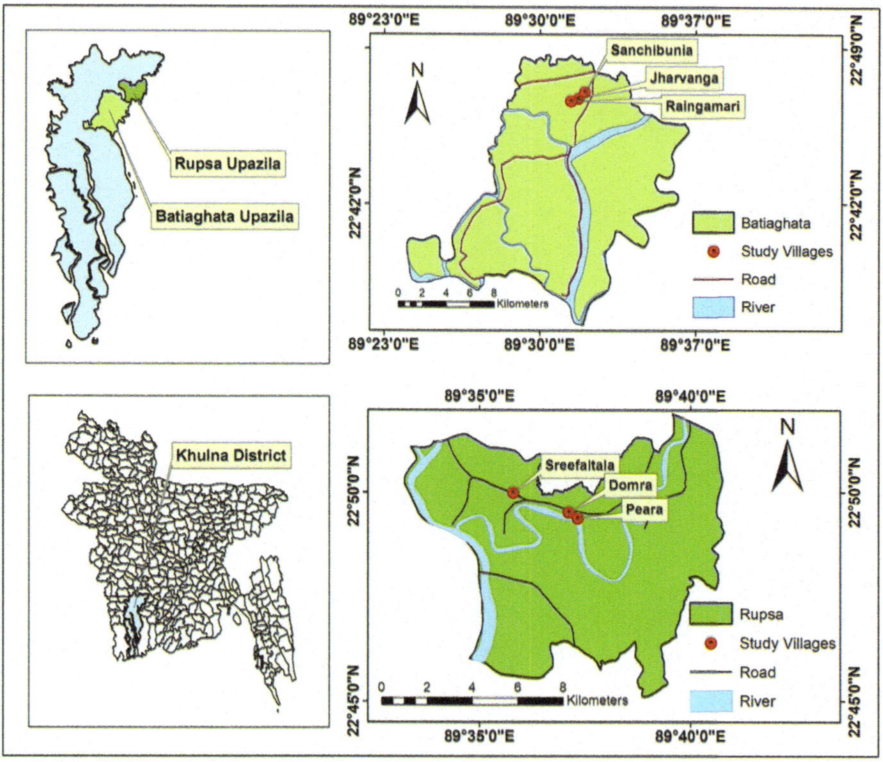

Fig. 1 Location map of the study area

complete the survey, we used online free version survey tool, KoBoToolbox (https://www.kobotoolbox.org/). To complete the study, we also conducted six focus group discussions (FGD) with both male and female participants from the sampled farmers. Numbers of participants in each FGD was 10 to 12. Also Arc GIS 10.5 software was applied for study area map preparation.

The totals of 200 sample farmers were calculated with 95% confidence (z = 1.96) and 10% error margin (e = 0.1) by the following equation (Berenson and Levine 1992):

$$n_o = z^2 p(1 - p)/e^2$$

Where no = the sample size not corrected with the finite household correction factor; z = the level of confidence; p = the proportion of the households; e = the accepted sampling error. Then the finite correction factor was applied to correct the calculated sample size (Berenson and Levine 1992).

$$n = n_o N/(n_o + (N - 1))$$

Where, n = the sample size considering the finite household correction factor; N = the total number of households in the study area.

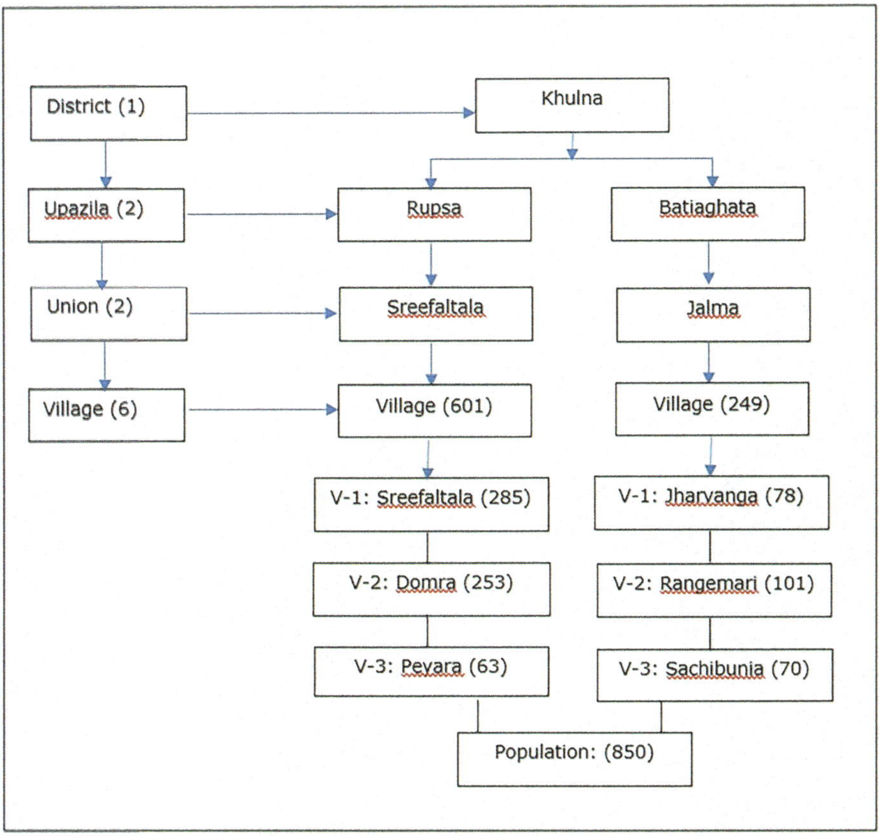

Fig. 2 Schematic diagram and population size at the study villages

2.3 *Data Analysis*

Online survey data was exported in excel environment. Then, we cleaned the dataset to overcome irregularities and missing data. We did post-coding for the open-ended responses. We used Microsoft Excel, version 2016, for analysis of the data mainly descriptive statistics and frequency analysis.

3 Result

This research was conducted to analyse agricultural information services available in Bangladesh and to decipher farmers' information needs in agricultural production. The available information sources for farmers are previous experiences, workshops and seminars, training, peer-farmers, farmer group meetings, print media such as agriculture magazines, newspapers, as well as websites, television, and radio. Government and non-Government organizations in Bangladesh provide many kinds of information services, for instance, call centre, SMS service, websites, mobile application, monthly magazine, community radio, television programs, bulletin, and leaflets. Major information sources and farmers' information needs is given in the following sections.

3.1 Agricultural Information Through Traditional Platforms

Farmers of peri-urban Khulna use their previous experiences from agricultural fields to make crop choices, and indigenous knowledge to predict weather conditions for cropping practices. They use several indicators to predict local weather conditions. Major indicators followed by the local farmers are wind direction and velocity, colour of the sky and clouds, lunar calendar and phases of the moon, rainbow formation, star density, behaviour of insects and plants, and blossoming of flowering plants. By using these indicators, farmers can predict the probability of temperature, rainfall and storm phenomena. A total of 71% of the interviewed farmers use these indicators to understand upcoming weather phenomena in the study area.

3.2 Weather Information Through ICT Platforms

Major ICT tools available in Bangladesh are television, radio, mobile phone and computer. These ICT tools provide information through different services like video, audio, telephone calls and SMS service, websites and android and iOS applications. Many helplines and SMS services are developed by the Government and private organizations for weather information services and dissemination. The mobile phone operators of our country also have some existing information services, which are given in Table 1.

From the desk research and personal interviews with officials of the agricultural extension department, we found that government and non-government organizations of Bangladesh already have several mobile apps to support local farmers for agricultural information. The popular apps are namely Agriculture Info Service, Farmer's Window, Pesticide Prescriber, BMD Weather App, and *Ajker Krishi*. However, common limitations for using these existing mobile apps by farmers are due to illiteracy, economic insufficiency, lack of ICT skills and lack of user-friendly

Table 1 Agricultural information service provided by mobile operators

Agricultural information service provided by mobile operators	
Mobile operators	Service
Banglalink	Krishijigyasha 7676
Grameenphone	Krishi Tothya Jiggasa 27676
ROBI	Robi Krishibarta: IVR and call, SMS, helpline 27676

mobile apps. In addition to that, strong internet connection is also necessary to run the mobile apps and update information on a regular basis.

3.3 ICT Tool Usage by Farmers

From the interviews, we found that the major ICT tools which are in use by the local farmers are radio, television and mobile phones. From the 200 interviews, we also found that, about 67% of farmers watch television, 19% of farmers listen to the radio and 14% of farmers use mobile phones to get weather information in the study area. Majority of farmers watch agricultural programs on television such as *'Krishi Dibanishi'* and *'Ridoy-e-Mati-O-Manus'*. During FGDs, male and female farmers groups confirmed that only female farmers get the opportunity to watch TV programs or emergency news, as most of the female farmers stay at home. Male farmers have less scope to watch television in daytime as they work mostly in the fields. However, male farmers get weather information using traditional and modern ICT platforms such as radio and mobile phone. The interview results depict that among 200 farmers, 85% of the farmers already have access to mobile phone. However, only 26% of farmers use mobile phones to collect agricultural information. Few farmers also use voice call, helpline service, radio, internet, website, mobile app to get information through mobile phones.

3.4 Agricultural Information Requirement by the Farmers

Study results show that access to weather and climate information is one of the major constraints for sustainable agricultural development in peri-urban Khulna. Also, the currently available weather and agricultural information is not location and time specific. Farmers cannot take key agricultural decisions based on the available weather and climate information through existing traditional platforms such as radio, television and newspaper. Local extension workers also indicated that lack of access to location and time specific weather information is a major limitation for providing agricultural advisory services to farmers. As a result, a majority local farmer suffers from frequent hydro-climatic hazards such as erratic rainfall and extreme rainfall, drought, cyclonic storm surge and waterlogging problems. The use of modern ICT tools such as mobile phone based information services could provide better information services at present context. However, only the modern ICT technology cannot safeguard entire agricultural production system. Building awareness on using such technologies and farmers capacity building for using ICT technologies is an urgent problem window in the present context, especially, for local farmers who are the most vulnerable group in the study area. During interviews, majority of the farmers expressed their needs for location and time specific weather and climate information services using modern ICT platform (Table 2).

Decision making according to the forecast information would help farmers to take better decisions and reduce their agricultural costs. This includes fertilizer cost,

Table 2 Farmer's weather information need

Evaluation of farmers' weather and climate information needs	
Strongly agree	44%
Agree	43%
Undecided	10%
Disagree	2%
Strongly disagree	1%

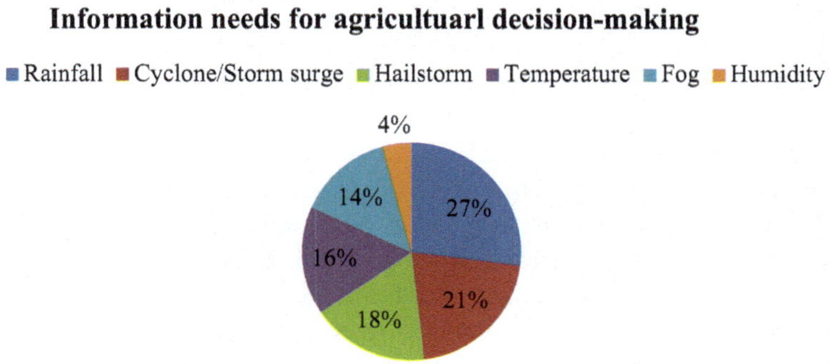

Fig. 3 Information need for crop related decision making

irrigation cost and labour cost, as well as, minimization of the production loss in cases of diverse weather hazards. For agricultural decision-making, farmers indicated their interests in different types of weather information in Fig. 3. Total 27% of farmers were interested in rainfall forecast, 21% of farmers were interested in cyclone and storm surge forecast, 18% of farmers were interested in hailstorm forecast, 16% of farmers were interested in temperature forecast including sunshine and heat stress, 14% of farmers were interested in fog forecast and only 4% were interested on humidity forecast. Humidity forecast is a very important factor to predict disease outbreak, as said by 4% of farmers during the interviews.

Most of the interviewed farmers were interested in their access to short-term (up to 1-week) and medium-term (up to 1-month) weather forecast for their agricultural decision-making. Among 200 farmers, 34% of them told that 1-week led-time forecast information is enough for crop decisions such as fertilizer application, irrigation and harvest. Additionally, 32% of them said that a minimum one-month led-time forecast is necessary for extra preparation such a variety selection, alternative crop selection and early or late plating based on seasonal weather phenomena. In total, 34% of the farmers said that one-week weather information service would be okay to prepare before any weather and climatic hazard. Whereas, 32% of the farmers claimed that monthly scale forecast would be necessary to take precautionary steps in agricultural processes. The farmers who wanted forecast information 1 month in advance, stated that it would be helpful for them to ready their mind and prepare for early sowing, harvesting or irrigation, if needed, as weather also relates with their economic facts and other factors. The farmers who wanted seasonal scale forecast also demanded a real-time weather forecast alongside the short-term and medium term forecasts.

4 Discussion

Farmers are mostly engaged with various climate-response strategies, among which, irrigation in the crop field is the most common. Other adaptation options to adverse climatic conditions are use of additional fertilizer, early or late cropping, early or late harvesting, use of high yielding seed, cropping pattern change and crop change.

Local farmers from the study area could understand the on-going weather trends a few years back, however, it is now confusing for them. They mentioned that weather and seasons are not as they used to be. They cannot predict the upcoming weather phenomenon based on their indigenous knowledge and indicators. It was also revealed in study that the available weather information from the news or electronic media are not based on specific locality. Bangladesh Meteorological Department (BMD) provides the regional weather forecast, which is not always accurate for all areas. Thus, the information gap creates misperception and farmers are not able to use this information available in the agricultural application. Therefore, most of the farmers strongly agreed that they need location and time specific weather information services so that they will be able to use the information in agricultural decision-making.

Access to information in peri-urban Khulna is one of the major constraints to agricultural development due to inadequate extension and information service. The use of ICT tools can develop the present scenario; however, only the technology itself cannot push the whole agricultural development. In addition to this, awareness on using information and skill of information usage is necessary to maintain the development. Farmers cannot identify key decisions based on water and weather, when they are asked. They are not aware about the available technology and suitable method for cultivation. Training is needed to teach the farmers about the available information services and to develop their understanding on weather parameters and their consequences. Therefore, the development process of weather information services must include the progress of farmers' skills to understand and use weather knowledge.

So, from the overall study it is deciphered that a user-friendly information tool development would be a very appropriate to address the present agricultural practices and farmers' decision-making in order to achieve the optimum production.

5 Conclusion

Despite the availability of some agricultural information, farmers do not use existing information for agricultural decision-making. A key reason is that very few farmers are conscious of the available information services and their proper usage for agricultural production. Thus, considering the present situation of peri-urban Khulna, an information service tool may have potential for sustainable agricultural practices by local farmers. In addition, to enhance uptake and use of the existing weather information, farmers' capacity building would bring the maximum usage output and impact through informed agricultural decision-making.

Questionnaire

1a. Are you agreed to participate in the interview.
1c. Name of the village.
1d. Name of the Upazila.
101. Name of the respondent.
102. Contact/Mobile No.
103. Gender.
104. Age of respondent.
105. Education.
106. Type of family.
107. Household population.
108. How long have you been involved in agricultural activities?
109. Monthly farm income of the household.
110. Do you have income source other than agriculture?
111. Name of alternative income sources?
117. Amount of personal land for agriculture (decimal).
501. How much are you dependent on water and weather information for agricultural decision-making?
502. Which information is needed for your crop related decision-making?
503. How do you access weather and water information at present situation?
503a. If other, specify.
504. Quality of the available weather and water information at present situation?
505. How existing weather and water information services could be improved?
601. Do you think that weather and water information are important for decision making?
602. What kind of information is needed for crop related key decision-making?
602a. If others, then tell about specific weather information which you need?
603. What kind of water information is needed for crop related key decision-making?
603a. If others, then tell about specific water information which you need?
604. What other information is needed for your crop related key decision-making?
605. How advance information is needed for crop related decision-making?
606. Will you be interested to join us for developing information platform in future?
607. What changes have you been observed in weather and climate condition?
608. How local farmers are knowledgeable to understand local weather phenomenon in advance?
609. Do they use any local indicators?
610. What are the local indicators used by farmers?
701. Do you use mobile phone?
702. Do you have smartphone (or in the family)?
703. Do you use mobile phone for agricultural information?
704. What type of information is received by mobile phone?
705. How do you get agri information by mobile phone?

706. If mobile app, then which type of the apps?

Don't use mobile app for information
Agriculture news apps
Disease and pest apps
Farm management apps
Weather/climate apps

707. How easy to find specific agriculture information using mobile apps?

Don't know about mobile apps
Don't use apps for information
Easy
Difficult
Very difficult
Very easy

708. What factors limit use of mobile apps by farmers?

Lack of ICT knowledge
Don't have smartphone
Unsuitable
Economic
Internet unavailability
Incompatible design
Others (specify)

708a. if others, specify.
709. Would you be interested to use agricultural apps, if we initiate for you?
710. If yes, which format will you prefer?

Audio-Video
Text format
Photograph/image/diagram
Others (specify)

710a. If other, specify

711. What would be the best way to communicate weather/water information with farmers?

712. How frequently?

Thanks for your kind patience!

References

Alam, M., & Haque, M. (2014). Contribution of television channels in disseminating agricultural information for the agricultural development of Bangladesh: A case study. *Library Philosophy and Practice* (e-journal).

Bertolini, R. (2004). *Making information and communication technologies work for food security in Africa.* 2020. IFPRI Brief- Africa Conference Brief II, International Food Policy Research Institute, Washington, USA, October 2004.

Bhatnagar, S. (2008). *Benefits from rural ICT applications in India: Reducing transaction costs and enhancing transparency.* Presentation at public lecture on ICT in Agriculture, Colombo, Sri Lanka.

Parkinson, J., & Tayler, K. (2003). Decentralized wastewater Management in peri-urban areas in low-income countries. *Journal of Environment and Urbanization, 15*(1), 75–90.

Silva, D. H., & Ratnadiwakara, D. (2008). *Using ICT to reduce transaction costs in agriculture through better communication: A case study from Sri Lanka.*

Assessment of the Impact of Agriculture on Livelihood of the Farmers in a Polder: A Case Study in the South-West Coastal Area of Bangladesh

Swarnali Mahmood and Abul Fazal M. Saleh

1 Introduction

The coastal zone of Bangladesh covers 32% of the total land area and constitutes of 28% of the population (PDO-ICZMP 2004). Apart from tidal and storm surges, increasing salinity and scarcity of fresh water for irrigation are the general problems for agricultural development in the coastal areas (Planning Commission 2015; Gain et al. 2017). These constraints restrict crop cultivation in dry (Rabi) season (i.e., between November and May) (Clarke et al. 2018), resulting in the enhancement of brackish water shrimp farming in the coastal areas, which then lead to the loss of agricultural land, contraction of traditional livelihood opportunities and growth of new non-agricultural work (Pokrant 2014). Despite these problems, the farmers bring in changes to their traditional agricultural practice by introducing short-period cash crops such as sesame, mungbean, sunflower, watermelon, maize, chili, etc. Besides growing Boro rice during the dry season, they thus follow an Aman-Rabi-Fallow cropping pattern instead of the Aman-Fallow-Fallow pattern to ensure increased food accessibility, economic improvement and ultimately food security (Kibria et al. 2015). During the dry season, the existing canals, on-farm reservoirs (OFRs) and ponds are the sources of irrigation water in the coastal area in which the farmers conserve non-saline river water or rainwater for irrigation (Rahman et al. 2000). Then the farmers apply irrigation water by the low lift pumps (LLPs) in Boro

S. Mahmood (✉)
Department of Coastal Studies and Disaster Management, University of Barishal, Barishal, Bangladesh
e-mail: smahmood@bu.ac.bd

A. F. M. Saleh
Institute of Water and Flood Management (IWFM), Bangladesh University of Engineering and Technology (BUET), Dhaka, Bangladesh

A. Haque, A. I. A. Chowdhury (eds.), *Water, Flood Management and Water Security Under a Changing Climate*, https://doi.org/10.1007/978-3-030-47786-8_24

rice fields during dry periods, but the farmers who do not have this facility, instead make use of residual soil water for non-rice Rabi crops after harvesting Aman rice.

Construction of polders was initiated during the 1960s to protect the agricultural lands against tidal flooding and saltwater intrusion. In the 1960s and 1970s, earthen embankments called 'polders' were put up to protect the low lands from tidal action and the uncontrolled flooding of the rivers in large parts of coastal Bangladesh (Chowdhury et al. 2010). This significant landscape change shaped the water management of the polders for irrigated crop agriculture (PDO-ICZMP 2004). At present the coastal zone of Bangladesh includes approximately 139 polders (Khan 2014; Kibria et al. 2015). In Bangladesh out of the 2.8 million ha of coastal zones almost half of the area (1.2 million ha) is now poldered (BWDB 2013). The polders provided the farmers with the necessary infrastructure to improve the productivity of their lands, mainly through irrigation and drainage (Chowdhury et al. 2010). The main impact of polders has been seen on rice cultivation, providing improved control over irrigation through protection from salinity (PDO-ICZMP 2004; Chowdhury et al. 2010).

Livelihood security consists of capabilities, assets (including both materials and social resources) and activities required for a living (Scoones 1998; Mutahara et al. 2016). A livelihood is sustainable only when it can maintain its capability and assets, and provide livelihood opportunities for the next generation (Chambers and Conway 1992; Mutahara et al. 2016), while also dealing with and recovering from both external and internal shocks and stresses (Charvet et al. 2014; Mutahara et al. 2016. The farmers' livelihoods depend on agriculture, which is why livelihood is closely linked with the cropping decisions, water management practice, crop yield, and income, this is because farmers have less ability to diversify their livelihood activities in coastal areas. For the integration of social, economic and environmental issues into all stages of farmers' livelihood security assessment, different dimensions need to be considered, such as availability of water for irrigation, irrigation and drainage infrastructure, proper training and information, and access to the market. Several livelihood vulnerability indices and livelihood security indices were developed and applied in the past, including the Social Vulnerability Index (Vincent 2004), Human Development Index (UNDP 2007), Livelihood Vulnerability Index for Mozambique (Hahn et al. 2009), Climate Change Vulnerability Index (Khajuria and Ravindranath 2012), Livelihood Vulnerability Index for Trinidad and Tobago (Shah et al. 2013), Livelihood Vulnerability Index for Vietnam (Can et al. 2013), Sustainable Livelihood Security Index (Mutahara et al. 2016), Farmers' Livelihood Security Index (Naher et al. 2017), etc. Successful application of some of these indices have been observed in some regions, but all of these are not entirely applicable to other regions, therefore, using appropriate index is necessary for conducting a study in a specific region.

While there have been previous attempts for assessing the impacts of polders on environmental and socio-economic characteristics (Chowdhury et al. 2010; CEGIS 2015), studies on the impacts of polders on agriculture-based livelihood have been scarce. Salinity intrusion in the non-poldered coastal areas not only decrease crop production (Sarwar and Khan 2007) through the unavailability of fresh water as well as increasing soil salinity leading to soil degradation, (Mondal et al. 2001) but also create severe problems in crop diversity (Gupta 1990; Gain et al. 2007). Studies reported that increased salinity from a 0.3 m sea-level rise would alone reduce the net production of rice by 0.5 million metric tons in the coastal areas of Bangladesh (World Bank 2000; Mahtab and Zahi 2018). Significant shifts have been identified in the participatory water management system in the coastal area of Bangladesh due to polderization (Dewan et al. 2015).

Moreover, there are multiple physical and socio-economic drivers, such as climatic change, upstream development, geologic process, and land use change (Roy et al. 2017), which impact ecology and society (Kibria and Haroon 2017) both inside and outside of the polders. Therefore, the cropping pattern, water management practice for crops, crop yield, production cost, and income related to the livelihood of the farmers, are expected to vary because of polderization. With these views in mind, this study attempts to assess the impact of polders on farmers' livelihood, and to understand, evaluate and compare the agriculture-based livelihoods in both polder and non-polder areas through a modified version of the Farmers' Livelihood Security Index (Naher et al. 2017) through participatory approaches using multi-criteria analysis.

2 Methodology

2.1 Selection of Study Area

To assess the changes resulting from the polder, a non-polder area was selected for comparison purposes. This area was similar to Polder 30 except it was not protected by a polder yet. The villages were purposively selected from each of the polder and non-polder areas so that they represent low elevation, this is because polders are supposed to have maximum impact on low lands since areas are exposed to soil salinity when kept unprotected. Polder 30, constructed during 1967–72, is in Batiaghata, Gangarampur, and part of Surkhali unions in Batiaghata upazila of Khulna district. An area adjacent to the polder, but not yet protected with an embankment, a small part of Jalma union, also located in Batiaghata upazila, Khulna, was selected as the non-polder area for comparison purposes (Fig. 1).

2.2 Selection of Farmers in the Study Area

For livelihood analysis, three common Rabi crop (i.e., Boro rice, mungbean, and sesame) producing farmers were selected in the study area during the dry season in 2017. Separate FGDs were conducted with the different groups of farmers that were producing different crops. On the other hand, Aman rice was found to be the main crop extensively produced during the monsoon (Kharif 1) to post-monsoon (Kharif 2) season. Thus, separate FGDs with Aman rice producing farmers were also

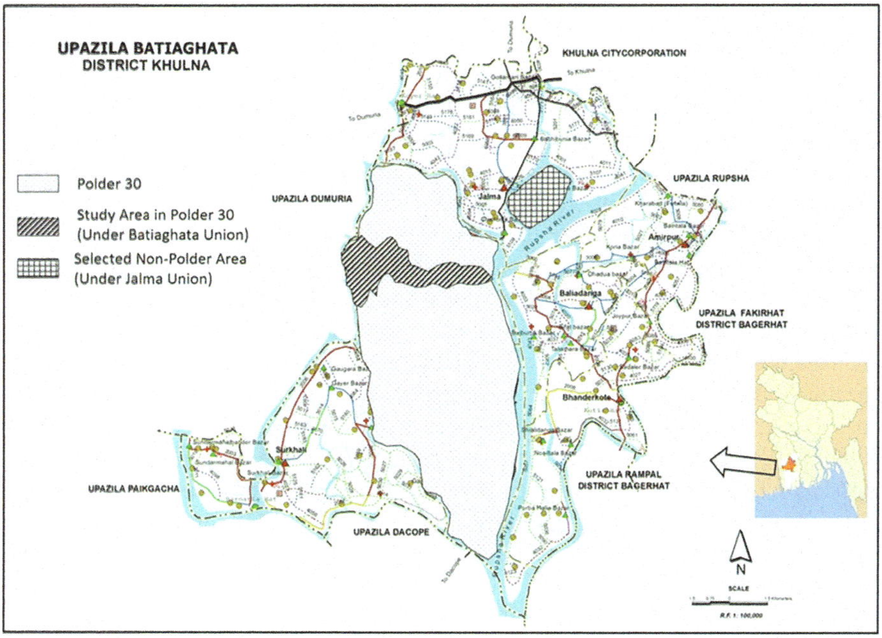

Fig. 1 Location of the study area

conducted at several locations of the selected polder and non-polder areas during Aman season (2017–2018). The farmers in the study area were selected by a non-probability purposive sampling method. All the farmers in this study were owners of their agricultural lands, on which they produced their crops.

2.3 Development of Indicators for Livelihood Index

Indicators for the security of coastal crop producing farmers' livelihoods were identified from different extents of their livelihood assets. Specific questions were modelled during data collection to obtain information on each of the indicators. In addition, background criteria for each of the questions were also observed from the data collection process through the FGDs (Boateng 2012). A total of 13 indicators were selected (Table 1) based on the FGDs with the selected crop producing farmers following DFID's sustainable livelihood framework (Ashley and Carney 1999). According to the DFID livelihood asset model, each livelihood group has five types of assets – (i) natural asset, (ii) financial asset, (iii) human asset, (iv) physical asset, and (v) social asset (DFID 1999). In the process of indicator development, all indicators of the DFID framework were not considered. Only the simple indicators were considered as only one livelihood group (i.e., crop producing farmers) was chosen in the study area.

Table 1 DFID framework indicators for developing the crop producing farmers' indicators

Assets	DFID framework indicators (DFID 1999)	Crop producing farmers' indicators (Source: Field Survey 2017-18)
Social	Networks and connectedness, access to wider institutions of society	Access to soil information, availability of effective extension services, availability of new varieties and availability of fertilizer and pesticides in the market
Human	Skills, knowledge, ability, education/training personnel	Training of the farmers and motivation of the farmers
Natural	Land, forests, marine/wild resources	Canal irrigation water and on-farm reservoir (OFR)/pond/rainfall
Financial	Availability of cash or equivalent, productivity	Profit and yield
Physical	Infrastructure, access to water supply	Irrigation water facility through low lift pump (LLP) and/or shallow tube well (STW), drainage facility through drainage sluice, amount of irrigation water, and access to marketing network

2.4 *Standardization and Weighting of Selected Indicators*

The actual values of the selected indicators used for the FLSI were standardized following the UNDP procedure of standardization of indicators for the sub-index value (UNDP 2007) as shown in Eq. 1.

$$\text{Standardized value of subindex} = \frac{Actual\ Value - Minimum\ Value}{Maximum\ Value - Minimum\ Value} \quad (1)$$

Before standardization, the actual values were collected in response to Boro rice, mungbean, sesame, and Aman rice producing farmers individually both in Polder 30 and the non-polder areas through conducting FGDs. The weight of each criterion was determined based on the field response. The number of times an indicator was cited as important was then used to generate the weighting system. A total of 18 FGDs were conducted in the selected study area. First, FGDs were conducted to understand the situation of crop farmers' livelihood security and indicators were developed from that. Second, more FGDs were conducted to find out the actual, minimum, and maximum values of indicators by full, half, one-third, and two-third ranking of indicators, and then the standardization was done. Finally, all indicators were normalized to have a relative position between 0 and 1 and these values were used to illustrate the pentagon of the five components of livelihood assets of crop producing farmers in both Polder 30 and the non-polder areas.

2.5 Farmers' Livelihood Security Index Calculation

The Farmers' Livelihood Security Index (FLSI) for producing Boro rice, mungbean, sesame, and Aman rice was individually calculated modifying the model as described by (Naher et al. 2017) as shown in Eq. 2.

$$\text{FLSI}\ (\%) = [\{(\text{Ssv} \times \text{Wi}) + (\text{Hsv} \times \text{Wii}) + (\text{Nsv} \times \text{Wiii}) + (\text{Fsv} \times \text{Wiv}) + (\text{Psv} \times \text{Wv})\} \times 100] \tag{2}$$

Where Ssv, Hsv, Nsv, Fsv, and Psv refer to the standardized values of the social, human, natural, financial, and physical assets, respectively. The term W refers to the weighting applied to each standardized value.

3 Results

3.1 Farmers' Livelihood Security Weightage Value for Crop Producing Farmers

Based on the local farmers' perception, the weighting and ranking of indicators for Rabi crops are presented in Tables 2 and 3 for both Polder 30 and the non-polder areas, respectively. In terms of financial assets, profit and yield from the crops are the indicators. The farmers of Polder 30 gave more importance to yield of crops than profit, whereas the farmers of the non-polder area gave more importance to the profit of crops than yield. In terms of physical assets, the farmers in Polder 30 gave more importance to the amount of irrigation water whereas in non-polder areas irrigation facility was given more importance as STWs were available in medium high lands of the non-polder areas for dry season irrigation. In terms of natural assets, during dry season crop agriculture canal irrigation was given more importance instead of OFR for both polder and non-polder areas. The social assets comprised of the availability of new varieties, fertilizers, pesticides in the market, effective extension services, and access to soil information. The farmers of Polder 30 gave more importance to access to soil information, whereas the farmers of the non-polder area gave more importance to the availability of new varieties, fertilizer, and pesticides in the market. The indicators of human assets were the training and motivation of the farmers. Training improves the agricultural knowledge of farmers. When experts motivate the farmers, the farmers become interested in taking a new challenge in an adverse situation. When the farmers get training they can know about the new techniques and the available variety of crops. Thus, the extension service is helpful to the farmers for the production of crops.

Based on the local farmers' perception, the weighting and ranking of indicators are shown in Tables 4 and 5 for Aman rice both in Polder 30 and the non-polder

Table 2 Weighting system for Rabi crop producing farmers in polder area

Component	Indicator	Times cited as most important	Relative importance	Weightage (indicators %)	Rank	Weightage (components %)
Social assets	Access to soil information	6	6.82	7	5	18
	Availability of effective extension services	2	2.27	2	12	
	Availability of new varieties in the market	3	3.41	3	10	
	Availability of fertilizer and pesticides in the market	5	5.68	6	7	
Human assets	Training of the farmers	7	7.95	8	4	10
	Motivation of the farmers	2	2.27	2	13	
Natural assets	Canal irrigation water	8	9.09	9	3	12
	OFR/pond irrigation water	3	3.41	3	11	
Financial assets	Profit	16	18.18	18	2	42
	Yield	21	23.86	24	1	
Physical assets	Access to marketing network	4	4.55	5	9	18
	Irrigation water facility (LLP)	5	5.68	6	8	
	Amount of irrigation water	6	6.82	7	6	
	Total	88	100.00	100		100

areas, respectively. In the case of financial assets, profit and yield from the crop were the indicators. It was observed that profit was given more importance than yield by the Aman rice producing farmers in both Polder and non-polder areas. In this case, better access to the market network was necessary for making a profit. It was found that Polder 30 had better road communication due to polder construction. Therefore, more profit for Aman rice producing farmers might be possible in Polder 30. For natural assets, in both cases, rainfall was the most important source of irrigation water for Aman rice. This natural irrigation also cut down the input cost of irrigation during Aman rice production everywhere. Moreover, freshwater can be conserved in ponds, OFRs, and canals during the rainy season which can be further used as a

Table 3 Weighting system for Rabi crop producing farmers in non-polder area

Component	Indicators	Times cited as most important	Relative importance	Weightage (indicators %)	Rank	Weightage (component %)
Social assets	Access to soil information	4	5.26	5	7	16
	Effective extension services	1	1.32	1	13	
	Availability of new varieties in the market	5	6.58	7	6	
	Availability of fertilizer and pesticides in the market	2	2.63	3	11	
Human assets	Training of the farmers	7	9.21	9	4	12
	Motivation of the farmers	2	2.63	3	12	
Natural assets	Canal irrigation water	4	5.26	5	8	9
	OFR/pond irrigation water	3	3.95	4	9	
Financial assets	Profit	17	22.37	22	1	40
	Yield	14	18.42	18	2	
Physical assets	Access to marketing network	3	3.95	4	10	23
	Irrigation water facility (LLP, STW)	8	10.53	11	3	
	Amount of irrigation water	6	7.89	8	5	
	Total	76	100.00	100		100

source for dry season irrigation. Among the indicators of physical assets, access to the market network was given more emphasis in both Polder 30 and the non-polder area for Aman producing farmers. Better access to the market network was closely related to better road communication, which was found in Polder 30. The indicators of human assets were the training and motivation of the farmers. The training improves the agricultural knowledge of farmers. In both Polder 30 and the non-polder area, the farmers gave more importance to training. When the farmers get training they can know about the new varieties and salt-tolerant varieties of crops. The extension service is helpful for the farmers for crop production. The social assets were comprised of the availability of new varieties, fertilizers,

Table 4 Weighting system for Aman rice producing farmers in polder area

Component	Indicator	Times cited as most important	Relative importance	Weightage (indicators %)	Rank	Weightage (component %)
Social assets	Access to soil information	2	3.33	3	10	25
	Availability of effective extension services	6	10.00	10	4	
	Availability of new varieties in the market	3	5.00	5	8	
	Availability of fertilizer and pesticides in the market	4	6.67	7	7	
Human assets	Training of the farmers	7	11.67	11	3	16
	Motivation of the farmers	3	5.00	5	9	
Natural assets	Canal irrigation water	0	–	0	11	2
	Rainfall	1	1.67	2	12	
Financial assets	Profit	13	21.67	22	1	42
	Yield	12	20.00	20	2	
Physical assets	Access to marketing network	5	8.33	8	5	15
	Irrigation water facility (drainage sluice)	4	6.67	7	6	
	Amount of irrigation water	0	–	0	13	
	Total	60	100.00	100		100

pesticides in the market, effective extension services, and access to soil information. The farmers of Polder 30 gave more importance to access to extension service. On the other hand, the farmers of the non-polder area gave equal importance to the availability of new varieties as well as fertilizer and pesticides in the market.

Table 5 Weighting system for Aman rice producing farmers in non-polder area

Component	Indicator	Times cited as most important	Relative importance	Weightage (indicator %)	Rank	Weightage (component %)
Social assets	Access to soil information	2	3.39	3	9	26
	Availability of effective extension services	3	5.08	5	8	
	Availability of new varieties in the market	5	8.47	9	5	
	Availability of fertilizer and pesticides in the market	5	8.47	9	6	
Human assets	Training of the farmers	7	11.86	12	3	20
	Motivation of the farmers	5	8.47	8	7	
Natural assets	Canal irrigation water	0	–	0	13	2
	Rainfall	1	1.69	2	10	
Financial assets	Profit	14	23.73	23	1	38
	Yield	9	15.25	15	2	
Physical assets	Access to marketing network	6	10.17	10	4	14
	Irrigation water facility	1	1.69	2	11	
	Amount of irrigation water	1	1.69	2	12	
	Total	59	100.00	100		100

3.2 *Farmers' Livelihood Security Standardized Value for Crop Producing Farmers*

After collecting data of actual value, on a scale ranging from zero to one, through FGDs with individual crop producing groups of farmers in Polder 30 and the non-polder area, standard values were achieved through standardization. The sub-index standardized values for Boro rice, mungbean, and sesame producing farmers in Polder 30 and non-polder areas are shown in Table 6.

For Rabi crops, significant differences in the standardized values of sub-index were found in case of natural, financial, and physical assets (Table 6). It can be explained that canal irrigation water was an essential asset to the Boro rice producing farmers in Polder 30. On the other hand, irrigation facility (i.e., STWs), as well as more irrigation

Table 6 Farmers' livelihood security standardized values of crop producing farmers

Assets	Indicators	Standardized value in Polder 30				Standardized value in non-polder area			
		Boro rice	Mungbean	Sesame	Aman rice	Boro rice	Mungbean	Sesame	Aman rice
Social	Access to soil information	0.04	0.50	0.50	0.02	0.03	0.50	0.50	0.02
	Availability of effective extension services	0.01	0.67	1.00	0.10	0.01	0.67	0.67	0.03
	Availability of new varieties in the market	0.01	0.50	0.50	0.03	0.07	0.67	0.67	0.09
	Availability of fertilizer and pesticides in the market	0.06	0.67	1.00	0.07	0.03	1.00	1.00	0.09
Human	Training of the farmers	0.00	0.33	0.00	0.00	0.02	0.00	0.00	0.03
	Motivation of the farmers	0.00	0.67	1.00	0.01	0.02	0.50	0.50	0.04
Natural	Canal irrigation water	0.00	0.33	0.00	0.00	0.03	0.67	0.67	0.00
	OFR/pond/rainfall	0.00	0.33	0.00	0.02	0.01	0.67	0.67	0.02
Financial	Profit	0.00	0.67	1.00	0.11	0.00	0.50	0.50	0.06
	Yield	0.06	0.67	1.00	0.10	0.09	0.50	0.50	0.04
Physical	Access to marketing network	0.00	0.33	0.00	0.08	0.02	0.67	0.67	0.03
	Irrigation water facility	0.00	0.33	0.00	0.07	0.11	1.00	1.00	0.02
	Amount of irrigation water	0.07	0.67	1.00	0.00	0.02	0.33	0.33	0.02

water availability, played an essential role in Boro rice production in the non-polder area. Thus, the yield was also higher for Boro rice in the non-polder area. As for mungbean and sesame, the standardized sub-index values slightly differed between Polder 30 and the non-polder area. In Polder 30, Boro rice, mungbean, and sesame producing farmers had better access to the marketing network.

For Aman rice, there was a significant difference in standardized values in access to extension service, profit, access to the market network, and irrigation water facility (Table 6). In Polder 30, these facilities were found to be better due to better road communication as well as drainage sluice that protected the polder area from flooding.

The social, human, natural, financial, and physical asset values of Boro rice, mungbean, sesame, and Aman rice producing farmers in Polder 30 and the non-polder area are shown in Table 7. From the viewpoint of Boro rice producing farmers, the social asset values in Polder 30 and the non-polder area were almost the same (Table 7). Human, natural, and financial asset values were found to be slightly better in the non-polder area. The main difference was found in physical asset values. The physical asset value was found to be double in the non-polder area than that in the polder area. In this case, irrigation facility (i.e., STWs), as well as more irrigation water availability, played an essential role in Boro rice production in the non-polder area. From the viewpoint of both mungbean and sesame producing farmers, financial asset value was found to be almost double in Polder 30 due to better road communication and market network. However, the physical asset value was found to be almost double in the non-polder area due to better water availability.

From Table 7, it can be seen that the financial asset value of Aman rice producing farmers in Polder 30 was 0.21, which was found to be double of that of the value found in the non-polder area (0.10). The natural asset values were the same in both Polder 30 and the non-polder area as the same source of irrigation water (i.e., rainfall) was used in both cases. The social asset values did not differ much in the two cases. The physical asset value of Aman rice producing farmers in Polder 30 was 0.15, which was also found to be double of that of the non-polder area (0.07). The better road communication, thus better access to the market network in Polder 30, resulted in overall higher asset values of Aman rice producing farmers in Polder 30.

Table 7 Comparison of each asset between crop producing farmers

	Boro rice producing farmers		Mungbean producing farmers		Sesame producing farmers		Aman rice producing farmers	
Assets	Polder 30	Non-polder area	Polder 30	Non-polder area	Polder 30	Non-polder area	Polder 30	Non-polder area
Social	0.12	0.13	0.13	0.11	0.06	0.07	0.21	0.22
Human	0.00	0.04	0.02	0.02	0.02	0.01	0.01	0.07
Natural	0.00	0.04	0.00	0.06	0.00	0.05	0.02	0.02
Financial	0.06	0.09	0.42	0.20	0.33	0.15	0.21	0.10
Physical	0.07	0.15	0.07	0.16	0.08	0.14	0.15	0.07

3.3 *Individual Security of an Asset by the Area of Pentagon*

The areas of livelihood assets of the Rabi crop producing farmers in Polder 30 and the non-polder area are shown in Fig. 2. Again, the areas of livelihood assets of Boro rice, mungbean, sesame, and Aman rice producing farmers in Polder 30 and the non-polder area are individually shown in Fig. 3.

From the comparative analysis, the livelihood asset area of Boro rice occupies more area than mungbean and sesame in the non-polder area (Fig. 2b), whereas, in Polder 30, the livelihood assets of Boro rice occupies less area than mungbean and sesame (Fig. 2a). It was observed that the overall livelihood of Boro rice producing farmers in the non-polder area was better than mungbean and sesame producing farmers due to availability and better facility of irrigation water in the non-polder area during Rabi season. The main differences in Polder 30 and the non-polder area were in natural and financial assets since yield and profitability of sesame and mungbean were found higher in Polder 30 due to better access to market (because of better road communication) as well as low requirement of irrigation for these crops (Fig. 3b, c).

It was observed from Fig. 3d that more livelihood asset area was occupied by Aman rice in Polder 30 than in the non-polder area. The noticeable change was found in physical assets as well as financial assets. Due to better protection from seasonal flooding through drainage sluice the polder protected the Aman rice farming area. On the other hand, Aman rice farming area in the medium low lands of the non-polder area was damaged due to heavy rainfall and seasonal flooding. Moreover, Aman rice, both in Polder 30 and the non-polder area, did not suffer constraints of irrigation water in terms of its availability. This is because the cultivation of Aman rice in both cases occurred in rain-fed conditions. The road

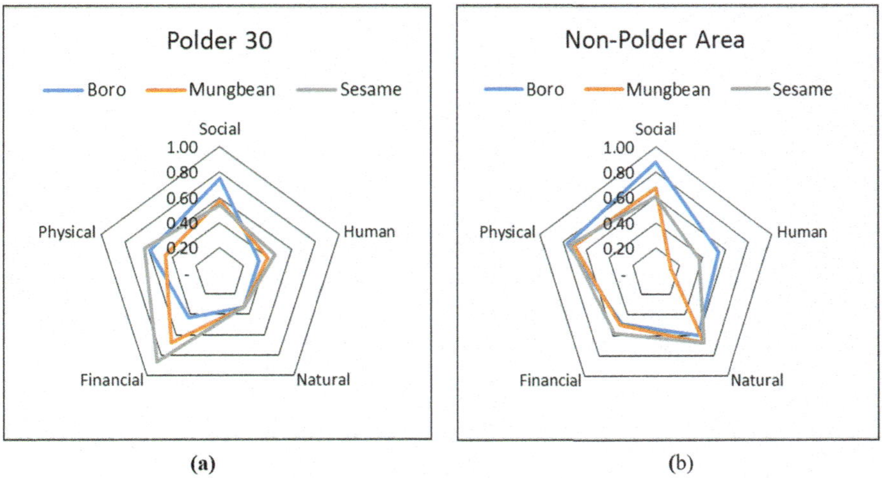

Fig. 2 The pentagon illustrating the five components of livelihood assets for the Rabi crop producing by farmers. (**a**) Polder area. (**b**) non-polder area

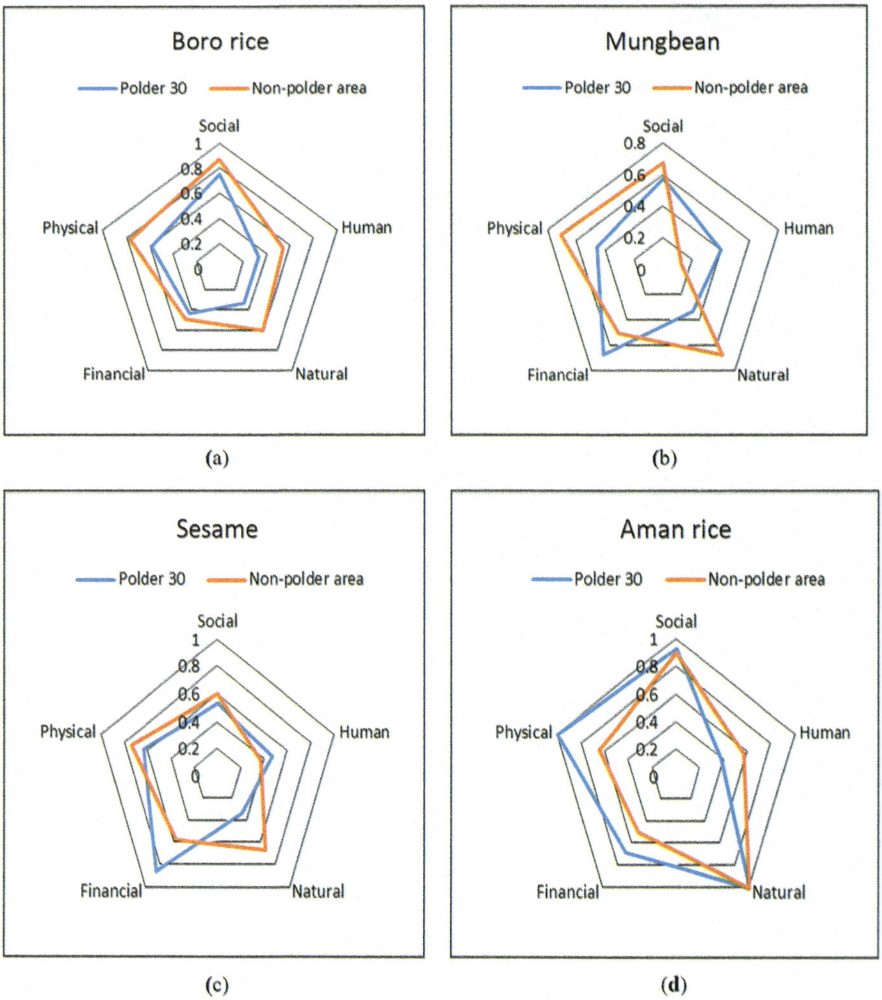

Fig. 3 The pentagon illustrating the five components of livelihood assets for farmers in polder and non-polder area. (**a**) Boro rice. (**b**) mungbean. (**c**) sesame. (**d**) Aman rice

communication in Polder 30 was found to be better, which attributed to better extension service, better availability of variety-fertilizer-pesticide, and provided better market network than those in the non-polder area. Therefore, the yield and profit were also found to be higher in Polder 30.

Table 8 Comparison of livelihood security among crop producing farmers with area cultivated

	Polder area		Non-polder area	
Crop producing farmer	FLSI (%)	Area cultivated (%)[a]	FLSI (%)	Area cultivated (%)[a]
Boro rice	24	28	45	76
Mungbean	64	–	55	–
Sesame	49	–	41	–
Aman rice	61	84	47	63

[a]Collected by personal communication with the Upazila Agriculture Office, Batiaghata upazila, Khulna during 2017–2018

3.4 Farmers Livelihood Security Index for Crop Producing Farmers

The Farmers' Livelihood Security Index (FLSI) of Boro rice, mungbean, sesame and Aman rice producing farmers were calculated for both Polder 30 and the non-polder area. The FLSI values of Boro rice, mungbean, and sesame producing farmers, along with the corresponding area (%) being cultivated, are shown in Table 8.

4 Discussion

4.1 Application of the Index

In this index value analysis, higher index value indicates higher security of that crop in the area for that season. From Table 8, the livelihood of Boro rice farmers was more secure in the non-polder area because of the availability of more irrigation water through STWs, but Boro rice was not found to be as extensively cultivated in Polder 30 as in the non-polder area. On the other hand, the livelihood of mungbean and sesame producing farmers were found to be slightly more secure in Polder 30 than in the non-polder area due to lower salinity and better market network through better road communications in Polder 30. Finally, the livelihood of Aman rice producing farmers in Polder 30 was found to be more secure than that in the non-polder area. The impact of polder had been significant for Aman rice cultivation as Aman rice was the single crop being cultivated in both polder and non-polder areas during that season (Table 8). Better road communication, better availability of extension service, and more yield and profit in Polder 30 during the Aman season can be attributed to the polder protection against seasonal flooding. Moreover, it was evident from the FGDs conducted with the farmers in the non-polder area that they were more inclined towards being in the inside of the polder due to flood protection and better water management provided by polders.

Some studies reported that the practical assessment of livelihood is very complicated as there are many related factors, aspects, and dimensions to a particular case or situation (Carney 1998; Sullivan 2002). This case study only focuses on some major components that influence crop production of the farmers inside and outside a polder. The sub-components used to construct the FLSI in this study were based on the current conditions of the study area as well as available data from FGDs. Thus, this study can be used as a reference for other cases with different conditions and can be updated or improved for different situations.

FLSI could be applied as a practical tool to identify the farmers' level of security for producing different crops. It could also be used to understand different factors contributing to farmers cropping decisions and crop water management in an area. It would also provide vital information to the policy and decision makers, and relevant officials for agricultural planning and interventions towards the improvement of the farmers' livelihood.

5 Conclusion

Initially when only Aman rice was cultivated in this region, the polders were constructed for the prevention of salinity and the tidal surge in the coastal region. Nowadays, as two to three crops are being cultivated in the area, adequate water supply and storage should be ensured in the polders for dry season crops. This can be done through re-excavation of existing canals and construction of more ponds and OFRs, which will then lead to better Rabi crop production and thus better livelihood of the farmers. In this study, the impact of polder on farmers' livelihood was found to be positive, particularly for Aman rice. However, this study was conducted for a period of a year (with one Boro and one Aman season) in one polder with an adjacent non-polder area in the south-west coastal district, and this might be inadequate to represent the overall impact of polder on the livelihood of all coastal farmers in this area. Moreover, conducting the same study in a different part of the coastal zone might achieve a different result. Looking at this situation through dimensions of gender or farmers' land ownership might also achieve different results. Therefore, emphasis is given on further research on the different dimensions of this issue.

Acknowledgements The authors acknowledge the IDRC-SAWA Fellowship Project, IWFM, BUET, for providing financial support for this research. The authors would also like to thank the farmers of the study area for their cooperation in providing the necessary information without which this work would not have been possible.

References

Ashley, C., & Carney, D. (1999). *Sustainable livelihoods: Lessons from early experience DFID*. London: Department for International Development (DFID).

Boateng, W. (2012). Evaluating the efficacy of focus group discussion (FGD) in qualitative social research. *International Journal of Business and Social Science, 3*, 156–175.

BWDB. (2013). *Environmental impact assessment. Coastal Embankment Improvement Project Phase-I* (Vol. 1). Dhaka: Bangladesh Water Development Board (BWDB).

Can, N. D., Tu, V. H., & Hoanh, C. T. (2013). Application of livelihood vulnerability index to assess risks from flood vulnerability and climate variability—A case study in the Mekong Delta of Vietnam. *Journal of Environmental Science and Engineering, pp, A*(2), 476–486.

Carney, D. (1998). *Sustainable rural livelihoods: What contribution can we make?* London: Department for International Development (DFID).

CEGIS. (2015). *Final report on environmental impact assessment (EIA) on rehabilitation of polder 30*. Center for Environmental and Geographic Information Services (CEGIS): Dhaka.

Chambers, R., & Conway, G. (1992). *Sustainable rural livelihoods: Practical concepts for the 21st century* (IDS Discussion Paper No. 296). Brighton: Institute of Development Studies (IDS), University of Sussex.

Charvet, I., Suppasri, A., & Imamura, F. (2014). Empirical fragility analysis of building damage caused by the 2011 Great East Japan Tsunami in Ishinomaki City using ordinal regression, and influence of key geographical features. *Stochastic Environmental Research and Risk Assessment, 28*, 1853–1867. https://doi.org/10.1007/s00477-014-0850-2.

Chowdhury, A. K. M., Jenkins, S. A. M., & Hossain, M. (2010). Assessing the impact of small scale coastal embankments: A case study of an LGED polder. In C. T. Hoanh, B. Szuster, S. P. Kam, A. Noble, & A. M. Ismail (Eds.), *Tropical deltas and coastal zones community, environment and food production at the land-water interface* (Comprehensive Assessment of Water Management in Agriculture Series) (pp. 421–435). Oxfordshire: CABI Publishing. https://doi.org/10.1079/9781845936181.0422.

Clarke, D., Lázár, A. N., Saleh, A. F. M., & Jahiruddin, M. (2018). Prospects for agriculture under climate change and soil salinisation. In R. Nicholls, C. Hutton, W. Adger, S. Hanson, M. Rahman, & M. Salehin (Eds.), *Ecosystem services for well-being in deltas* (pp. 447–467). Cham: Macmillan. https://doi.org/10.1007/978-3-319-71093-8_24.

Dewan, C., Mukherji, A., & Buisson, M. C. (2015). Evolution of water management in coastal Bangladesh: From temporary earthen embankments to depoliticized community-managed polders. *Water International, 40*(3), 401–416. https://doi.org/10.1080/02508060.2015.1025196.

DFID. (1999). *Sustainable livelihoods guidance sheet 1999*. London: Department for International Development (DFID).

Gain, A. K., Aryal, K. P., Sana, P., & Uddin, M. N. (2007). Effect of river salinity on crop diversity: A case study of south west coastal region of Bangladesh. *Nepal Agricultural Research Journal, 8*, 29–37. https://doi.org/10.3126/narj.v8i0.11576.

Gain, A. K., Benson, D., Rahman, R., Datta, D. K., & Rouillard, J. J. (2017). Tidal river management in the South West Ganges-Brahmaputra Delta in Bangladesh: Moving towards a transdisciplinary approach? *Environmental Science and Policy, 75*, 111–120. https://doi.org/10.1016/j.envsci.2017.05.020.

Gupta, I. C. (1990). *Use of saline water in agriculture: A case study of arid and semi-arid zones of India*. New Delhi: Oxford and IBH Pub. Pvt. Ltd..

Hahn, M. B., Riederer, A. M., & Foster, S. O. (2009). The livelihood vulnerability index: A pragmatic approach to assessing risks from climate variability and change – A case study in Mozambique. *Global Environmental Change, 19*(1), 74–88. https://doi.org/10.1016/j.gloenvcha.2008.11.002.

Khajuria, A., & Ravindranath, N. H. (2012). Climate change vulnerability assessment: Approaches DPSIR framework and vulnerability index. *Journal of Earth Science and Climate Change, 3*, 109. https://doi.org/10.4172/2157-7617.1000109.

Khan, Z. H. (2014). *Tidal River Management (TRM) in the coastal area of Bangladesh*. Dhaka: Institute of Water Modeling (IWM).

Kibria, G., & Haroon, A. K. Y. (2017). Climate change impacts on wetlands of Bangladesh, its biodiversity and ecology, and actions and programs to reduce risks. In A. Prusty, P. Chandra, & P. A. Azeez (Eds.), *Wetland science* (pp. 189–204). New Delhi: Springer. https://doi.org/10.1007/978-81-322-3715-0_10.

Kibria, M. G., Saha, D., Kabir, T., Naher, T., Maliha, S., & Mondal, M. S. (2015). Achieving food security in storm surge-prone coastal polders of South-West Bangladesh. *SAWAS Journal, 5*(1), 26–43.

Mahtab, M. H., & Zahi, A. (2018). Coastal surface water suitability analysis for irrigation in Bangladesh. *Applied Water Science, 8*, 28. https://doi.org/10.1007/s13201-018-0650-9.

Mondal, M. K., Bhuiyan, S. I., & Franco, D. T. (2001). Soil salinity reduction and prediction of salt dynamics in the coastal ricelands of Bangladesh. *Agricultural Water Management, 47*(1), 9–23. https://doi.org/10.1016/S0378-3774(00)00098-6.

Mutahara, M., Haque, A., Khan, M. S. A., Warner, J. F., & Wester, P. (2016). Development of a sustainable livelihood security model for storm-surge hazard in the coastal areas of Bangladesh. *Stochastic Environmental Research and Risk Assessment, 30*(5), 1301–1315. https://doi.org/10.1007/s00477-016-1232-8.

Naher, T., AFM, S., & MSA, K. (2017). Development of livelihood security index for farmers in the coastal areas of Bangladesh. In *Proceedings of the 6th international conference on Water & Flood Management, ICWFM 2017, 4–6 March 2017, Dhaka, Bangladesh* (pp. 427–436). Dhaka: Institute of Water and Flood Management (IWFM), Bangladesh University of Engineering and Technology (BUET).

PDO-ICZMP. (2004). Where land meets the sea. In M. R. Islam (Ed.), *Program Development Office for Integrated Coastal Zone Management Plan Project (PDO-ICZMP)*. Water Resources Planning Organization (WARPO), Ministry of Water Resources: Dhaka.

Planning Commission. (2015). *Coast and polder issues: Bangladesh delta plan 2100 formulation project, general economic division, planning commission*. Dhaka: Government of the People's Republic of Bangladesh.

Pokrant, B. (2014). Brackish water shrimp farming and the growth of aquatic monocultures in coastal Bangladesh. In J. Christensen & M. Tull (Eds.), *Historical perspectives of fisheries exploitation in the Indo-Pacific* (MARE Publication Series) (Vol. 12, pp. 107–132). Amsterdam: Amsterdam University Press. https://doi.org/10.1007/978-94-017-8727-7_6.

Rahman, M. M., Hassan, M. Q., Islam, M. S., & Shamsad, S. Z. K. M. (2000). Environmental impact assessment on water quality deterioration caused by the decreased Ganges outflow and saline water Intrusion in South-Western Bangladesh. *Environmental Geology, 40*, 31–40. https://doi.org/10.1007/s002540000152.

Roy, K., Gain, A. K., Mallick, B., & Vogt, J. (2017). Social, hydro-ecological and climatic change in the southwest coastal region of Bangladesh. *Regional Environmental Change, 17*, 1895–1906. https://doi.org/10.1007/s10113-017-1158-9.

Sarwar, G. M., & Khan, M. H. (2007). Sea level rise a threat to the coast of Bangladesh. *International Asienforum, 38*(3–4), 375–397.

Scoones, I. (1998). *Sustainable rural livelihoods: A framework for analysis* (IDS Working Paper 72). Brighton: Institute of Development Studies (IDS), University of Sussex.

Shah, K. U., Dulal, H. B., Johnson, C., & Baptiste, A. (2013). Understanding livelihood vulnerability to climate change: Applying the livelihood vulnerability index in Trinidad and Tobago. *Geoforum, 47*, 125–137. https://doi.org/10.1016/j.geoforum.2013.04.004.

Sullivan, C. (2002). Calculating a water poverty index. *World Development, 30*(7), 1195–1210. https://doi.org/10.1016/S0305-750X(02)00035-9.

UNDP. (2007). *Human development report 2007/2008: Fighting climate change: Human solidarity in a divided world*. New York: United Nations Development Programme (UNDP).

Vincent, K. (2004). *Creating an index of social vulnerability to climate change for Africa* (Working Paper 56). Cork: Tyndall Center for Climate Change Research.

World Bank. (2000). *Climate change and sustainable development* (Report No. 21104 BD). Dhaka: World Bank.

Climate Smart Agriculture and Intelligent Irrigation System for Management of Water Resources in Arid and Semi-Arid Regions – A Review

Zakaria Fouad Fawzy and Shaymaa I. Shedeed

1 Introduction

In order to support alternative soil water management solutions in vulnerable dry lands, to optimize available water resources and increase water use efficiency – a pilot study treating deficit irrigation in vulnerable dry lands of Egypt was targeted. Deficit irrigation is an optimization strategy in which irrigation is applied during drought-sensitive growth stages of a crop, while water usage is reduced during drought-tolerant stages. By reducing the water usage, the technique aims to optimize the water use efficiency.

1.1 Climate Change and Its Impact on Agriculture

Driven mainly by population and economic growth, total world food consumption is expected to increase over 50% by 2030 and may double by 2050 (Barker et al. 2007). Most of the increase in food production in the next decades is expected to occur through further intensification of current cropping systems rather than through freeing up new lands for agricultural production. Intensification of cropping systems has been a highly successful strategy for increasing food production. The most relevant example would be the renowned success of the Green Revolution, where in the adoption of modern varieties, irrigation, fertilizers and agrochemicals resulted in dramatic increases in food production. However, this strategy also resulted in unexpected environmental consequences, one of them being the emission of greenhouse gases into the atmosphere. Therefore, future strategies that promotes further

Z. F. Fawzy (✉) · S. I. Shedeed
Agriculture and Biological Research Division, National Research Centre, Giza, Egypt

A. Haque, A. I. A. Chowdhury (eds.), *Water, Flood Management and Water Security Under a Changing Climate*, https://doi.org/10.1007/978-3-030-47786-8_25

intensification of agriculture should aim at the development of sustainable cropping systems that not only consider increasing food production but also consider minimizing environmental impact (Ortiz-Monasterio et al. 2010).

1.2 The Complexities of Climate Change and the Role of CSA

Climate change has wide-ranging impacts on agriculture and food security both in the present and in the future. Changes in rainfall patterns, high average temperatures, increased variation in rainfall and temperature, changes in water availability, sea level rise, salinity increase, changes in the frequency and intensity of adverse weather events all have negative consequences on the agriculture and fisheries sectors. Uncertainties related to the effects of climate change, in terms of timing and severity, coupled with the consequences of many interrelated sectors other than agriculture (health, energy, economy, migration), may result in a very complex challenge.

Climate Smart Agriculture (CSA) is an integrated approach that helps guide actions to transform and reorient agricultural systems to effectively support development and ensure food security in a changing climate. Climate smart agriculture aims at addressing three main objectives – i) sustainable increase in agricultural productivity and income, ii) adaptation and capacity-building to adapt to climate change and iii) reducing and / or removing greenhouse gas emissions, where possible.

Of course, achieving these goals require changes in agricultural behavior, strategies and practices employed by rural households. This can be achieved by improving their access to climate resilience, knowledge and information technologies to help increase productivity as well as, access to inputs, market information and information that contribute to the diversification of sources of income.

Climate smart agriculture is a new conceptual framework that addresses both the issue of food security and the challenges of climate change. On this basis, climate smart agriculture helps translate and increase the impact of Sustainable Development Goals (SDG) on agriculture, food security and livelihoods in rural areas. It also contributes to the changes required to improve the capacity of food systems to cope with difficulties in growing cities. Climate smart agriculture supports adapting agricultural sectors to expected impacts of climate change by enhancing the capacity of production systems and communities to cope with adverse climate events. Climate smart agricultural practices are defined on the basis of agro-ecological and socio-economic conditions and constitute enhancing resilience and reducing greenhouse gas emissions by providing targeted support to countries to increase productivity. This calls for: identification of appropriate systems, practices and technologies, establishing an enabling institutional structure according to their respective social, economic, environmental and climatic conditions and providing capacity, methodologies and tools to conduct the required assessments and analyses. Small enterprises need to be engaged to demonstrate the potential of climate smart agriculture. Governments and the broader society also have a critical role to play for achieving required transformational changes. The future of food security is closely linked with the future of the environment as well as social development, and therefore an integrated approach is imperative.

1.3 What Is Different about Climate-Smart Agriculture?

What is new about CSA is an explicit consideration of climatic risks that happening which have been increasing in frequency and intensity over the last few decades. New climate risks require changes in agricultural technologies and approaches to improve the lives of those still locked in food insecurity and poverty and to prevent the loss of development gains already achieved. CSA approaches entail greater investments in

1. Managing climate risks.
2. Understanding and planning for adaptive transitions that may be needed, for example into new farming systems or livelihoods.
3. Exploiting opportunities for reducing or removing greenhouse gas emissions where feasible.

1.4 What Are the Main Elements of Climate-Smart Agriculture?

CSA is not a set of practices that can be universally applied, but rather an approach that involves different elements embedded in local contexts. CSA relates to actions both on-farm and beyond the farm, and incorporates technologies, policies, institutions and investments. Different elements which can be integrated in climate-smart agricultural approaches include:

1. Management of farms, crops, livestock, aquaculture and capture fisheries to manage resources better and produce more with less while increasing resilience.
2. Ecosystem and landscape management to conserve ecosystem services that are key to for simultaneously increasing resource efficiency and resilience.
3. Services for farmers and land managers to enable them to implement the necessary changes.

1.5 Principles of Climate Smart Agriculture(CSA)

As mentioned earlier, CSA has three focal areas i.e. sustainably increasing agricultural productivity and incomes, adapting and building resilience to climate change, reducing and/or removing greenhouse gases emissions, where possible. Adapting and building resilience to climate change is especially important since it ensures food sufficiency despite unsuitable conditions. Resilience is defined as the ability of a social or ecological system to absorb disturbances while retaining the same basic structure and ways of functioning, the capacity for self-organization, and the capacity to adapt to stress and change (IPCC 2007). Above all, the practices must enhance the natural resource base. Therefore, the most important premise of CSA is the building of healthy soils (Magdoff 2007; Stabinsky and Ching 2012) through increasing the soil organic matter (SOM) status of the soil (Blanco-Canqui and Lal 2004). Soil management practices for CSA include: direct seeding under no/reduced-tillage (Zheng et al. 2014), improved protective soil cover through cover crops, crop residues or mulch (Muzangwa et al. 2013), and crop diversification through rotations (Davis et al. 2012). Moreover, integrated soil fertility management, which includes both inorganic and organic sources and considers combining inputs of organic matter i.e. mulch, compost, crop residues and green manure with fertilizers to address or prevent macro- and micro-nutrient deficiencies should be carefully considered (FAO 2013).

1.6 Evaluating the Future of Climate Smart Agriculture

This section discusses how to measure the impact of CSA, the barriers for the industry to adopt CSA and strategies to overcome them. Critical views on the concept are explored and the principles of CSA with other agricultural practices are compared.

CSA is an approach that helps to guide actions needed to transform and reorient agricultural systems to effectively support development and ensure food security in a changing climate. CSA aims to tackle three main objectives: sustainably increasing agricultural productivity and incomes; adapting and building resilience to climate change; and reducing and/or removing greenhouse gas emissions, where possible.

CSA is an approach for developing agricultural strategies to secure sustainable food security under climate change. CSA provides the means to help stakeholders from local to national and international levels identify agricultural strategies suitable to their local conditions. CSA is one of the 11 Corporate Areas for Resource Mobilization under the FAO's Strategic Objectives "FAO 2013". It is in line with FAO's vision for Sustainable Food and Agriculture and supports FAO's goal to make agriculture, forestry and fisheries more productive and more sustainable".

CSA constitutes using smart farm technologies such as the Internet of Think, GIS Big Data as well as the Expert Agriculture System. Smart farm technology exposes farmers to a wealth of new possibilities that could increase their production. The information available to them through wireless sensors could contain data on livestock, crops, soil, environmental conditions, machinery and many more factors.

These elements constitute very critical components of production in every farm. A better understanding of this information puts farmers at an advantage compared to farmers without access to similar technologies.

Things board platform provides out-of-the-box components and APIs to dramatically reduce time to market and effort to develop smart farming solutions and projects. The platform is device-agnostic. This solution can feed and analyze the data from any sensor, connected device or application and help save up to 90% of development time for smart farming projects.

2 Methodology

There is a long list of tools based on agro-meteorology, which in turn is the key to the application of CSA. These include – engaging the agro-meteorological experts from these organizations to conduct research on many agricultural meteorological topics, to offer farmer awareness programmers, and to provide related publications. These experts are crucial in developing the future work plan of WMO's Commission for Agricultural Meteorology (CAgM) which will further assist National Meteorological Hydrological Service (NMHS) and other national institutions in providing better agro-meteorological advisory services (ICRAF 2015). In WMO, about 120 of its members or 62%, provide agro-meteorological services within their country. NMHS provide farmers with a range of weather / climate information (such as real-time weather data - short, medium and long term weather forecasts), with different levels of detail. All of these provide the basis for tactical and strategic adaptation, especially when tailored to farmers' needs, including an indicator of usage and results. There is a lot of skepticism about the ability of CSA to mitigate the impact of climate change by fostering soil resilience and let alone feed communities. This is, in spite of the potential benefits of the system, especially to small holder farmers who bear the brunt of climate change impacts. Most of the farmers are resource poor and they usually own lands in marginal areas in some zones of arid regions such as Egypt. According to Ortmann and Machethe (2003), there are more than three million small holder famers owning less than 1.3 hectares in marginal with low yields. CSA would be the most appropriate system for such farmers since it uses locally available resources and does not rely on the use of external inputs (Magdoff 2007). Nonetheless, in a review on adapting crops and cropping systems to future climates, Matthews et al. (2013) highlighted that most recommended adaptations will involve several trade-offs. For example, Pretty (2009) points out that farmers cannot simply cut their existing use of such inputs as fertilizer or pesticides and hope to maintain outputs and neither can they introduce a new productive element into their farming systems, and hope it succeeds. Instead the transition costs arise for several reasons, which include the following:

- Farmers must first invest in learning about a greater diversity of practices and measures.
- Farmers need to acquire information and management skills.
- During the transition and learning period, farmers must experiment more, and incur the costs of making mistakes as well as of acquiring new knowledge and information.
- New technologies often require more labor.

3 Results and Discussion

Intelligent water-saving irrigation system driven by complementary wind-solar energy is a precision irrigation system, which uses wind power or solar power to drive the intelligent control system and transports the on-demand supplied water and fertilizer to the root of the plant by underground micro seepage piping. Such green equipment system is essential for normal growth of plants in different arid conditions, which breakthroughs the traditional concept of drought resistance with intensive energy, high strength, and temporary drought relief measures (Fig. 1).

3.1 Water Management with Smart Irrigation

In Egypt, agricultural educators are working with farmers to install subsurface drip irrigation systems, which supply controlled amounts of water to crops with little waste. Drip is especially suitable for arid, hot and windy areas. Subsurface application of water to the root zone also has the potential to improve yields by reducing the incidence of disease and weeds.

Sub-surface drip irrigation is not subject to the amount of evaporation or runoff that occurs in the more common flood-furrow systems. Furthermore, some of the more sophisticated systems feature computer-programmed controllers with an option to apply agro-chemicals.

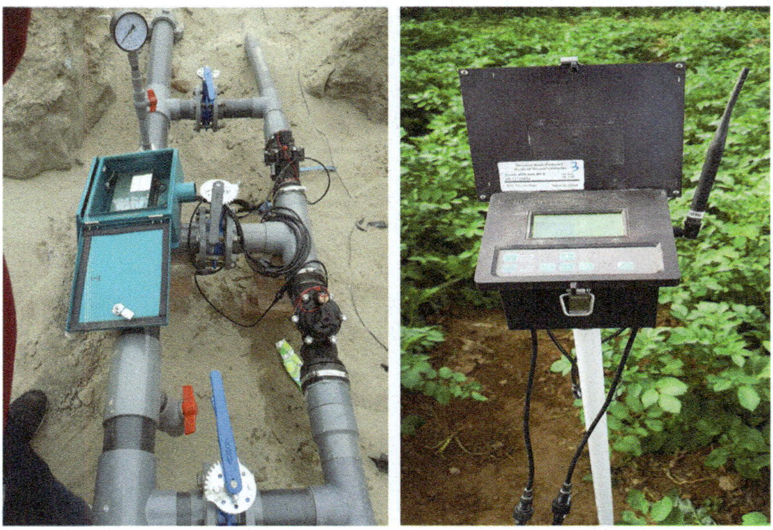

Fig. 1 Equipment of smart irrigation system titled "Intelligent Control System of Water Saving Irrigation"

Drip systems also avoid increasing salinity downstream, a serious problem with flood irrigation in arid environments. Irrigation water discharging from furrows carries dissolved solids picked up from the soil. Farmers near a river's headwaters who divert part of the water into furrows multiply the amount of solids and salinity for each farm that follows.

3.2 *Irrigation Systems*

With proper design and installation, a center pivot sprinkler system can achieve high irrigation efficiency and uniform application. A variety of packages and operating methods need to be employed including contact extension or engaging appropriate state agency to learn about furrow irrigation strategies, such as tail water recovery, irrigating every other row and polyacrylamide application.

3.3 *Irrigation Management*

Determining the amount and timing of irrigation for efficient water use can play a huge role in water conservation. Monitoring water application for crop needs and soil moisture content remains a key strategy. It considers a variety of low-cost, user-friendly electronic devices.

3.4 *The most Related Smart Irrigation Software*

Smart irrigation software allows farmers and others involved with harvesting crops to optimize their water usage throughout the crop lifecycle. These platforms provide precise, live updates on usage levels—both in specific segments of land and as a whole—along with the ability to enable or disable irrigation systems remotely. Many smart irrigation systems provide scheduling features in which users can map out water cycles, as well as options to automatically increase or decrease sprinkler output based on weather and other environmental factors. With the data collected from this software, businesses can make informed decisions about irrigation to maximize water conservation and quality.

Smart irrigation software often communicates with specially designed controllers or sensors from the vendor, or with commonly used hardware on the market. Used in conjunction with other agriculture software, such as precision agriculture and farm management systems, these tools can streamline farming processes while reducing environmental impact and labor costs.

4 Recommendations

Following are the key recommendations for CSA as a strategy to address the impacts of climate change:

- Promoting the culture of climate smart agriculture in Egypt and other countries.
- Working to reduce the potential negative impacts of climate change in the agricultural field, especially in relation to combating desertification.
- Development of varieties resistant to high temperatures and the development of varieties resistant to drought and salinity.
- Introducing new crops such as improved cassava varieties for Egyptian crops, extracting starch and flour and adding it to wheat flour to increase the self-sufficiency rate of flour as well as quinoa and other crops.
- Cultivation of desert areas with crops with minimal water needs and those that can tolerate the conditions of drought and salinity.
- Intensifying the cultivation of the Jojoba crop in the desert areas due to their economic and environmental benefits.
- Monitor and reduce greenhouse gas emissions and identify carbon footprint and water footprint for agricultural products.
- Work on increasing agricultural production and improving its quality through good agricultural practices and climate smart agriculture.
- Maintaining the ecosystem and biological diversity in agriculture, especially in the modern agricultural areas, the major agricultural projects areas in the new lands, the one million feddan project and the 100 thousand-Greenhouses project and benefiting from the other experience in this field.
- Promote activate cooperation and joint activities between the Egypt and other agro – based countries, especially in the field of climate change and agricultural productivity.
- Develop agricultural development programs that work through expected climate changes and avoid their negative impacts and adaptation.
- Develop mechanisms for the development and establishment of fish farms, taking into account the lack of available water.
- Coordination between the Egyptian and other countries to establish a database and a regional network for climate change to encourage research, studies and training in this field, with attention to the institutional systems of monitoring and early warning, which helps in formulating future agricultural policies.
- Increase awareness and adaptation of climate change issues and reduce their expected negative impacts.
- Develop better understanding in water resources and agriculture which are affected by different factors of climate changes, and monitor these changes.
- Develop plans and policies on climate change in the planning stages, based on accurate forecasting of weather and climate pattern.
- Expand the use of modern irrigation technologies, particularly in the desert areas and newly reclaimed lands, and also in the lands of the valley and the delta.

- Employ cropping patterns that maximize productive yield compared to the water unit used.
- Incorporate modern technology and remote sensing in the field of climate and agricultural environment.
- Joint exchange of experiences between Egypt and China in the field of climate change, agricultural production and combating desertification. This can be accomplished via a Memorandum of Understanding (MoU) between the National Research Center in Egypt and Institute of Geographic and Natural Resources Research and Beijing Institute of Technology in China.

5 Conclusion

It may be concluded that CSA by employing resilient agriculture practices could help address climate change both in terms of mitigation and adaptation. Within the current and projected situation of climate change globally, climate change mitigation alone will not be enough, and therefore long term solutions that combine climate change adaptation in the agriculture sector are imperative. Such practices could include smart irrigation, organic agriculture, manure management, smart agroforestry practices etc. There is a need to focus on new irrigation systems such as "surface and sub-surface irrigation system", and how these can be applied. Furthermore, CSA should also consider the adoption of solar energy to feed practices and strategies.

References

Barker, T., Bashmakov, I., Bernstein, L., Bogner, J. E., Bosch, P. R., Dave, R., Davidson, O. R., Fisher, B. S., & S.Gupta S. (2007). Technical summary. In B. Metz, O. R. Davidson, P. R. Bosch, R. Dave, & L. A. Meyer (Eds.), *Climate change 2007: Mitigation. Contribution of working group III to the fourth assessment report of the intergovernmental panel on climate change* (pp. 620–690). Cambridge: Cambridge University Press.

Blanco-Canqui, H., & Lal, R. (2004). Mechanisms of carbon sequestration in soil aggregates. *Critical Reviews in Plant Science, 23*, 481–504. https://doi.org/10.1080/07352680490886842.

Davis, A. S., Hill, J. D., Chase, C. A., Johanns, A. M., & Liebman, M. (2012). Increasing cropping system diversity balances productivity, profitability and environmental Health. *PLoS One, 7*, e47149. https://doi.org/10.1371/journal.pone.0047149.

FAO. (2013). *Climate-smart agriculture sourcebook.* Rome: Food and Agriculture Organization of the United Nations. http://www.fao.org/docrep/018/i3325e/i3325e04.pdf.

ICRAF. (2015). *Spatially Characterised Agroforestry Database (SCAF).* Available at http://worldagroforestry.org/regions/so

IPCC. (2007). Climate change: synthesis report. Contribution of working groups I, II and III to the fourth assessment report of the Intergovernmental Panel on Climate Change. In R. K. Pachauri & A. Reisinger (Eds.), *Core writing team* (p. 104). Geneva: IPCC. http://www.ipcc.ch/ipccreports/ar4-syr.htm.

Magdoff, F. (2007). Ecological agriculture: Principles, practices, and constraints. *Renewable Agricultural Food Systems, 22*, 109–117. https://doi.org/10.1017/S1742170507001846.

Matthews, R. B., Rivington, M., Muhammed, S., Newton, A. C., & Hallet, P. D. (2013). Adapting crops and cropping systems to future climates to ensure food security: The role of crop modelling. *Global Food Security, 2*, 24–28. https://doi.org/10.1016/j.gfs.2012.11.009.

Muzangwa, L., Chiduza, C., & Muchaonyerwa, P. (2013). Feasibility of winter cover crop production under rainfed conditions in the Eastern Cape Province of South Africa. *African Crop Science Journal, 21*, 73–184.

Ortiz-Monasterio, I., Wassmann, R., Govaerts, B., Hosen, Y., Katayanagi, N., & Verhulst, N. (2010). Greenhouse gas mitigation in the main cereal systems: Rice, wheat and maize. In M. P. Reynolds (Ed.), *Climate change and crop production* (CABI series in climate change) (Vol. 1). Wallingford/Oxfordshire/Cambridge: CAB International.

Ortmann, G., & Machethe, C. (2003). Problems and opportunities in south African agriculture. In L. Nieuwoudt & J. Groenewald (Eds.), *The challenge of change: Agriculture, land and the South African economy* (pp. 47–62). Pietermaritzburg: University of Natal Press.

Pretty, J. N. (2009). Can ecological agriculture feed nine billion people? *Monthly Review: An Independent Socialist Magazine, 61*, 46–58. http://monthlyreview.org/2009/11/01/can-ecological-agriculture-feed-nine-billionpeople/.

Stabinsky, D., & Ching, L. (2012). *Ecological agriculture, climate resilience and a roadmap to get there* (TWN Environment & Development Series No. 14). Penang: Third World Network.

Zheng, C., Jiang, Y., Chen, C., Sun, Y., Feng, J., Deng, A., & Song, Z. (2014). The impacts of conservation agriculture on crop yield in China depend on specific practices, crops and cropping regions. *The Crop Journal, 2*, 289–296. https://doi.org/10.1016/j.cj.2014.06.006.

The manufacturer's authorised representative in the EU is Springer
Nature Customer Service Centre GmbH, Europaplatz 3, 69115 Heidelberg,
Germany. If you have any concerns regarding our products, please
contact ProductSafety@springernature.com

Printed and bound by CPI Group (UK) Ltd, Croydon, CR0 4YY
30/09/2026
02225491-0012